全国高等职业教育医疗器械类专业
国家卫生健康委员会"十三五"规划教材

供医疗器械类专业用

医疗器械专业英语

第 2 版

主　审　郑彦云
主　编　陈秋兰
副主编　黄敏菊　杨　娴　景　然　严家来

编　者　（以姓氏笔画为序）

陈秋兰　广东食品药品职业学院　　　　严家来　安徽医学高等专科学校
孙延宁　山东医学高等专科学校　　　　黄敏菊　广东省医疗器械质量监督检验所
张涌萍　广东食品药品职业学院　　　　景　然　大庆医学高等专科学校
杨　娴　湖北中医药高等专科学校　　　葛　露　辽宁医药职业学院

人民卫生出版社

图书在版编目（CIP）数据

医疗器械专业英语／陈秋兰主编．—2版．—北京：人民卫生出版社，2018

ISBN 978-7-117-25805-0

Ⅰ.①医… Ⅱ.①陈… Ⅲ.①医疗器械－英语－高等职业教育－教材 Ⅳ.①TH77

中国版本图书馆CIP数据核字（2018）第132248号

| 人卫智网 | www.ipmph.com | 医学教育、学术、考试、健康、购书智慧智能综合服务平台 |
| 人卫官网 | www.pmph.com | 人卫官方资讯发布平台 |

版权所有，侵权必究！

医疗器械专业英语
第 2 版

主　　编：陈秋兰
出版发行：人民卫生出版社（中继线 010-59780011）
地　　址：北京市朝阳区潘家园南里 19 号
邮　　编：100021
E - mail：pmph @ pmph.com
购书热线：010-59787592　010-59787584　010-65264830
印　　刷：中农印务有限公司
经　　销：新华书店
开　　本：850×1168　1/16　印张：23
字　　数：541 千字
版　　次：2011 年 7 月第 1 版　2018 年 12 月第 2 版
　　　　　2024 年 5 月第 2 版第 8 次印刷（总第 13 次印刷）
标准书号：ISBN 978-7-117-25805-0
定　　价：59.00 元

打击盗版举报电话：010-59787491　E-mail：WQ @ pmph.com
（凡属印装质量问题请与本社市场营销中心联系退换）

全国高等职业教育医疗器械类专业
国家卫生健康委员会"十三五"规划教材
出版说明

《国务院关于加快发展现代职业教育的决定》《高等职业教育创新发展行动计划(2015—2018年)》《教育部关于深化职业教育教学改革全面提高人才培养质量的若干意见》等一系列重要指导性文件相继出台,明确了职业教育的战略地位、发展方向。同时,在过去的几年,中国医疗器械行业以明显高于同期国民经济发展的增幅快速成长。特别是随着《关于深化审评审批制度改革鼓励药品医疗器械创新的意见》的印发、《医疗器械监督管理条例》的修订,以及一系列相关政策法规的出台,中国医疗器械行业已经踏上了迅速崛起的"高速路"。

为全面贯彻国家教育方针,跟上行业发展的步伐,将现代职教发展理念融入教材建设全过程,人民卫生出版社组建了全国食品药品职业教育教材建设指导委员会。在指导委员会的直接指导下,经过广泛调研论证,人卫社启动了全国高等职业教育医疗器械类专业第二轮规划教材的修订出版工作。

本套规划教材首版于2011年,是国内首套高职高专医疗器械相关专业的规划教材,其中部分教材入选了"十二五"职业教育国家规划教材。本轮规划教材是国家卫生健康委员会"十三五"规划教材,是"十三五"时期人卫社重点教材建设项目,适用于包括医疗设备应用技术、医疗器械维护与管理、精密医疗器械技术等医疗器类相关专业。本轮教材继续秉承"五个对接"的职教理念,结合国内医疗器械类专业领域教育教学发展趋势,紧跟行业发展的方向与需求,重点突出如下特点:

1. **适应发展需求,体现高职特色**　本套教材定位于高等职业教育医疗器械类专业,教材的顶层设计既考虑行业创新驱动发展对技术技能型人才的需要,又充分考虑职业人才的全面发展和技术技能型人才的成长规律;既集合了我国职业教育快速发展的实践经验,又充分体现了现代高等职业教育的发展理念,突出高等职业教育特色。

2. **完善课程标准,兼顾接续培养**　本套教材根据各专业对应从业岗位的任职标准优化课程标准,避免重要知识点的遗漏和不必要的交叉重复,以保证教学内容的设计与职业标准精准对接,学校的人才培养与企业的岗位需求精准对接。同时,本套教材顺应接续培养的需要,适当考虑建立各课程的衔接体系,以保证高等职业教育对口招收中职学生的需要和高职学生对口升学至应用型本科专业学习的衔接。

3. **推进产学结合,实现一体化教学**　本套教材的内容编排以技能培养为目标,以技术应用为主线,使学生在逐步了解岗位工作实践、掌握工作技能的过程中获取相应的知识。为此,在编写队伍组建上,特别邀请了一大批具有丰富实践经验的行业专家参加编写工作,与从全国高职院校中遴选出的优秀师资共同合作,确保教材内容贴近一线工作岗位实际,促使一体化教学成为现实。

4. **注重素养教育,打造工匠精神**　在全国"劳动光荣、技能宝贵"的氛围逐渐形成,"工匠精神"在各行各业广为倡导的形势下,医疗器械行业的从业人员更要有崇高的道德和职业素养。教材更加强调要充分体现对学生职业素养的培养,在适当的环节,特别是案例中要体现出医疗器械从业人员的行为准则和道德规范,以及精益求精的工作态度。

5. **培养创新意识，提高创业能力**　为有效地开展大学生创新创业教育,促进学生全面发展和全面成才,本套教材特别注意将创新创业教育融入专业课程中,帮助学生培养创新思维,提高创新能力、实践能力和解决复杂问题的能力,引导学生独立思考、客观判断,以积极的、锲而不舍的精神寻求解决问题的方案。

6. **对接岗位实际，确保课证融通**　按照课程标准与职业标准融通、课程评价方式与职业技能鉴定方式融通、学历教育管理与职业资格管理融通的现代职业教育发展趋势,本套教材中的专业课程,充分考虑学生考取相关职业资格证书的需要,其内容和实训项目的选取尽量涵盖相关的考试内容,使其成为一本既是学历教育的教科书,又是职业岗位证书的培训教材,实现"双证书"培养。

7. **营造真实场景，活化教学模式**　本套教材在继承保持人卫版职业教育教材栏目式编写模式的基础上,进行了进一步系统优化。例如,增加了"导学情景",借助真实工作情景开启知识内容的学习;"复习导图"以思维导图的模式,为学生梳理本章的知识脉络,帮助学生构建知识框架。进而提高教材的可读性,体现教材的职业教育属性,做到学以致用。

8. **全面"纸数"融合，促进多媒体共享**　为了适应新的教学模式的需要,本套教材同步建设以纸质教材内容为核心的多样化的数字教学资源,从广度、深度上拓展纸质教材内容。通过在纸质教材中增加二维码的方式"无缝隙"地链接视频、动画、图片、PPT、音频、文档等富媒体资源,丰富纸质教材的表现形式,补充拓展性的知识内容,为多元化的人才培养提供更多的信息知识支撑。

本套教材的编写过程中,全体编者以高度负责、严谨认真的态度为教材的编写工作付出了诸多心血,各参编院校为编写工作的顺利开展给予了大力支持,从而使本套教材得以高质量如期出版,在此对有关单位和各位专家表示诚挚的感谢!教材出版后,各位教师、学生在使用过程中,如发现问题请反馈给我们(renweiyaoxue@163.com),以便及时更正和修订完善。

人民卫生出版社
2018年3月

全国高等职业教育医疗器械类专业
国家卫生健康委员会"十三五"规划教材
教材目录

序号	教材名称	主编	单位
1	医疗器械概论(第2版)	郑彦云	广东食品药品职业学院
2	临床信息管理系统(第2版)	王云光	上海健康医学院
3	医电产品生产工艺与管理(第2版)	李晓欧	上海健康医学院
4	医疗器械管理与法规(第2版)	蒋海洪	上海健康医学院
5	医疗器械营销实务(第2版)	金 兴	上海健康医学院
6	医疗器械专业英语(第2版)	陈秋兰	广东食品药品职业学院
7	医用X线机应用与维护(第2版)*	徐小萍	上海健康医学院
8	医用电子仪器分析与维护(第2版)	莫国民	上海健康医学院
9	医用物理(第2版)	梅 滨	上海健康医学院
10	医用治疗设备(第2版)	张 欣	上海健康医学院
11	医用超声诊断仪器应用与维护(第2版)*	金浩宇	广东食品药品职业学院
		李哲旭	上海健康医学院
12	医用超声诊断仪器应用与维护实训教程(第2版)*	王 锐	沈阳药科大学
13	医用电子线路设计与制作(第2版)	刘 红	上海健康医学院
14	医用检验仪器应用与维护(第2版)*	蒋长顺	安徽医学高等专科学校
15	医院医疗设备管理实务(第2版)	袁丹江	湖北中医药高等专科学校/荆州市中心医院
16	医用光学仪器应用与维护(第2版)*	冯 奇	浙江医药高等专科学校

说明:*为"十二五"职业教育国家规划教材,全套教材均配有数字资源。

全国食品药品职业教育教材建设指导委员会成员名单

主 任 委 员：姚文兵　中国药科大学
副主任委员：刘　斌　天津职业大学　　　　　　　马　波　安徽中医药高等专科学校
　　　　　　冯连贵　重庆医药高等专科学校　　　袁　龙　江苏省徐州医药高等职业学校
　　　　　　张彦文　天津医学高等专科学校　　　缪立德　长江职业学院
　　　　　　陶书中　江苏食品药品职业技术学院　张伟群　安庆医药高等专科学校
　　　　　　许莉勇　浙江医药高等专科学校　　　罗晓清　苏州卫生职业技术学院
　　　　　　昝雪峰　楚雄医药高等专科学校　　　葛淑兰　山东医学高等专科学校
　　　　　　陈国忠　江苏医药职业学院　　　　　孙勇民　天津现代职业技术学院

委　　　员（以姓氏笔画为序）：

于文国	河北化工医药职业技术学院	杨元娟	重庆医药高等专科学校
王　宁	江苏医药职业学院	杨先振	楚雄医药高等专科学校
王玮瑛	黑龙江护理高等专科学校	邹浩军	无锡卫生高等职业技术学校
王明军	厦门医学高等专科学校	张　庆	济南护理职业学院
王峥业	江苏省徐州医药高等职业学校	张　建	天津生物工程职业技术学院
王瑞兰	广东食品药品职业学院	张　铎	河北化工医药职业技术学院
牛红云	黑龙江农垦职业学院	张志琴	楚雄医药高等专科学校
毛小明	安庆医药高等专科学校	张佳佳	浙江医药高等专科学校
边　江	中国医学装备协会康复医学装备技术专业委员会	张健泓	广东食品药品职业学院
		张海涛	辽宁农业职业技术学院
师邱毅	浙江医药高等专科学校	陈芳梅	广西卫生职业技术学院
吕　平	天津职业大学	陈海洋	湖南环境生物职业技术学院
朱照静	重庆医药高等专科学校	罗兴洪	先声药业集团
刘　燕	肇庆医学高等专科学校	罗跃娥	天津医学高等专科学校
刘玉兵	黑龙江农业经济职业学院	郏枝花	安徽医学高等专科学校
刘德军	江苏省连云港中医药高等职业技术学校	金浩宇	广东食品药品职业学院
孙　莹	长春医学高等专科学校	周双林	浙江医药高等专科学校
严　振	广东省药品监督管理局	郝晶晶	北京卫生职业学院
李　霞	天津职业大学	胡雪琴	重庆医药高等专科学校
李群力	金华职业技术学院	段如春	楚雄医药高等专科学校

全国食品药品职业教育教材建设指导委员会成员名单

袁加程	江苏食品药品职业技术学院	**黄美娥**	湖南食品药品职业学院
莫国民	上海健康医学院	**晨　阳**	江苏医药职业学院
顾立众	江苏食品药品职业技术学院	**葛　虹**	广东食品药品职业学院
倪　峰	福建卫生职业技术学院	**蒋长顺**	安徽医学高等专科学校
徐一新	上海健康医学院	**景维斌**	江苏省徐州医药高等职业学校
黄丽萍	安徽中医药高等专科学校	**潘志恒**	天津现代职业技术学院

前　言

《医疗器械专业英语》第2版是针对高等职业教育医疗器械类专业学生和从业人员,为提高专业英语水平,熟练运用医疗器械行业英语的需求而编写。为适应行业发展和"互联网+"环境下教学改革发展的趋势,根据高等职业教育医疗器械类专业师生和行业人员的意见和建议,在第一版的基础上进行了修订。

本版教材修订的原则:

1. 在内容选择和结构设计上,以教育部"工学结合、突出实践能力培养、以就业为导向"的现代职业教育理念为依据,注重培养学生对医疗器械专业领域的行业应用文章的阅读、翻译能力。

2. 努力贯彻"实用、可用、好用"的原则,加强实践性教学,注重培养学生分析和解决实际问题的能力。

和上版相比,本版教材将医疗器械相关的产品知识、技术与管理和本专业特色有机结合,扩大了题材范围,兼顾职业性、应用性、趣味性和前瞻性。主要有以下特点:

1. 教材内容的选择上,突出职业行业特色,内容涉及面广,涵盖了专业英语文体特点与翻译技巧、专业基础知识介绍、典型产品介绍、监管法规、产品标准与应用案例、市场营销和行业交流等七方面的内容。结构上力求突出实践性和职业性,强化职业能力的培养,以满足高职高专学生在学习和将来从业中对英语素养的需求。

2. 在编写体例上,以纸质为基本载体,融合数字资源。将导学情景、学习目的、知识要求置于每个项目内容之首,知识导入、课文、词汇、课文注释位于每个项目中间部分,目标检测、练习和文化沙龙置于每个项目的末尾。数字资源部分提供了同步练习,阅读理解和听力训练等内容。

本教材由郑彦云主审,各章执笔为:陈秋兰(绪论、第三章)、严家来(第一章)、孙延宁(第二章第一、二节)、张涌萍(第二章第三、四节)、葛露(第二章第五、六节)、黄敏菊(第二章第七节、第四章)、杨娴(第五章)、景然(第六章)。

通过本教材的学习可以掌握大量的专业英语词汇、句法和表达方法等,从而提高医疗器械专业英语的阅读能力、理解能力、翻译能力,为从事医疗器械相关工作打下坚实的基础。

在编写过程中,有幸得到了广东省医疗器械质量监督检验所李继彦博士和百特医疗用品贸易(上海)有限公司的大力支持,并参考了有关专著和资料,同时得到了全国食品药品职业教育教材建设指导委员会、人民卫生出版社和有关学校的大力支持,谨在此向有关行业专家、企业和作者致以崇高的敬意和感谢！对各位参编同仁的辛勤付出表示衷心的感谢，在此一并致谢！

由于编者水平有限、时间仓促，不妥之处在所难免，敬请读者不吝赐教。

陈秋兰

2018 年 9 月

Table of contents

绪论　专业英语文体特点与翻译技巧　　1

Chapter 1　Fundamentals of medical devices　　12

Section 1　Basic introduction to medical devices　　12

　Text 1　Introduction to medical devices　　12

　Text 2　FDA: definition and classification of medical devices　　18

　Text 3　Global medical device market　　23

Section 2　Fundamentals of medical devices　　31

　Text 1　A brief introduction to electronic basics　　31

　Text 2　A brief introduction to biological basics　　37

　Text 3　A brief introduction to microorganisms　　43

Chapter 2　Introduction to typical products　　51

Section 1　Consumer products　　51

　Text 1　Contact lenses　　51

　Text 2　Blood pressure meters　　57

　Text 3　Glucose meters　　63

　Text 4　Hearing aids　　70

Section 2　Electrocardiograph (ECG) monitors　　76

　Text 1　Introduction to electrocardiograph (ECG) monitors　　76

　Text 2　How an ECG monitor works　　82

Section 3　Ultrasound imaging　　90

　Text 1　Introduction to ultrasound imaging　　90

　Text 2　Fundamentals of ultrasound imaging　　96

Section 4　X-ray imaging　　103

　Text 1　Introduction to medical X-ray imaging　　103

　Text 2　Radiography (Plain X-rays)　　109

　Text 3　Computed tomography (CT)　　116

Section 5　Magnetic resonance imaging (MRI)　　123

　Text 1　Introduction to magnetic resonance imaging (MRI)　　123

Table of contents

    Text 2 Physics of magnetic resonance imaging (MRI)   129

  Section 6 Clinical laboratory equipment   135

    Text 1 Hematology analyzer   135

    Text 2 Biochemistry analyzer   142

    Text 3 Urine analyzer   149

  Section 7 Biomaterials   156

    Text 1 Medical dressing materials   156

    Text 2 Dental materials   162

    Text 3 Metallic biomaterials in orthopaedic surgery   169

Chapter 3 Medical devices regulations 176

  Section 1 Global medical device regulatory harmonization   176

    Text 1 Global regulations of medical devices   176

    Text 2 Global harmonization of medical devices   182

  Section 2 US FDA regulations   189

    Text 1 How to market your medical devices in U.S.?   189

    Text 2 FDA: Premarket notification (PMN) 510 (k)   195

  Section 3 NMPA regulations   202

    Text 1 Regulations on supervisory management of medical devices   202

    Text 2 Provisions for medical device registration   208

  Section 4 Quality management system (QMS) for medical device   215

    Text 1 Quality management system (QMS) for medical devices   215

    Text 2 ISO 13485: 2003 Medical devices—quality management systems—requirements for regulatory purposes   221

Chapter 4 Medical device standards and practices 228

  Section 1 Medical device standards   228

    Text 1 ISO 10993-5: Test for in vitro cytotoxicity   228

    Text 2 Heavy metals impurities test methods in United States pharmacopeia   235

    Text 3 IEC 60601-1-2-2014: Medical electrical equipment-General requirements for basic safety and essential performance   240

  Section 2 Application cases   246

Text 1	Classes and types of medical electrical equipment	246
Text 2	The AK 96 dialysis machine	252
Text 3	Maintenance manual of the AK 96 dialysis machine	258

Chapter 5 Medical device marketing and management — 265

- Section 1 Business etiquette — 265
  - Text 1 Business etiquette tips everyone should follow — 265
  - Text 2 Customer service etiquette — 271
- Section 2 Marketing — 276
  - Text 1 Medical device sales jobs — 276
  - Text 2 How medical device companies can strategically sell products to customers? — 282
- Section 3 Business negotiation — 288
  - Text 1 How to win price negotiations? — 288
  - Text 2 Meet goals in negotiation — 293

Chapter 6 Industrial communication — 299

- Section 1 Job hunting — 299
  - Text 1 Resume for employment — 299
  - Text 2 Job-hunting letter — 305
  - Text 3 Interview skills — 311
- Section 2 Technical communication — 317
  - Text 1 How to become a successful medical device sales representative? — 317
  - Text 2 The internet-where the customers are? — 322

参考文献 — 329

目标检测参考答案（Learning Test 部分） — 331

医疗器械专业英语课程标准 — 349

绪 论

专业英语文体特点与翻译技巧

专业英语是自然科学和技术人员从事专业活动时所使用的一种文体,涉及各行各业的众多领域,是英语的一种变体形式。科学著作、学术论文、实验报告、产品说明书等都属于专业英语。广义上来说,专业英语泛指一切涉及科学或技术的书面语和口头语。具体包括:①科技著述、科技论文(报告)、实验报告(方案)等;②各类科技情报及其他文字资料;③科技实用手册,包括仪器、仪表、机械和工具等的结构描述和操作规程的叙述;④科技问题的会议、会谈及交谈用语;⑤科技影片或录像等有声资料的解说词等。

专业英语是科技人员在世界范围内用以科学技术交流的重要工具,在其使用和发展过程中形成了自己鲜明的特色,与新闻报刊、日常交流用语的文体不一样,目前已发展成为一种重要的英语语体。总体来说,专业英语不追求语言的艺术美,而是讲究逻辑清晰、条例清楚、叙述准确严密,具有行文简洁、陈述客观、描述准确、结构严密、强调事实、专业性强等特点。

专业英语的翻译是一门科学而不是艺术,不需要大量的再创作,需要的是译者能够准确、客观地翻译所表达的专业知识。总体来说,专业英语的文体特点可通过构词、句法、修辞等层面表现出来。

一、专业英语的词汇特点

科技文章要求概念清楚,避免含糊不清和一词多义。为了准确、科学地表达观点,描述客观事实,遣词就显得尤为重要。专业英语最显著的特点体现在词汇的使用及构词法上。

(一) 多用术语、专业性强

每一学科、行业或特定领域内都有其相应的专业术语,用于正确表达科学概念,与具体的科技领域息息相关,具有丰富的内涵和外延。为了揭示自然科学和客观事物的现象和发展规律,专业英语必须使用表达准确的专业术语。例如,magnetic resonance imaging(MRI,磁共振成像),biochemistry(生物化学),voltage(电压)等,都是专业性极强的词汇。此外,随着科技知识的普及,有些科技词汇已为大众所熟知,并成为普通语言的一部分,如 cardiovascular disease(心血管疾病),blood pressure(血压)等。

另一方面,即使是来自英语中的普通词,也被赋予了新的词义。例如:Work is the transfer of energy expressed as the product of a force and the distance through which its point of application moves in the direction of the force. 这句话中,work(功),energy(能),product(乘积),force(力)都是来源于普通词汇的物理术语。

在科技文章中,同一词语的词义也和专业紧密相关。有时同一个英语常用词不仅被多个专业采用,而且含义也各不相同,区别很大。如 field 一词,一般英语中指"野外、田间";在物理领域常称为

"场",如 magnetic field(磁场);在光学方面可作"视场"解,如 sight field(视场);在科研领域可作"领域"解,如 research field(研究领域)等。对于这一类词语,只有通过上下文和具体的语言环境才能确定其真正的含义。

(二) 大量源于希腊语或拉丁语

大量的科技词汇都源于希腊语和拉丁语,因为这两种语言不会由于社会的发展而引起词义的变化,也不因词的多义引起歧义。例如:physics(物理)、radius(半径)、parameter(参数)、thesis(论文)等。这些来源于希腊语或拉丁语的词汇的复数形式有些仍按原来的形式,如,vertebra(椎骨)的复数形式是 vertebrae,medium(媒介)的复数形式是 media,thrombus(血栓)的复数形式是 thrombi。也有些词汇由于在英语里使用时间较长,不仅保留了原来的复数形式,也采用了英语的复数形式。例如,formula(公式,拉丁语)的复数形式可以是 formulae,也可以是 formulas,stratum(层,拉丁语)的复数形式可以是 strata,也可以是 stratums。

(三) 新词不断、构词灵活

随着科技的日新月异,新事物、新概念和新现象大量涌现,因而需要构造新的词汇,这些新词记录着时代脉搏的跳动。到了信息与科技爆炸的21世纪,情况更是如此。据统计,每年都能出现5000以上的新英语词汇。一般来说,科技新词的产生与其所属学科关系密切,例如,材料领域的 stereo-lithography(立体光刻),生物化学领域的 stereo-chemistry(立体化学)、reprogenetics(生殖遗传学),物理领域的 superconducting material(超导材料)、microfluidics(微流控),以及一些交叉学科领域的 bioinformatics(生物信息学)、optical tweezer(光镊)、biophotonics(生物光子学)等。

为了更准确地表达这些新生事物,专业词汇的构词法也灵活多样,常见的构词法有合成法、派生法、缩略法等。

1. 合成法 由两个或两个以上独立的词合在一起,构成一个新词。例如:

base+line=baseline(基线)

water+proof=waterproof(防水的)

power+meter=powermeter(功率计)

power+plant=powerplant(发电站、发电厂)

work+shop=workshop(车间)

clean+room=cleanroom(洁净室)

over+night=overnight(整夜)

safe+guard=safeguard(保护)

有的合成词的两个成分之间有连字符"-"。例如:

heat+resistant=heat-resistant(耐热的)

photo+lithography=photo-lithography(光刻)

wet+etching=wet-etching(湿蚀刻)

state+of+the+art=state-of-the-art(最先进的)

专业英语中有很多专业术语由两个或更多的词组成,它们的构成成分虽然看起来是独立的,但实际上合起来才是一个完整的概念,因此也把它们看成一个术语。例如:

computer language(计算机语言)

liquid crystal(液晶)

red blood cell(红细胞)

biomedical engineering(生物医学工程)

signal-to-noise ratio(信噪比)

2. 派生法 派生词是由词根加上前缀和/或后缀来构成新词,也叫缀合,有些词缀和词根之间有连接符。加前缀构成新词一般只改变词义,不改变词性。英语中有大量的前缀,并且是有固定意义的。例如:

前缀	前缀词义	派生词
un- non- in- im- dis- mal- il- ir- a-	否定前缀	unknown 未知的 uncomfortable 不舒适的 unbelievable 难以置信的 non-active 不活跃的 inefficient 低效的 inaccurate 不准确的 immobilize 固定 impossible 不可能的 irregular 不规则的 disorder 失调 disagree 不同意 illegal 非法的 malnutrition 营养不良 asymmetry 不对称
anti-	反-,抗-	antibody 抗体 anti-interference 抗干扰 antigen 抗原 antibiotic 抗生素 antioxidant 抗氧化剂
auto-	自动-	auto-focusing 自动聚焦
counter-	相反,反对-	counterclockwise 逆时针
bio-	生物-	biomedical 生物医学
sub-	亚-,子-	submicron 亚微米 subsystem 分系统
pre-	在…之前,预-	pre-warning 预警 pre-mix 预混合
post-	在…之后	post-war 战后
re-	重新-	redo 重做
mono-	单-	monolayer 单层

续表

前缀	前缀词义	派生词
multi-	多-	multilayer 多层
inter-	相互-	interaction 相互作用
bi-	双-	bi-directional 双向的
super-	超-	superconductor 超导体
ultra-		ultrafast 超快的
tri-	三-	triangle 三角形
semi-	半-	semiconductor 半导体

加后缀的作用和前缀有所不同，它们主要用来改变词性，只有少数会改变词义。记住常用的一些词缀对于记忆生词和猜测词义很有帮助。例如：

后缀	后缀词义	派生词
-able	形容词后缀	avoidable 可避免的
-ible		audible 可听见的
-uble		soluble 可溶的
-al		fundamental 基本的
-ant		abundant 大量的
-ent		efficient 有效的
-ed		limited 有限的
-ful		powerful 强大的
-ic		electric 电气的
-ical		economical 经济的
-less		useless 无用的
-ous		numerous 众多的
-tive		conductive 导电的
-sive		invasive 侵入性的
-tion	名词后缀	invention 发明
-sion		permission 许可
-ment		treatment 治疗
-ence		existence 存在
-ance		maintenance 维护
-ity		possibility 可能性
-er/or		employer 雇主
		conductor 导体
-logy		biology 生物学
-logist		biologist 生物学家
-ness		darkness 黑暗
-ize	动词后缀	liquidize 液化
-fy		simplify 简化
-en		strengthen 加强
-ly	副词后缀	independently 独立地
-ward		upward 向上

3. 缩合法 将两个或多个单词进行省略或简化,组合成新词。可取多个词汇的首字母进行组合,或直接去掉词汇的某一部分。例如:

acquired immune deficiency syndrome 缩合成 AIDS(获得性免疫性缺陷综合征)

unidentified flying object 缩合成 UFO(不明飞行物)

signal-to-noise ratio 缩合成 SNR(信噪比)

magnetic resonance image 缩合成 MRI(磁共振成像)

digital radiography 缩合成 DR(数字 X 线摄影)

radio detection and ranging 缩合成 radar(无线电探测和测距)

technology 简写成 tech(技术)

semiconductor 简写成 semicon(半导体)

influenza 简写成 flu(流行性感冒)

telecommunication 简写成 telecom(电信)

二、专业英语的句法特点

(一) 经常使用被动语态,而且没有行为的被动主体

被动语态在专业英语中用得非常频繁,这是因为科技文章的主要目的是讲述客观现象,描写行为或状态本身,如介绍科技成果、仪器操作等,使用被动句比使用主动句更少主观色彩,能够突出论证主体本身。其次,行为或状态的主体或者没必要指出,或者根本指不出来。因此在科技英语中,凡是在不需要或不可能指出行为主体的场合,或者在需要突出行为客体的场合都使用被动语态。在很多情况下,被动结构也更简洁。例如:

1. The working temperature of the machine should be measured regularly.

2. The signal-to-noise ratio can be increased by the use of high-sensitivity cooled CCD camera.

3. The absorption spectrum is determined by the substances in the solution.

(二) 多用长难句

专业英语用于表达科学理论、技术原理、仪器维护方法等,常涉及各事物之间错综复杂的关系,而复杂的科学思维常无法用简单语句来表达,所以语法结构复杂的长句较多地应用于专业英语,而这种严谨周密、层次分明、重点突出的语言手段也就成了专业英语文体又一重要特征。为了保证表达内容的严密性、准确性和逻辑性,往往使用大量的复合句和有附加成分的长句。一般来说,长难句的特点是后置定语、非谓语动词、宾语从句、状语从句等成分多。例如:

1. Because Chinese companies have limited R&D and design budgets, imitations of foreign medical devices are common, but they occur mostly in product categories in which technology is easily obtained, such as biochemical analysis instruments.

2. In addition, medical devices include in vitro diagnostic products, such as general-purpose lab equipment, reagents, and test kits, which may include monoclonal antibody technology.

3. To find the classification of your device, as well as whether any exemptions may exist, you need to

find the regulation number that is the classification regulation for your device.

(三) 大量使用名词化结构

名词化是指把动词通过加缀、转化等变成有动作含义的名词。科技英语要求行文简洁、表达客观、内容准确、信息量大，常强调存在的事实，而非某一行为，所以大量使用名称化结构。如果是动词短语或句子，则把其变成名词短语，例如：

1. earth rotates 可以转换成 the rotation of earth
2. monitor heart diseases 可以转换成 monitoring of heart diseases
3. to examine specimens 可以转换成 the examination of specimens
4. Biomedical engineers develop the equipment. 可以转换成 the development of the equipment by biomedical engineers.

为了客观地表达事物的本质，避免主观性，一个方法是采用被动语态来突出事物本身并实现无人称性，另一个方法就是使用名词词组。例如：

We can normally regulate the temperature by using a thermo-coupler. 可以转化成 Regulation of the temperature can normally be realized by the use of a thermo-coupler.

此外，主语在英语中所处的位置很重要，名词化可以把最重要的信息放在句首，起到突出作用，并且这样做可以使句子变得更加简练。例如：

If the thickness of the insulation layer is increased, it will reduce the heat losses. 转化成 The increase of the thickness of the insulation layer will reduce the heat losses. 后者更加突出"绝缘层厚度的增加"这一重点。

(四) 多用非谓语动词(动词非限定形式)

每个简单的英文句子中，只能有一个谓语动词，作谓语的动词称为限定动词，不作谓语的动词或动词短语由于其结构不受主语的限定，没有人称和数的变化，因而称为非限定动词。非谓语动词包括不定式(to+V)、V-ing、和过去分词(V-ed)。它们在专业英语中的使用频率比一般文体要高得多。例如：

This module enables the student to program a simple microcontroller to perform typical industrial task. (这个模块使学生能对一个简单的微处理器进行编程以执行典型的行业任务)。这里的 to program, to perform 都是非谓语动词，描述动作。

非谓语动词的作用大致有两个：

(1)用非谓语动词作定语，对描述的对象加以限定，使描述更加准确。

(2)用非谓语动词短语代替从句，使语句更加简练。

1. 非谓语动词不定式(to+V)　　动词不定式短语在句中常用作主语、后置定语、表语、宾语、状语和宾语补足语。它们多以短语的形式出现，即以动词不定式为中心，再加上其他的状语宾语。例如：

(1)The most important aim of safety management is to maintain and promote workers' health and safety at work. (作表语)

(2)The first step to solve the problem is to analyze the possible causes. (前者为后置定语，后者作

表语)

(3) To pass the examination is his current goal. (作主语)

(4) The electromotive force creates the electric pressure that causes the current to flow through a conductor. (作宾语补足语)

专业英语中,不定式短语还常常单独放在句尾并用逗号隔开,用以补充说明作用或目的,或者放在句首并以逗号隔开,用以突出主句中所述事件的目的或作用。

(1) You need to find the regulation number that is the classification regulation for your device, to find the classification of your device, as well as whether any exemptions may exist.

(2) To find the classification of your device, as well as whether any exemptions may exist, you need to find the regulation number that is the classification regulation for your device.

2. 非谓语动词 V-ing V-ing 形式的动词在有的语法书中分为动名词与现在分词,但现在常不加以区分。V-ing 在句中可作表语、主语、宾语、定语和宾语补足语等成分。同时保留了动词性,因此可根据词性带上宾语和状语。例如:

(1) Physical, mental and emotional health is considered when treating a patient. (作状语)

(2) A direct current is a current always flowing in the same direction. (作宾语补足语)

(3) While taking up any home medical equipment you need to take consultation from your doctor, nurses, or the specialist. (作状语)

(4) Having your blood pressure checked regularly indicates important fluctuations. (作主语)

(5) Organizational culture is a major component affecting its performance and behavior. (作后置定语)

在专业英语中,with/without/for+(名词)+V-ing 结构常用作补充说明,在这种结构中,with/without 没有词义,表示一种伴随情况,而 for 则多用于表示一些目的性动作,有时还用逗号隔开。例如:

(1) An MRI scan is the best way to see inside the human body, without cutting it open.

(2) As of 2009, the medical equipment industry is worth more than \$200 billion globally, with the United States leading the market.

(3) An X-ray is very effective for showing doctors a broken bone.

3. 非谓语动词 V-ed 动词的 V-ed 形式与 be 结合构成被动语态,与 have 连用构成完成时,因此,V-ed 形式本身含有被动与完成的意思。在句中可作后置定语、状语等,保留了动词性,表示这一动作是已完成的或所修饰的名词所(被动)接受的。V-ed 做定语常放在名词后面,含有被动的意思。在专业英语中,"with/without/for+(名词)+V-ed"结构常用作补充说明。在这种结构中,with/without 没有词义,表示一种伴随情况,而 for 则多用于表示一些目的性动作,有时还用逗号隔开。例如:

(1) In today's unpredictable era, it is practically impossible to stay away from the unexpected accidents. (作定语)

(2) The volunteers would come into a room where there was a row of five cubicles, with their doors shut. (作后置定语)

(3) The results <u>obtained</u> must be checked. (作后置定语)

(4) All metals are fairly hard, <u>compared</u> to polymers. (作状语)

(5) With an estimated 14 percent of the world's population <u>expected</u> to be 65 or older by 2030, the need for medical devices and qualified technicians to fix them will continue to grow. (作后置定语)

(五) 多采用后置短语 (从句) 结构

大量使用后置短语 (从句), 尤其是后置定语, 是科技文章的特点之一。常用的句子结构形式有: 介词短语后置, 形容词及形容词短语后置, 副词、分词、定语从句后置。其中, V-ed 形式的过去分词作后置定语时带有强烈的被动动词意义。例如:

1. In small and medium-sized companies, the safety manager and the safety representative often have other duties <u>besides their health and safety tasks</u>. (介词短语后置)

2. The safety manager's role is to act as an expert <u>who is aware of the health and safety legislation and other obligations concerning the company</u>. (定语从句后置)

3. The heat <u>produced</u> is equal to the electrical energy <u>wasted</u>. (分词后置)

4. The efforts <u>necessary to assure that sufficient emphasis is placed on system safety</u> are often organized into formal programs. ("形容词短语+从句"后置)

5. If you face certain illness and require urgent medicare, the home medical equipment is the best possible option <u>available</u>. (形容词后置)

三、专业英语翻译技巧

英、汉两种语言在语言结构与表达形式方面各有其自身的特点。专业英语中大量使用名词化结构、被动语态、复杂长句等。而汉语中动词、主动态和简单句用得多, 此外, 汉语中没有冠词、关系代词、关系副词、分词、不定式、动名词等。因此, 在专业英语翻译中, 要使译文既忠实于原意又顺畅可读, 就不能局限于逐词对等, 必须采用适当的语态转换、句子成分顺序转换, 乃至句型转换等翻译技巧。

(一) 根据专业领域确定词义

大量的词汇在专业英语中有其独特的词义, 和专业领域有关, 同一个单词在不同的专业领域可能具有完全不同的词义, 甚至在同一个领域中也可能不同。例如:

1. cell: 在生物学中, 常译为"细胞", 如 blood cell (血细胞); 在化学领域, 常译为"电池", 如 fuel cell (燃料电池)。

2. power: 在数学中, 常译为"幂", 如 The fourth power of two is 16. (2 的 4 次幂为 16); 在电学中, 常译为"电源、功率"等意思, 如: power meter (功率计)。

在阅读专业文献时, 应注意扩大自己的专业词汇量, 许多在日常英语中的单词, 在专业领域也可能有其特定的词义。在确定词义时, 可根据文章涉及的专业内容和上下文来确定。

(二) 英语名词化结构及非谓语动词一般均可译成汉语的动词

专业科技英语中大量使用名词化结构, 而汉语中动词用得多, 翻译时上述词可以予以转换处理。

例如：

1. A change of state from a solid to a liquid form requires heat energy. (从固态变为液态需要热能。)

2. Light from the sun is a mixture of light of many different colors. (太阳光是由许多不同颜色的光混合成的。)

（三）被动语态转主动表达

科技英语往往强调如何去做，而不介意谁去做，所以常采用被动语态。英语的许多被动句不需要或无法讲出动作的发出者，因此往往可译成汉语的无主句，而把原句中的主语译成宾语。例如：

1. Now the heart can be safely opened and its valves can berepaired. (现在可以安全地打开心脏，并对心脏瓣膜进行修复。)

2. Much progress has been made in electrical engineering in less than a century. (不到一个世纪，电气工程就取得了很大的进步。)

3. The earth was formed from the same kind of materials that makes up the sun. (构成地球的物质与构成太阳的物质是相同的。)

4. In the watch making industry, the tradition of high precision engineering must be kept. (在钟表制造业中，必须保持高精度工艺的传统。)

当英语被动句中的主语为无生命的名词，又不出现由介词 by 引导的行为主体时，往往可译成汉语的主动句，原句的主语在译文中仍为主语。有时候，可把原主语译成宾语，而把行为主体或相当于行为主体的介词宾语译成主语。例如：

1. Materials may be grouped in several ways. (材料可以按几种方式分类。)

2. Sand casting is used to make large parts. (砂型铸造用于制造大型零件。)

3. Friction can be reduced and the life of the machine prolonged by lubrication. (润滑能减少摩擦，延长机器寿命。)

4. A new kind of magnifying glass is being made in that factory. (那个工厂正在制造一种新的放大镜。)

如果原句中未包含动作的发出者，译成主动句时可从逻辑主语出发，适当增添不确定主语，如"人们""大家""我们"等。凡是着重描述事物的过程、性质和状态的英语被动句，与系表结构相近，往往可译成"是……"的结构。例如：

1. Salt is known to have a very strong corroding effect on metals. (大家知道，盐对金属有很强的腐蚀作用。)

2. Magnesium is found to be the best material for making flash light. (人们发现，镁是制造闪光灯的最好材料。)

3. The volume is not measured in square millimeters. It is measured in cubic millimeters. (体积不是以平方毫米计量的。它是以立方毫米计量的。)

4. The first explosive in the world was made in China. (世界上最早的炸药是在中国制造的。)

(四) 专业英语长难句翻译方法

专业英语长难句的结构特点是后置定语、非谓语动词、同位语、宾语从句、状语从句等修饰性成分多,这些成分相对复杂,可以是单词、短语、也可以是从句,甚至有时一个句子能包含所有这些成分。一般情况下,简单句子中的某些主干成分(主语、谓语、宾语)会放在一起。长句表达虽然有条理性、周密性和严谨性,但是由于修饰性成分的大量存在,这些主干成分常常被修饰性成分隔开了,这就给翻译带来很大困难。

值得注意的是,这些长难句一般都是由一些基本的句型扩展或变化而来,因此翻译的关键在于抓住全句的主干成分,理解句子的中心意思。在此基础上,再去弄清句子的组织构成以及相互间的连接关系,补足附加成分的内容。一般来说,翻译时,可采用下列步骤:

1. 弄清句子的逻辑关系和连接词汇;
2. 辨别句子的主从结构,分切句子内容;
3. 省略句子的修饰性成分,找出句子的主干(主语、谓语、宾语);
4. 根据上下文和全句内容领会句子主要意思;
5. 根据上下层次和前后联系,补齐修饰性成分的内容。

例如:

1. Most MRI systems use a superconducting magnet, which consists of many coils or windings of wire through which a current of electricity is passed, creating a magnetic field of up to 2.0 tesla. (大多数的 MRI 系统使用超导磁体,由许多线圈组成,线圈中有电流通过,产生了一个 2.0 特斯拉的磁场。)

本句的主干为 Most MRI systems use a superconducting magnet, 后面的为定语从句和后置动名词短语,补充说明 superconducting magnet(超导磁体)的构成和作用。

2. When patients slide into an MRI machine, they take with them the billions of atoms that make up the human body. (当患者滑入 MRI,他们带上了构成人体的数以亿计的原子。)

本句的主干为 they take with them the billions of atoms, 前面部分为 when 开头的状语从句,后面部分为定语从句,修饰 atoms.

3. With an estimated 14 percent of the world's population expected to be 65 or older by 2030, the need for medical devices and qualified technicians to fix them will continue to grow. (截至 2030 年,预计世界将有 14% 的人口在 65 岁以上,对于医疗器械及从事医疗器械维修的合格工程师的需求也将继续攀升。)

本句的主干为 the need will continue to grow, 前面部分为表示伴随状况的介词短语。

4. These measures are formulated according to the Law on the Inspection of Imported and Exported Commodities of the People's Republic of China(hereinafter referred to as the Inspection Law) and its implementing regulation and other relevant laws and administrative regulations for the purpose of strengthening the administration of inspection and supervision of the imported medical instruments and safeguarding the human health and life safety.[为加强进口医疗器械检验监督管理,保障人体健康和生命安全,根据《中华人

民共和国进出口商品检验法》(以下简称商检法)及其实施条例和其他有关法律法规规定,制定本办法。]

本句的主干为These measures are formulated.后面部分为以according to, for开头的介词短语,补充说明办法制定的依据和目的。

Chapter 1

Fundamentals of medical devices

Learning objective

In this chapter, basic knowledge of medical devices, such as FDA definition and classification of medical devices, global medical device market, fundamentals of medical devices, will be introduced. Read the following texts, you should meet the following requirements:

(1) Grasp basic concepts, key points, involved professional terms and phrases.

(2) Acquire the knowledge of involved subjects of medical devices.

(3) Understand the basics and progress of related disciplines and their correlation.

Section 1　Basic introduction to medical devices

Chapter 1-
Section 1-PPT

Text 1　Introduction to medical devices

Lead-in questions:

(1) Are definitions of medical devices the same all over the world?

(2) Can you list some examples of medical devices?

(3) How many existing medical devices are there?

(4) What is the annual average growth rate of the global market from 2010 to 2020?

(5) Which company dominates the slow-growing orthopedics market in 2013 in the United States?

In China, as defined by National Medical produces Administration (NMPA), a medical device is any instrument, equipment, appliance, in vitro diagnostic reagent and calibrator, material, and other similar or relevant articles directly or indirectly contacting human body, including necessary computer software; the effectiveness are obtained mainly through physical means other than pharmacological, immunological or metabolic ways, or such ways are involved in but only play auxiliary roles; which is used to achieve the following intended objectives:

Chapter 1-
Section 1-
Text 1-听课文

(Ⅰ) Diagnosis, prevention, monitoring, treatment or alleviation of disease;

(Ⅱ) Diagnosis, monitoring, treatment, alleviation or functional compensation for an injury or handicap conditions;

(Ⅲ) Inspection, substitution, regulation or support of physiological structure or physiological process;

(Ⅳ) Support or sustainability of life;

(Ⅴ) Control of pregnancy;

(Ⅵ) Examination of the sample coming from human body to provide information for medical or diagnostic purpose.

Definitions of medical devices are slightly different all over the world, but main descriptions of medical devices are generally similar. Other than drugs, medical-related instruments, apparatuses, implements, machines, appliances, implants, in vitro reagents or calibrators, software, materials or other similarly related articles are considered as medical devices.

Examples of medical devices range from something as simple as bandages, medical thermometers, or disposable gloves to advanced devices such as equipment (e.g., MRI, ultrasound, CT), implants, and prostheses. There are approximately 1.5 million different devices. Medical devices vary greatly in complexity and application. Current medical devices can replace almost every part of the human body with partial to full functional support. The development of medical devices witnessed steady growth as increasing innovative factors became involved. For example, contact lenses that have been used widely as vision aid and for aesthetic purposes may have additional functions in the future. Google and Novartis are working on smart contact lenses, thus we foresee eyewear that will be able to monitor blood-sugar levels for diabetics.

The global medical device market reached approximately \$363.8 billion in 2013 with annual average growth of the global market around 18.41% from 2010 to 2020. The global market share mainly consisted of North America, Europe, and Asia-Pacific, the rest of the world only makes up 6% of the market share. The top 10 medical device markets by sales revenue in 2012 are United States (36.3%), Japan (9.9%), Germany (7.0%), China (4.3%), France (4.1%), UK (3.0%), Italy (2.6%), Russia (2.1%), Canada (2.1%), and Brazil (1.6%). In recent years, the four BRIC markets (Brazil, Russia, India, and China) have rapidly grown to a total market of US \$26.2 billion in 2012. While Chinese per capita spending remains low in absolute terms at US \$10.5, the Chinese market was the fourth largest market in the world during 2012 and it is forecasted to become the second largest by 2017.

The top 10 companies in the medical device market are Johnson & Johnson (USA), Siemens (Germany), Medtronic (USA), Roche (Switzerland), Covidien (USA), Abbott Laboratories (USA), Stryker (USA), General Electric (USA), Philips (the Netherlands), and Essilor International (France), respectively, in 2013. Orthopedics and cardiovascular are the two largest medical device market areas in the United States. J&J mainly dominates the slow-growing orthopedics market with sales of US \$8.9 billion and a 26.5% market share in 2013, while Medtronic was predicted to maintain in the top spot for cardiology from 2013 to 2020 with sales of US \$8.8 billion and a 22.2% market share in 2013.

The design of medical devices has developed into a major segment of the field of biomedical engineering, which is multidisciplinary, related to biology, medicine, materials and engineering, etc. Most universities have already established bioengineering departments to educate students in this field. Those students join biomedical companies to contribute to the development of the medical device industry upon graduation.

The driving forces behind the development of the medical device market are: ①longer life span with a growing aging population; ②higher quality of life and changing life styles; ③public awareness, all stakeholders are sharing the awareness and responsibility. In almost every country, the population aged 60 years and above is growing faster than any other age group, as a result of both longer life expectancy and declining birth rates. With the development of positive medical devices, patients are diagnosed at early stages and receive timely treatments, significantly improving their quality of life both in the hospital and at home even with permanent injury as compared to before.

For the safety issues, devices must be constructed from safe raw materials. The composition of the materials that make up the device should be clear as any potential leaching materials would present a hazard. We often see the bisphenol A (BPA) free label on water containers. Both BPA and phthalates (PAE) are suspected to be endocrine disruptors. Besides, the design and manufacture processes are also crucial to the safety aspect of the product. Risk analysis, human factor, validation tests, and other related issues have to be taken into considerations.

Words and phrases

instrument	[ˈɪnstrəmənt]	n.	仪器
equipment	[ɪˈkwɪpmənt]	n.	设备
appliance	[əˈplaɪəns]	n.	器具,器械
reagent	[riˈeɪdʒənt]	n.	试剂
calibrator	[ˈkælɪˌbreɪtə]	n.	校准器
material	[məˈtɪəriəl]	n.	材料
software	[ˈsɒftweə(r)]	n.	软件
pharmacological	[ˌfɑːməkəˈlɒdʒɪkl]	adj.	药理学的
immunological	[ˌɪmjʊnəˈlɒdʒɪkl]	adj.	免疫学的
metabolic	[ˌmetəˈbɒlɪk]	adj.	代谢的,新陈代谢的
intended	[ɪnˈtend]	adj.	预期的
diagnosis	[ˌdaɪəɡˈnəʊsɪs]	n.	诊断,检查
prevention	[prɪˈvenʃn]	n.	预防
alleviation	[əˌliːvɪˈeɪʃn]	n.	缓解,减轻
compensation	[ˌkɒmpenˈseɪʃn]	n.	补偿
injury	[ˈɪndʒəri]	n.	损伤,创伤
substitution	[ˌsʌbstɪˈtjuːʃn]	n.	替代,替换
regulation	[ˌreɡjuˈleɪʃn]	n.	调节
physiological	[ˌfɪziəˈlɒdʒɪkl]	adj.	生理的,生理学的

implant	[ɪmˈplɑːnt]	n.	移植物,植入物
bandage	[ˈbændɪdʒ]	n.	绷带
thermometer	[θəˈmɒmɪtə(r)]	n.	温度计,体温表
disposable	[dɪˈspəʊzəbl]	adj.	一次性的
advanced	[ədˈvɑːnst]	adj.	先进的
prosthesis	[prɒsˈθiːsɪs]	n.	假体(如假肢等)
vary	[ˈveəri]	vi.	变化,不同
complexity	[kəmˈpleksəti]	n.	复杂性
application	[ˌæplɪˈkeɪʃn]	n.	应用
replace	[rɪˈpleɪs]	vt.	代替,替换
innovative	[ˈɪnəveɪtɪv]	adj.	创新的,革新的
aesthetic	[iːsˈθetɪk]	adj.	审美的,美的,美学的
foresee	[fɔːˈsiː]	vt.	预见
diabetic	[ˌdaɪəˈbetɪk]	n.	糖尿病患者
market	[ˈmɑːkɪt]	n.	市场
reach	[riːtʃ]	vt.	达到,到达
revenue	[ˈrevənjuː]	n.	收益
forecast	[ˈfɔːkɑːst]	vt.	预测
orthopedics	[ˌɔːθəˈpiːdɪks]	n.	骨科
cardiovascular	[ˌkɑːdiəʊˈvæskjələ(r)]	adj.	心血管的
predict	[prɪˈdɪkt]	vt.	预言,预测
multidisciplinary	[ˌmʌltidɪsəˈplɪnəri]	adj.	多学科的
biology	[baɪˈɒlədʒi]	n.	生物学,生物
medicine	[ˈmedsɪn]	n.	医学
engineering	[ˌendʒɪˈnɪərɪŋ]	n.	工程(学)
university	[ˌjuːnɪˈvɜːsəti]	n.	大学
awareness	[əˈweənəs]	n.	认识,意识
decline	[dɪˈklaɪn]	vt.	下降
composition	[ˌkɒmpəˈzɪʃn]	n.	成分,组分,组成
endocrine	[ˈendəʊkrɪn]	adj.	内分泌(腺)的
manufacture	[ˌmænjuˈfæktʃə]	n.	制造,生产
in vitro diagnostic			体外诊断
physiological process			生理过程
human body			人体
be able to			能够

absolute term	常数项
biomedical engineering	生物医学工程
contribute to	贡献于,为……做出贡献
life span	寿命
as a result of	由于
birth rate	出生率
raw material	原材料,原料
make up	构成,组成

Key sentences

(1) Definitions of medical devices are slightly different all over the world, but main descriptions of medical devices are generally similar.

医疗器械的定义在世界各地略有不同,但医疗器械的主要描述大致相似。

(2) Examples of medical devices range from something as simple as bandages, medical thermometers, or disposable gloves to advanced devices such as equipment (e. g., MRI, ultrasound, CT), implants, and prostheses.

医疗器械的涉及范围包括从简单的绷带、医用温度计或一次性手套,到诸如设备(例如,MRI、超声、CT)、植入物和假体等先进器械。

(3) The global medical device market reached approximately $363.8 billion in 2013 with annual average growth of the global market around 18.41% from 2010 to 2020.

全球医疗器械市场在2013年(销售额)达到约3638亿美元,2010~2020年全球市场年平均增长率约为18.41%。

(4) The design of medical devices has developed into a major segment of the field of biomedical engineering, which is multidisciplinary, related to biology, medicine, materials and engineering, etc..

医疗器械的设计已发展成为生物医学工程领域的一个重要分支,是一个涉及生物学、医学、材料和工程等的多学科领域。

(5) For the safety issues, devices must be constructed from safe raw materials.

对于安全问题,器械必须由安全的原材料构成。

Learning test

Choose the following words below to fill in the blanks, change the form if necessary.

reagent	injury	revenue	disposable	prevention
instrument	diagnosis	reach	vary	innovative

(1) Advertising _____ is the biggest item in the company's earnings.

(2) The practitioners of China's medical device market has _____ 300 thousands people.

(3) This precise _____ can be used to detect extremely weak signal.

(4) The use of _____ syringes should also be strictly disinfected.

(5) The chemical _____ is used in cell toxicity test.

(6) Medical devices provide assistance in the _____ of diseases.

(7) This method is effective for the _____ of the occurrence of infectious diseases.

(8) He was out of work owing to a physical _____ .

(9) Mainland companies have developed plenty of _____ products in recent years.

(10) The series of MRI machines _____ in size and openness.

Translation

1. Translate the following words into Chinese.

(1) device   (2) instrument

(3) diagnosis   (4) software

(5) manufacture   (6) engineering

(7) treatment   (8) material

(9) prosthesis   (10) biology

2. Translate the following phrases into Chinese.

(1) human body   (2) biomedical engineering

(3) life span   (4) birth rate

(5) in vitro diagnostic   (6) physiological process

Practice

1. Translate the following sentences into Chinese.

(1) Medical devices vary greatly in complexity and application.

(2) Those students join biomedical companies to contribute to the development of the medical device industry upon graduation.

(3) In almost every country, the population aged 60 years and above is growing faster than any other age group, as a result of both longer life expectancy and declining birth rates.

(4) The composition of the materials that make up the device should be clear as any potential leaching materials would present a hazard.

(5) Besides, the design and manufacture processes are also crucial to the safety aspect of the product.

2. Translate the following sentences into English.

(1) 本产品属于医疗器械,包括报警装置和制动装置两部分。

(2) 药品不是医疗器械。

(3) 在丢弃医疗器械之前请先看一下制造商的使用说明书。

(4) 由于一些不可控因素,今年的销售额下降了。

> **Culture salon**
>
> ### Single-use medical device reprocessing
>
> Single-use medical device reprocessing is the disinfection, cleaning, remanufacturing, testing, packaging and labeling, and sterilization among other steps, of a used (or, in some cases, a device opened from its original packaging but unused) medical device to be put in service again. All reprocessed medical devices must meet strict cleaning, functionality, and sterility specifications prior to use. Currently, approximately 2% of all "Single Use Device" (SUDs) on the U.S. market are eligible for reprocessing by a qualified third-party vendor. The U.S. revenue for reprocessed devices (not SUDs) is estimated to be around $400 million annually.

Text 2　FDA: definition and classification of medical devices

Lead-in questions:

(1) Are electronic radiation emitting products medical devices?

(2) Does a medical device achieve its purpose through chemical action?

(3) Are drugs regulated as medical devices?

(4) What is the basis of medical device classification?

(5) Can you give some examples of Class Ⅲ devices according to the text?

Medical devices range from simple tongue depressors and bedpans to complex programmable pacemakers with micro-chip technology and laser surgical devices. In addition, medical devices include *in vitro* diagnostic products, such as general-purpose lab equipment, reagents, and test kits, which may include monoclonal antibody technology. Certain electronic radiation emitting products with medical application and claims meet the definition of medical device. Examples include diagnostic ultrasound products, X-ray machines and medical lasers.

The United States FDA has a clear and strict definition of medical equipment as follows. A medical device is a medical machine, contrivance, implant, in vitro reagent, or other similar or related article, including a component part, or accessory that is:

(1) Recognized in the official National Formulary, or the United States Pharmacopoeia, or any supplement to them.

(2) Intended for use in the diagnosis of disease or other conditions, or in the cure, mitigation, treatment, or prevention of disease, in man or other animals.

(3) Intended to affect the structure or any function of the body of man or other animals, and does not achieve any of its primary purpose through chemical action within or on the body of man or other animals

and does not depend on metabolic action to achieve its primary purpose.

This definition provides a clear distinction between a medical device and other FDA regulated products such as drugs. If the primary intended use of the product is achieved through chemical action or by being metabolized by the body, the product is usually a drug. FDA's Center for Devices and Radiological Health (CDRH) is responsible for regulating medical devices to protect and promote the public health. They assure that patients and providers have timely and continued access to safe, effective, and high-quality medical devices and safe radiation-emitting products. In cases where it is not clear whether a product is a medical device there are procedures in place to use DICE(Division of Industry and Consumer Education) staff directory to assist you in making a determination.

Federal law(Federal Food, Drug, and Cosmetic Act, section 513) established the risk-based device classification system for medical devices. Each device is assigned to one of three regulatory classes: Class Ⅰ, Class Ⅱ or Class Ⅲ, based on the level of control necessary to provide reasonable assurance of its safety and effectiveness.

Regulations differ by class based on their complexity or the potential hazards in the event of malfunction. As device class increases from Class Ⅰ, Class Ⅱ to Class Ⅲ, the regulatory controls also increase, with Class Ⅰ devices subject to the least regulatory control, and Class Ⅲ devices subject to the most stringent regulatory control. Class Ⅰ devices are the least likely to cause major bodily harm or death in the event of failure, and are subjected to less stringent regulations than are devices categorized as Class Ⅱ or Class Ⅲ.

The regulatory controls for each device class include:
- Class Ⅰ (low to moderate risk): general controls.
- Class Ⅱ (moderate to high risk): general controls and special controls.
- Class Ⅲ (high risk): general controls and premarket approval(PMA).

Class Ⅰ devices are not intended to help support or sustain life or be substantially important in preventing impairment to human health, and may not present an unreasonable risk of illness or injury. Class Ⅱ devices are those for which general controls alone cannot assure safety and effectiveness, and existing methods are available that provide such assurances. In addition to complying with general controls, Class Ⅱ devices are also subject to special controls. Special controls may include special labeling requirements, mandatory performance standards and postmarket surveillance. Devices in Class Ⅱ are held to a higher level of assurance than Class Ⅰ devices, and are designed to perform as indicated without causing injury or harm to patient or user. Examples of Class Ⅱ devices include acupuncture needles, powered wheelchairs, infusion pumps, air purifiers, and surgical drapes.

A Class Ⅲ device is one for which insufficient information exists to assure safety and effectiveness solely through the general or special controls sufficient for Class Ⅰ or Class Ⅱ devices. Such a device needs premarket approval, a scientific review to ensure the device's safety and effectiveness, in addition to the general controls of Class Ⅰ. Class Ⅲ devices are usually those that support or sustain human life, are of substantial importance in preventing impairment of human health, or present a potential, unreasonable risk of illness or injury. Examples of Class Ⅲ devices that currently require a premarket notification include implantable pacemaker, pulse generators, HIV diagnostic tests kits, automated external defibrillators, and endosseous implants.

Words and phrases

bedpan	[ˈbedˌpæn]	n.	便盆
pacemaker	[ˈpeɪsmeɪkə(r)]	n.	心脏起搏器
laser	[ˈleɪzə(r)]	n.	激光
equipment	[ɪˈkwɪpmənt]	n.	设备
reagent	[riˈeɪdʒənt]	n.	反应物,试剂
monoclonal	[ˌmɒnəʊˈkləʊnəl]	adj.	单克隆的
antibody	[ˈæntibɒdi]	n.	抗体
eletronic	[ɪˌlekˈtrɒnɪk]	adj.	电子的
radiation	[ˌreɪdiˈeɪʃn]	n.	辐射
ultrasound	[ˈʌltrəsaʊnd]	n.	超声,超声波
machine	[məˈʃiːn]	n.	机器,仪器
definition	[ˌdefɪˈnɪʃn]	n.	定义
contrivance	[kənˈtraɪvəns]	n.	设备,机器,装置,器件
article	[ˈɑːtɪkl]	n.	物质,材料
component	[kəmˈpəʊnənt]	n.	元件,零件,部件
accessory	[əkˈsesəri]	n.	附件
official	[əˈfɪʃl]	adj.	官方的,正式的
formulary	[ˈfɔːmjʊləri]	n.	处方集,药品集
pharmacopeia	[ˌfɑːməkəʊˈpeɪə]	n.	药典
supplement	[ˈsʌplɪmənt]	n.	增刊,副刊,补充
mitigation	[ˌmɪtɪˈgeɪʃn]	n.	缓解,减轻
treatment	[ˈtriːtmənt]	n.	治疗
affect	[əˈfekt]	vt.	影响
metabolic	[ˌmetəˈbɒlɪk]	adj.	新陈代谢的
drug	[drʌg]	n.	药物,药品
timely	[ˈtaɪmli]	adj.	及时的,适时的
quality	[ˈkwɒləti]	n.	质量
consumer	[kənˈsjuːmə(r)]	n.	消费者,顾客
regulation	[ˌregjuˈleɪʃn]	n.	管理,控制
harzard	[ˈhæzəd]	n.	风险,危险

malfunction	[ˌmælˈfʌŋkʃn]	n.	故障,功能障碍
requirement	[rɪˈkwaɪəmənt]	n.	要求,需求
mandatory	[ˈmændətəri]	adj.	强制性的
performance	[pərˈfɔːrməns]	n.	性能
surveillance	[sɜːˈveɪləns]	n.	监督
scientific	[ˌsaɪənˈtɪfɪk]	adj.	科学的
review	[rɪˈvjuː]	n.	审查,审核
pulse	[pʌls]	n.	脉冲
generator	[ˈdʒenəreɪtə(r)]	n.	发生器
defibrillator	[diːˈfɪbrɪleɪtə(r)]	n.	(电击)除颤器
micro-chip			微芯片
monoclonal antibody			单克隆抗体
electronic radiation			电子辐射
National Formulary			国家处方集
United States Pharmacopoeia			美国药典
chemical action			化学作用
depend on			取决于
Center for Devices and Radiological Health(CDRH)			器械和辐射健康中心
responsible for			负责
access to			获得,得到
high-quality			高质量
DICE(Division of Industry and Consumer Education)			工业和消费者教育分部
subject to			受到,遭受

Key sentences

(1) Medical devices range from simple tongue depressors and bedpans to complex programmable pacemakers with micro-chip technology and laser surgical devices.

医疗器械范围从简单的压舌板、便盆到复杂的采用微芯片技术的可编程起搏器和激光手术设备。

(2) FDA's Center for Devices and Radiological Health(CDRH) is responsible for regulating medical devices to protect and promote the public health.

FDA 器械和辐射健康中心负责医疗器械的监管,以保护和促进公众健康。

(3) Each device is assigned to one of three regulatory classes: Class Ⅰ, Class Ⅱ or Class Ⅲ, based on the level of control necessary to provide reasonable assurance of its safety and effectiveness.

根据为安全性和有效性提供合理保证所必需的控制水平,每种器械被归为三个监管类别之一:Ⅰ类,Ⅱ类或Ⅲ类。

(4) Class Ⅰ devices are not intended to help support or sustain life or be substantially important in pre-

venting impairment to human health, and may not present an unreasonable risk of illness or injury.

Ⅰ类器械不是用于帮助支持或维持生命,或在防止人体健康损害方面至关重要,并且不会产生不合理的疾病或损伤风险。

(5) A Class Ⅲ device is one for which insufficient information exists to assure safety and effectiveness solely through the general or special controls sufficient for Class Ⅰ or Class Ⅱ devices.

Ⅲ类设备是指没有足够的信息(证明),仅通过Ⅰ类或Ⅱ类设备的一般或特殊控制可确保其安全性和有效性的器械。

Learning test

Choose the following words below to fill in the blanks, change the form if necessary.

| include | complex | assure | regulate | electronic |
| affect | effective | treatment | assist | consumer |

(1) The doctor's _____ has played a great role in the recovery of his illness.

(2) This circuit contains thousands of _____ components.

(3) This project involves a lot of _____ technical problems.

(4) I can _____ you that the temperature meter is high-quality.

(5) Medical staff in hospitals mainly _____ doctors, nurses, technologist.

(6) Alcohol greatly _____ the functioning of the stomach.

(7) He will _____ you to complete the task.

(8) The reduced _____ demand is also affecting company profits.

(9) This medicine is _____ if used within three years.

(10) Contact lens is _____ as medical device.

Translation

1. Translate the following words into Chinese.

(1) definition (2) performance

(3) drug (4) pharmacopeia

(5) malfunction (6) ultrasound

(7) mandatory (8) surveillance

(9) radiation (10) quality

2. Translate the following phrases into Chinese.

(1) access to (2) high-quality

(3) micro-chip (4) responsible for

(5) electronic radiation (6) depend on

Practice

1. Translate the following sentences into Chinese.

(1) This definition provides a clear distinction between a medical device and other FDA regulated prod-

ucts such as drugs.

(2) They assure that patients and providers have timely and continued access to safe, effective, and high-quality medical devices and safe radiation-emitting products.

(3) Regulations differ by class based on their complexity or the potential hazards in the event of malfunction.

(4) Class Ⅱ devices are those for which general controls alone cannot assure safety and effectiveness, and existing methods are available that provide such assurances.

(5) Examples of Class Ⅱ devices include acupuncture needles, powered wheelchairs, infusion pumps, air purifiers, and surgical drapes.

2. Translate the following sentences into English.

(1)体温计是一种常见的家用医疗器械。

(2)性能和价格是家用医疗器械销售的决定性因素。

(3)MRI 的辐射剂量对患者是安全的。

(4)扫描仪的核心部件是一个微芯片。

(5)在中国,药品的监管比医疗器械更严格。

Culture salon

Risk assessment management for a new medical device

Safety of a medical device is of the highest priority for any regulatory body and is the basis for product categorization and control. While a medical device being safe is the desired outcome, definition of what constitutes safe, and the procedures and actions to ensure safety require analyzing risk and its management. Thus identification and management of risk is a vital component in meeting regulatory requirements. Risk assessment is about balancing the product's benefit against risk. Such benefit-risk consideration may also influence the chosen path for medical device regulatory approval. Safety and risk arising from the various stages of the product's life cycle is not just limited to medical device regulations but may also be required from other related regulations such as environment and electronic devices.

Chapter 1-
Section 1-
Text 2-看译文

Text 3　Global medical device market

Lead-in questions:

(1) Why is the globalization of medical device industry intensified?

(2) What is the total sales of medical devices in China in 2014?

(3) Which countries are the top three leading suppliers of medical equipment in the world?

(4) What are the diagnostic imaging devices?

(5) Can you list five famous medical device companies according to the text?

Demand for medical devices grows faster than expected in recent years. World market totalled $209 billion in 2006, and grew with an average annual rate of 8%~9% through 2010. Medical device is a highly dynamic industry, where dramatic innovations and developments are taking place every day. Driven by both the increasing demand in overseas markets and companies' ambition to pursue profit globally, the globalization of medical device industry is intensified. So is the need to understand the global market.

Chapter 1-Section 1-Text 3-听课文

The sales of medical devices in China have increased rapidly over the last decade, reaching a total value of CNY 255.6 billion (EUR 36 billion) at the end of 2014. On account of an ageing population, increased awareness of diseases and growing expenditure on healthcare, the sector is expected to continue to expand in the coming years. Competition in the market is strong, with foreign companies concentrated in the high-end segment and Chinese companies occupying the mid-to-low end section.

Regulation-wise, even though China's healthcare system and medical device regulations can often be challenging to understand, the PRC's government has recently implemented a set of new measures aimed at increasing the standardization of the market and promoting its development. Such trends will provide SMEs (Small and Medium Enterprises) of EU (European Union), North American and Japan etc.. With a new set of opportunities, provided that the right preparation is made to face existing challenges.

The U.S., Japan, and Germany are the top three leading suppliers of medical equipment in the world. There are a significant number of companies that have established a presence by setting up their regional headquarters there in an effort to be closer to and better serve their customers. The U.S. enjoys a good reputation and is recognized by the industry as technologically superior, providing high quality, advanced and reliable equipment. However, the U.S. is not price competitive in the lower-end consumables, typically supplied by manufacturers in more countries throughout Asia.

Medical devices market is the fastest increasing market worldwide due to its extensive use in hospitals, research and development centers, educational institutes and clinical laboratories. Medical devices market is majorly classified into seven categories including diagnostic imaging devices, drug delivery devices, molecular diagnostic, surgical devices, monitoring devices, bioimplants and neurostimulation, and others. Following are the highlights of the various segments of the medical devices market.

(1) Diagnostic imaging devices

Diagnostic imaging devices are the devices used for diagnosing disease condition. Diagnostic devices market is further differentiated into:

• Magnetic resonance imaging (MRI)

• Computed tomography (CT) scan

• X-ray systems

• Ultrasound imaging systems

• Nuclear imaging systems [Positronemission tomography (PET), Single photon emission computed tomography(SPECT)]

  • Electrocardiography(ECG)

(2) Drug delivery devices

Drug delivery devices are the devices and accessories helpful in performing drug administration. This market consists of:

  • Inhaler devices(dry powder inhalers and metered dose inhalers)

  • Nebulizers

  • Microneedles

  • Pre-filled syringes(PFS)(glass PFS and plastic PFS)

  • Pen injectors

  • Auto injectors

  • Needle free injectors

(3) Molecular diagnostic devices

Molecular diagnostic devices are specialized to test nucleic acid or proteins. This category of medical devices includes:

  • Polymerase chain reaction(PCR)

  • Microarrays

(4) Surgical devices

Surgical devices are devices majorly used during both invasive and non-invasive surgery. This category of medical devices includes:

  • Handheld instruments(forceps, spatula, dialator)

  • Guiding devices(guiding catheters and guidewires)

  • Auxiliary devices(cannula, staplers, clamp, closure)

  • Balloon inflation systems

  • Mechanical cutters(trocar, rasp, scissors)

(5) Monitoring devices

Monitoring devices are the devices used for keeping the record of patient after the treatment. This category of devices consists of:

  • Heart rate monitors(holter monitors, event monitors, implantable loop monitors)

  • Blood pressure monitoring system

  • Blood glucose monitors

  • Thermometers

  • Body mass index analyzers

(6) Bioimplants and neurostimulation devices

Bioimplants and neurostimulation market include the devices which are implanted inside the body for treatment of the disease condition. This segment of market is differentiated into:

- Cardiovascular implants(pacing devices, stents, structural cardiac implants)
- Spinal implants(thoracolumbar implants, intervertebral spacers, machined allograft spacers, cervical implants)
- Orthopedics and trauma(reconstructive joint replacements, orthobiologics, trauma implants)
- Dental implants(plate form and root form dental implants)
- Ophthalmic implants(intraocular lens, glaucoma and other lenses)
- Neurostimulator implants(cortical stimulators, deep brain stimulators, sacral nerve stimulators, spinal stimulators)
- Hearing aids
- Cosmetic implants

(7) Others:

This category of medical devices include all the accessories and materials which are helpful in performing medical procedures such as:

- Syringes
- Gloves
- Medical and surgical sterilizers
- Needles

The medical devices market is showing continuous growth driven by various factors such as rise in ageing population, innovations in medical devices, changing lifestyles and increasing awareness about the medical conditions and available treatments. Diagnostic imaging devices segment is the major revenue generating market for medical devices market due to extensive research and development. Other major revenue segment for medical devices market is bioimplants and neurostimulation specifically in markets of North America and Europe due to increasing demand for cosmetic implants and economic stability in these regions.

North America leads in the medical devices market followed by the Europe majorly because of ageing population. According to the data of Centers for Disease Control and Prevention(CDC), in 2010 40 million people aged 65 and above in United States accounting for 13% of the total population and CDC projected that around 20% of the population in the country would be under this category by 2030. Market of Asia-Pacific is also growing at faster pace for medical devices majorly driven by the growing economies of Japan, China and India. The key players in the market include Stryker Corporation, Johnson & Johnson, Philips Healthcare, Siemens Healthcare, GE Healthcare, Toshiba Medical Systems Corporation, Boston Scientific Corporation, St. Jude Medical, Inc., Biomet, Inc. and Smith and Nephew. Among all of the above Stryker Corporation and Johnson & Johnson are currently dominating medical devices markets.

Words and phrases

demand	[dɪˈmænd]	n.	需求,要求
expect	[ɪkˈspekt]	vt.	期望,预期
market	[ˈmɑːkɪt]	n.	市场
dynamic	[daɪˈnæmɪk]	adj.	充满活力的,朝阳的
dramatic	[drəˈmætɪk]	adj.	巨大的,惊人的
innovation	[ˌɪnəˈveɪʃən]	n.	创新,改革
ambition	[æmˈbɪʃən]	n.	抱负,志向
pursue	[pəˈsjuː]	vt.	追求
profit	[ˈprɒfɪt]	n.	利润,收益
globalization	[ˌgləʊbəlaɪˈzeɪʃn]	n.	全球化
intensify	[ɪnˈtensɪfaɪ]	vt.	增强,加剧
expenditure	[ɪkˈspendɪtʃə(r)]	n.	花费,支出
healthcare	[ˈhelθkeə]	n.	医疗保健
sector	[ˈsektə(r)]	n.	行业,产业
promote	[prəˈməʊt]	vt.	促进,推进,提升
extensive	[ɪkˈstensɪv]	adj.	广泛的
imaging	[ˈɪmɪdʒɪŋ]	n.	成像
molecular	[məˈlekjələ(r)]	adj.	分子的
bioimplant	[biːəʊɪmplɑːnt]	n.	生物植入物
neurostimulation	[nˈjʊərəʊˌstɪmjʊˈleɪʃn]	n.	神经刺激
highlight	[ˈhaɪˌlaɪt]	n.	亮点,重点
nuclear	[ˈnjuːklɪə(r)]	adj.	原子核的
photon	[ˈfəʊtɒn]	n.	光子
accessory	[ækˈsesəri]	n.	附件
inhaler	[ɪnˈheɪlə(r)]	n.	吸入器
nebulizer	[ˈnebjəˌlaɪzə]	n.	喷雾器,雾化器
microneedle	[maɪkrəʊˈniːdl]	n.	微针
syringe	[sɪˈrɪndʒ]	n.	注射器
injector	[ɪnˈdʒektə]	n.	注射器
protein	[ˈprəʊtiːn]	n.	蛋白质
microarray	[ˌmaɪkrəʊəˈreɪ]	n.	微阵列
invasive	[ɪnˈveɪsɪv]	adj.	创伤性的,侵害性的
mechanical	[məˈkænɪkl]	adj.	机械的
monitor	[ˈmɒnɪtə(r)]	n. & vt.	监护仪,监测器,监测

thermometer	[θərˈmɑːmɪtə(r)]	n.	温度计,体温计
cardiovascular	[ˌkɑːdiəʊˈvæskjələ(r)]	adj.	心血管的
stent	[stent]	n.	支架
spinal	[ˈspaɪnl]	adj.	脊柱的
orthopedics	[ˌɔːθəˈpiːdɪks]	n.	骨科
trauma	[ˈtrɔːmə]	n.	损伤
joint	[dʒɔɪnt]	n.	关节
dental	[ˈdentl]	adj.	牙科的,牙齿的
ophthalmic	[ɒfˈθælmɪk]	adj.	眼的,眼科的
revenue	[ˈrevənuː]	n.	收益,收入
take place			发生
high-end			高端,高档
mid-to-low end			中低端
electrocardiography (ECG)			心电图
labor-abundant			劳动力资源丰富的
drug delivery			给药
magnetic resonance imaging (MRI)			磁共振成像
computed tomography (CT) scan			计算机断层扫描
ultrasound imaging system			超声成像系统
nuclear imaging system			核(医学)成像系统
positron emission tomography (PET)			正电子发射计算机断层扫描
single photon emission computed tomography (SPECT)			单光子发射计算机断层扫描
pre-filled syringe (PFS)			预填充注射器
nucleic acid			核酸
polymerase chain reaction (PCR)			聚合酶链反应
body mass index (BMI)			体重指数
heart rate			心率
blood pressure monitoring system			血压监测系统
blood glucose			血糖
hearing aid			助听器

Key sentences

(1) Driven by both the increasing demand in overseas markets and companies' ambition to pursue profit globally, the globalization of medical device industry is intensified.

在海外市场需求增长和公司追求全球利润的抱负的推动下,医疗器械行业的全球化加剧。

(2) On account of an ageing population, increased awareness of diseases and growing expenditure on

healthcare, the sector is expected to continue to expand in the coming years.

由于人口老龄化、对疾病的认识增强,医疗保健支出不断增长,未来几年本行业将继续发展。

(3) The U.S., Japan and Germany are the top three leading suppliers of medical equipment in the world.

美国、日本和德国是世界上前三名医疗设备主要供应方。

(4) Medical devices market is majorly classified into seven categories including diagnostic imaging devices, drug delivery devices, molecular diagnostic, surgical devices, monitoring devices, bioimplants and neurostimulation, and others.

医疗器械市场主要分为七类,包括影像诊断器械、给药器械、分子诊断、手术器械、监护器械、生物植入物和神经刺激、其他器械。

(5) The medical devices market is showing continuous growth driven by various factors such as rise in aging population, innovations in medical devices, changing lifestyles and increasing awareness about the medical conditions and available treatments.

受诸如人口老龄化、医疗器械创新、生活方式等改变,以及对医疗条件和现有治疗方法的认识的提高等因素驱动,医疗器械市场呈现持续增长。

Learning test

Choose the following words below to fill in the blanks, change the form if necessary.

| market | invasive | healthcare | expenditure | imaging |
| dynamic | demand | ambition | protein | highlight |

(1) The _____ of this article is to describe the future direction of economic development.

(2) Excluding water, half of the body's weight is _____ .

(3) Private enterprises usually have the _____ to chase profits.

(4) The monthly _____ of our family is around 3000 RMB.

(5) The two big companies control 50% of the salt _____ .

(6) China's _____ is open to the world in compliance with the relevant laws of China.

(7) Affordable _____ is the major problem in rural area.

(8) The _____ principles of CT and MRI are different.

(9) Stent implantation is a minimally _____ surgery.

(10) Economy is _____ in Guangzhou.

(11) The _____ of the doctors is increasing in China.

Translation

1. Translate the following words into Chinese.

(1) innovation   (2) profit

(3) healthcare   (4) imaging

(5) dental   (6) mechanical

(7) trauma　　　　　　　　　(8) orthopedics

(9) joint　　　　　　　　　　(10) ophthalmic

2. Translate the following phrases into Chinese.

(1) heart rate　　　　　　　　(2) high-end

(3) hearing aid　　　　　　　(4) magnetic resonance imaging (MRI)

(5) electrocardiography　　　 (6) blood glucose

Practice

1. Translate the following sentences into Chinese.

(1) Demand for medical devices grows faster than expected in recent years.

(2) Following are the highlights of the various segments of the medical devices market.

(3) Diagnostic imaging devices are the devices used for diagnosing disease condition.

(4) Monitoring devices are the devices used for keeping the record of patient after the treatment.

(5) Surgical devices are devices majorly used during both invasive and non-invasive surgery.

2. Translate the following sentences into English.

(1) 2020年全球医疗器械市场将达到4775亿美元。

(2) 心电图机则是记录人体生理电信号的仪器。

(3) 戴有心脏起搏器的患者不能作MRI检查。

(4) 血管支架可通过微创手术置入血管内。

(5) 本公司是专业从事医疗器械销售及售后服务的公司。

Culture salon

Import and export of medical devices in China

According to customs data, in 2014, the import and export value of medical devices was approximately USD 35.7 billion (EUR 31.2 billion). In 2014, China's medical device total import value was USD 15.7 billion (EUR 13.7 billion), the growth rate is around 15%. Joint ventures (JV) and wholly foreign-owned enterprises (WFOE) were the main importers of medical devices, making up almost 40% of China's total import value of medical devices in 2014. Diagnostic medical devices totalled approximately 70% of the medical devices import value, the highest amongst all imported medical devices.

Chapter 1-Section 1-Text 3-看译文

Chapter 1-Section 1-习题

Section 2 Fundamentals of medical devices

Text 1 A brief introduction to electronic basics

Chapter 1-
Section 2-PPT

Lead-in questions:

(1) What is Electronics?
(2) What are the applications of Electronics?
(3) How many types of electronic circuits?
(4) What is the feature of a parallel circuit?
(5) Can you list some examples of mixed-signal circuits?

Chapter 1-
Section 2-
Text 1-听课文

Electronics is the science of controlling electrical energy, in which the electrons have a fundamental role. Electronics deals with electrical circuits that involve active electrical components such as vacuum tubes, transistors, diodes, integrated circuits, associated passive electrical components, and interconnection technologies. Commonly, electronic devices contain circuits consisting primarily or exclusively of active semiconductors supplemented with passive elements, such a circuit is described as an electronic circuit.

Electronics is widely used in information processing, telecommunication, and signal processing. The ability of electronic devices to act as switches makes digital information processing possible. Interconnection technologies such as circuit boards, electronic packaging technology, and other varied forms of communication infrastructure complete circuit functionality and transform the mixed components into a regular working system.

Electronics is distinct from electrical and electro-mechanical science and technology, which deal with the generation, distribution, switching, storage, and conversion of electrical energy to and from other energy forms using wires, motors, generators, batteries, switches, relays, transformers, resistors, and other passive components.

An electronic component is any basic discrete device or physical entity in an electronic system used to affectelectrons or their associated electric fields. Electronic components have a number of electrical terminals or leads as shown in Figure 1-1. These leads connect to create an electronic circuit with a particular function (for example an amplifier, radio receiver, or oscillator). Basic electronic components may be packaged discretely, as arrays or networks of like components, or integrated inside of packages such as semiconductor or integrated circuits, hybrid integrated circuits, or thick film devices.

An electronic circuit is composed of individual electronic components, such as resistors, transistors, capacitors, inductors and diodes, connected by conductive wires or traces through which electric current can

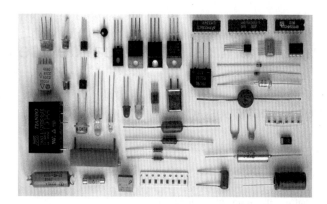

Figure 1-1　Various common electronic components.

flow. The combination of components and wires allows various simple and complex operations to be performed: signals can be amplified, computations can be performed, and data can be moved from one place to another.

An electronic circuit can usually be categorized as an analog circuit, a digital circuit, or a mixed-signal circuit(a combination of analog circuits and digital circuits).

(1) Analog electronic circuits

Analog electronic circuits are those in which current or voltage may vary continuously with time to correspond to the information being represented. The basic components of analog circuits are wires, resistors, capacitors, inductors, diodes, and transistors. Analog circuitry is constructed from two fundamental building blocks: series and parallel circuits.

In a series circuit, the same current passes through a series of components. A string of Christmas lights is a good example of a series circuit: if one goes out, they all do.

In a parallel circuit, all the components are connected to the same voltage, and the current divides between the various components according to their resistance.

Analog circuit analysis employs Kirchhoff's circuit laws: all the currents at a node(a place where wires meet), and the voltage around a closed loop of wires is 0. Wires are usually treated as ideal zero-voltage interconnections.

(2) Digital electronic circuits

Digital circuits are electric circuits based on a number of discrete voltage levels. Digital circuits are the most common physical representation of Boolean algebra, and are the basis of all digital computers.

Because of this discretization, relatively small changes to the analog signal levels due to manufacturing tolerance, signal attenuation or noise do not leave the discrete envelope, and as a result are ignored by signal state sensing circuitry.

Most digital circuits use a binary system with two voltage levels labeled "0" and "1". Often logic "0" will be a lower voltage and referred to as "Low" while logic "1" is referred to as "High".

Digital techniques are useful because it is easier to get an electronic device to switch into one of a number of known states than to accurately reproduce a continuous range of values. Computers, electronic clocks, and programmable logic controllers (used to control industrial processes) are constructed of digital circuits. Digital signal processors are another example.

Digital electronic circuits are usually made from large assemblies of logic gates, simple electronic representations of Boolean logic functions. A logic gate is generally created from one or more electrically controlled switches, usually transistors but thermionic valves have seen historic use. The output of a logic gate can, in turn, control or feed into more logic gates.

(3) Mixed-signal circuits

Mixed-signal or hybrid circuits contain elements of both analog and digital circuits. Examples include comparators, timers, phase-locked loops, analog-to-digital converters, and digital-to-analog converters. Most modern radio and communications circuitry uses mixed signal circuits. For example, in a receiver, analog circuitry is used to amplify and frequency-convert signals so that they reach a suitable state to be converted into digital values, after which further signal processing can be performed in the digital domain.

Words and phrases

fundamental	[ˌfʌndəˈmentl]	adj. & n.	基础的,基本的,基础
electronics	[ɪˌlekˈtrɒnɪks]	n.	电子学
electron	[ɪˈlektrɒn]	n.	电子
component	[kəmˈpəʊnənt]	n.	(电子)元件
transistor	[trænˈzɪstə(r)]	n.	晶体管
diode	[ˈdaɪəʊd]	n.	二极管
electronic	[ɪˌlekˈtrɒnɪk]	adj.	电子的
contain	[kənˈteɪn]	vt.	包含
consist	[kənˈsɪst]	vi.	包括,由…组成
circuit	[ˈsɜːkɪt]	n.	电路,线路
exclusively	[ɪkˈskluːsɪvlɪ]	adv.	完全
passive	[ˈpæsɪv]	adj.	被动的;无源的
semiconductor	[ˌsemikənˈdʌktə(r)]	n.	半导体
telecommunication	[ˌtelɪkəˌmjuːnɪˈkeɪʃn]	n.	电信,电通信,远程通讯
switch	[swɪtʃ]	n.	开关
digital	[ˈdɪdʒɪtl]	adj.	数字的
technology	[tekˈnɒlədʒi]	n.	技术
packaging	[ˈpækɪdʒɪŋ]	n.	封装
functionality	[ˌfʌŋkʃəˈnæləti]	n.	功能性,功能
transform	[trænsˈfɔːm]	vt.	转换,变换
generation	[ˌdʒenəˈreɪʃn]	n.	产生

storage	[ˈstɔːrɪdʒ]	n.	存储
conversion	[kənˈvɜːʃn]	n.	转换,变换
wire	[ˈwaɪə(r)]	n.	电线,导线
motor	[ˈməʊtə(r)]	n.	马达,发动机
battery	[ˈbætəri]	n.	电池
relay	[ˈriːleɪ]	n.	继电器
transformer	[trænsˈfɔːmə(r)]	n.	变压器
resistor	[rɪˈzɪstə(r)]	n.	电阻,电阻器
discrete	[dɪˈskriːt]	adj.	分离的,非连续的
terminal	[ˈtɜːmɪnl]	n.	(电路的)端子
amplifier	[ˈæmplɪfaɪə(r)]	n.	放大器
oscillator	[ˈɒsɪleɪtə(r)]	n.	振荡器
analog	[ˈænəlɔːg]	n.	模拟
hybrid	[ˈhaɪbrɪd]	adj.	混合的
capacitor	[kəˈpæsɪtə(r)]	n.	电容器,电容
inductor	[ɪnˈdʌktə]	n.	电感
current	[ˈkʌrənt]	n.	电流
voltage	[ˈvəʊltɪdʒ]	n.	电压
node	[nəʊd]	n.	节点
value	[ˈvæljuː]	n.	值,数值
output	[ˈaʊtpʊt]	n.	输出
vacuum tube			真空管
integrated circuit			集成电路
consist of			包括,由……组成
passive element			无源元件
interconnection technology			互联技术
circuit board			电路板
electronic packaging technology			电子封装技术
deal with			处理,处置
distinct from			有别于,不同于
electric field			电场
analog circuit			模拟电路
digital circuit			数字电路
series circuit			串联电路
parallel circuit			并联电路

Kirchhoff's circuit laws	基尔霍夫电路定律
Boolean algebra	布尔代数
binary system	二进制系统
logic gate	逻辑门

Key sentences

(1) Electronics is the science of controlling electrical energy, in which the electrons have a fundamental role.

电子学是控制电能的科学,其中电子起着根本性的作用。

(2) An electronic component is any basic discrete device or physical entity in an electronic system used to affect electrons or their associated electric fields.

电子元件是电子系统中用于影响电子或其相关电场的基础分立器件或物理实体。

(3) Analog electronic circuits are those in which current or voltage may vary continuously with time to correspond to the information being represented.

模拟电子电路是指电流或电压可根据所表示的信息随时间连续变化的电路。

(4) Digital circuits are electric circuits based on a number of discrete voltage levels.

数字电路是基于若干离散电压电平的电路。

(5) Mixed-signal or hybrid circuits contain elements of both analog and digital circuits.

混合信号或混合电路同时包含模拟和数字电路的元件。

Learning test

Choose the following words below to fill in the blanks, change the form if necessary.

current   resistor   contain   digital   passive

value   electronic   fundamental   transform   technology

(1) Most vegetables _____ a number of vitamins.

(2) The calculated _____ of 8 divided by 4 is 2.

(3) In an analogy circuit, there are a variety of _____ components.

(4) The _____ increased rapidly because of a sudden short circuit.

(5) Your body can _____ food into energy through metabolic action.

(6) Capacitors are _____ electronic components.

(7) Advanced artificial intelligence(AI) _____ is developing rapidly in recent years.

(8) The signal will be converted into _____ code.

(9) The physical _____ of MRI is quite complicated and difficult to understand.

(10) In a closed circuit, a voltage is generated across a _____ .

Translation

1. Translate the following words into Chinese.

(1) circuit   (2) analog

(3) voltage   (4) output
(5) wire   (6) capacitor
(7) value   (8) amplifier
(9) battery   (10) switch

2. Translate the following phrases into Chinese.

(1) integrated circuit   (2) electric field
(3) circuit board   (4) series circuit
(5) analog circuit   (6) consist of

Practice

1. Translate the following sentences into Chinese.

(1) Electronics is widely used in information processing, telecommunication, and signal processing.

(2) The basic components of analog circuits are wires, resistors, capacitors, inductors, diodes, and transistors.

(3) In a series circuit, the same current passes through a series of components.

(4) Most digital circuits use a binary system with two voltage levels labeled "0" and "1".

(5) Most modern radio and communications circuitry uses mixed signal circuits.

2. Translate the following sentences into English.

(1) 电池是主动元件。

(2) 电阻和电容是模拟电路中常用的元件。

(3) 在电子电路中，电阻器用于降低电流。

(4) 人体安全电流是10毫安。

(5) 心脏的跳动会产生微弱电流。

Culture salon

What is electronic noise?

Electronic noise is defined as unwanted disturbances superposed on a useful signal that tend to obscure its information content. Noise is not the same as signal distortion caused by a circuit. Noise is associated with all electronic circuits. Noise may be electromagnetically or thermally generated, which can be decreased by lowering the operating temperature of the circuit. Other types of noise, such as shot noise, cannot be removed due to their limitations in physical properties.

Chapter 1-Section 2-Text 1-看译文

Text 2 A brief introduction to biological basics

Lead-in questions:

(1) What is the definition of biology?

(2) What is the fundamental unit of life?

(3) Can you list some examples of the subdisciplines of biology?

(4) What is clinical biochemistry?

(5) What is electrophysiology?

Biology is a natural science concerned with the study of life and living organisms, including their structure, function, growth, evolution, distribution, identification and taxonomy. Modern biology is a vast field, composed of many branches and subdisciplines. However, despite the broad scope of biology, there are certain general concepts that govern all study and research, consolidating it into a single field. In general, biology recognizes the cell as the basic unit of life, genes as the basic unit of heredity, and evolution as the engine that propels the synthesis and creation of new species (Figure 1-2).

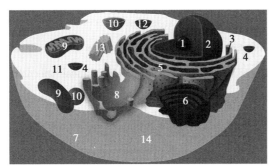

Figure 1-2 Components of a typical animal cell
1. Nucleolus; 2. Nucleus; 3. Ribosome; 4. Vesicle; 5. Rough endoplasmic reticulum; 6. Golgi apparatus; 7. Cytoskeleton; 8. Smooth endoplasmic reticulum; 9. Mitochondrion; 10. Vacuole; 11. Cytosol; 12. Lysosome; 13. Centrosome; 14. Cell membrane.

The cell is the fundamental unit of life, and all living things are composed of one or more cells. All cells arise from other cells through cell division. In multicellular organisms, every cell in the organism's body derives ultimately from a single cell in a fertilized egg. The cell is also considered to be the basic unit in many pathological processes. In addition, the phenomenon of energy flow occurs in cells in processes that are part of the function known as metabolism. Finally, cells contain hereditary information(DNA), which is passed from cell to cell during cell division. The structural and physiological properties of cells, including their behaviors, interactions, and environment can be studied on both the microscopic and molecular levels. Understanding the structure and function of cells is fundamental to all of the biological sciences.

Sub-disciplines of biology are defined by the scale at which organisms are studied, the kinds of organisms studied, and the methods used to study them. For example, biochemistry examines the rudimentary chemistry of life, and physiology examines the physical and chemical functions of tissues, organs, and organ systems of an organism. In this text, fundamentals of clinical biochemistry and electrophysiology relevant to medical devices will be briefly introduced.

(1) Clinical biochemistry

Biochemistry mainly deals with the structures, functions and interactions of biological macromolecules, such as proteins, nucleic acids, carbohydrates and lipids, which provide the structure of cells and perform many of the functions associated with life. The chemistry of the cell also depends on the reactions of smaller molecules and ions. These can be inorganic, for example water and metal ions, or organic, for example the amino acids, which are used to synthesize proteins. The main chemical elements involved in the study of biochemistry are carbon, hydrogen, nitrogen, oxygen, phosphorus, and sulfur, often referred to as CHNOPS.

Clinical biochemistry is an application of laboratory techniques to find out causes of diseases. It provides clinical data for diagnosis, prognosis and treatment of disease. Determination of chemical constituents of various body fluids such as blood, serum, urine, cerebrospinal fluid are useful in diagnosing various clinical conditions like diabetes mellitus, jaundice, gout etc. It is also useful in determining severity of diseases of organs, disorders of endocrine glands, disturbances in acid-base balance of body.

All the diseases are manifestation of abnormalities of molecules, chemical reactions or processes. All of these are caused by different factors, which affect biochemical mechanisms in the cell or in the body. Knowledge of biochemistry helps the physicians in diagnosis and prognosis of diseases. Clinicians must have knowledge of biochemical changes of various foodstuffs, hormones, vitamins, minerals, etc. to diagnose a disease properly and for its cure because certain diseases are caused due to alterations of these constituents in the body.

Thus knowledge of biochemistry is essential to determine cause of disease, to assist in monitoring process of particular treatment and for assessing response to therapy. Biochemistry is a subject linking chemistry and biology with medicine and pharmacy.

(2) Electrophysiology

When the heart beats too fast, too slow or doesn't beat with a regular rhythm, the condition is called an arrhythmia. In some cases, the heart no longer pumps blood effectively, and as a result, the arrhythmia can be life threatening. Electrophysiology is a branch of heart medicine that looks for the causes of arrhythmia and ways to prevent it.

The heart is made up of mostly muscle. The muscle forms the walls of the four chambers of the heart. These chambers fill with blood, and when the muscle that forms the chamber walls tightens (or contracts), the blood is pushed out of the chamber. The four chambers don't push blood all at once. Instead, they must contract in the right order to ensure that the blood is effectively collected from the rest of the body, moved through the lungs (where it picks up oxygen), then pushed back through the body. When the chambers are all working in the right order, they produce the lub-dub sound of a healthy heartbeat. Electrical signals (traveling along nerve fibers) control the order in which the chambers contract and push blood. Electrophysiologists look at these signals to diagnose heart problems and seek ways to overcome these problems.

To avoid open surgery, doctors will use small medical tools delivered to the heart by a catheter. A catheter is a thin tube that is inserted into a blood vessel through a tiny cut in the leg or arm and threaded through the bloodstream to the heart. Once in place, it acts as a tunnel, enabling the doctor to efficiently guide the tools to where they are needed.

In this procedure, the doctor inserts a catheter that runs from a small cut in the upper leg to the heart. The doctor then guides tiny electrodes (electrical connectors at the end of wires) up the catheter and into the heart. Using an X-ray machine to "see" inside the patient's body, the doctor can precisely position the electrodes within the heart. With the electrodes in place, the doctor can test the electrical pathways (the "wiring") within the heart to see what's causing the unusual heartbeat. The doctor can measure the electrical current that travels through the pathways as well as injecting electricity into the pathway to see how the heart responds. The actual procedure can take as little as 20 minutes or as long as several hours.

Words and phrases

biology	[baɪˈɒlədʒi]	n.	生物学,生物
science	[ˈsaɪəns]	n.	科学
organism	[ˈɔːgənɪzəm]	n.	有机体,生物体
structure	[ˈstrʌktʃə(r)]	n.	结构
function	[ˈfʌŋkʃn]	n.	功能
evolution	[ˌiːvəˈluːʃn]	n.	进化,演变
distribution	[ˌdɪstrɪˈbjuːʃn]	n.	分布
taxonomy	[tækˈsɒnəmi]	n.	分类,(生物)分类学
field	[fiːld]	n.	领域
subdiscipline	[ˈsʌbˈdɪsɪplɪn]	n.	(学科的)分支,分科
scope	[skəʊp]	n.	范围
cell	[sel]	n.	细胞
gene	[dʒiːn]	n.	基因
heredity	[həˈredəti]	n.	遗传,遗传性
propel	[prəˈpel]	vt.	推动,驱动
synthesis	[ˈsɪnθəsɪs]	n.	合成,综合
creation	[kriˈeɪʃn]	n.	创造
species	[ˈspiːʃiːz]	n.	物种,种类
phenomenon	[fəˈnɒmɪnən]	n.	现象

英文	音标	词性	中文
information	[ˌɪnfəˈmeɪʃn]	n.	信息
property	[ˈprɒpəti]	n.	特性,属性
microscopic	[ˌmaɪkrəˈskɒpɪk]	adj.	微观的,显微(镜)
rudimentary	[ˌruːdɪˈmentri]	adj.	基本的
tissue	[ˈtɪʃuː]	n.	组织
electrophysiology	[ɪˌlektrəʊˌfɪzɪˈɒlədʒɪ]	n.	电生理学
protein	[ˈprəʊtiːn]	n.	蛋白质
carbohydrate	[ˌkɑːbəʊˈhaɪdreɪt]	n.	碳水化合物
lipid	[ˈlɪpɪd]	n.	脂质
chemistry	[ˈkemɪstri]	n.	化学
molecule	[ˈmɒlɪkjuːl]	n.	分子
inorganic	[ˌɪnɔːˈgænɪk]	adj.	无机的
hydrogen	[ˈhaɪdrədʒən]	n.	氢,氢气
nitrogen	[ˈnaɪtrədʒən]	n.	氮,氮气
oxygen	[ˈɒksɪdʒən]	n.	氧,氧气
phosphorus	[ˈfɒsfərəs]	n.	磷
sulfur	[ˈsʌlfə]	n.	硫,硫黄
data	[ˈdeɪtə]	n.	数据
prognosis	[prɒgˈnəʊsɪs]	n.	预测,预后
blood	[blʌd]	n.	血液
serum	[ˈsɪərəm]	n.	血清
urine	[ˈjʊərɪn]	n.	尿
jaundice	[ˈdʒɔːndɪs]	n.	黄疸
gout	[gaʊt]	n.	痛风
endocrine	[ˈendəʊkrɪn]	adj.	内分泌(腺)的
mechanism	[ˈmekənɪzəm]	n.	机制,机能
foodstuff	[ˈfuːdstʌf]	n.	食物,食品
hormone	[ˈhɔːməʊn]	n.	荷尔蒙,激素
vitamin	[ˈvɪtəmɪn]	n.	维生素
mineral	[ˈmɪnərəl]	n.	矿物质
constituent	[kənˈstɪtjuənt]	n.	成分
rhythm	[ˈrɪðəm]	n.	节律
muscle	[ˈmʌsl]	n.	肌肉
catheter	[ˈkæθɪtə(r)]	n.	导管
respond	[rɪˈspɒnd]	vi.	回应,响应

composed of	由…组成
cell division	细胞分裂
fertilized egg	受精卵
deoxyribonucleic acid(DNA)	脱氧核糖核酸
pathological process	病理过程
clinical biochemistry	临床生物化学
nucleic acid	核酸
amino acid	氨基酸
diabetes mellitus	糖尿病
endocrine gland	内分泌腺
cerebrospinal fluid	脑脊液
blood vessel	血管

Key sentences

(1) In general, biology recognizes the cell as the basic unit of life, genes as the basic unit of heredity, and evolution as the engine that propels the synthesis and creation of new species.

一般来说,生物学认为细胞是生命的基本单位,基因是遗传的基本单位,进化是推动新物种合成和创造的动力。

(2) In this text, fundamentals of clinical biochemistry and electrophysiology relevant to medical devices will be briefly introduced.

本文将简要介绍与医疗器械有关的临床生物化学和电生理学的基础知识。

(3) Clinical biochemistry is an application of laboratory techniques to find out causes of diseases.

临床生物化学是应用实验室技术来找出疾病起因。

(4) All the diseases are manifestation of abnormalities of molecules, chemical reactions or processes.

所有的疾病都是分子、化学反应或过程异常的表现。

(5) Electrophysiology is a branch of heart medicine that looks for the causes of arrhythmia and ways to prevent it.

电生理学是心脏医学的一个分支,旨在寻找心律失常的起因和预防心律失常的方法。

Learning test

Choose the following words below to fill in the blanks, change the form if necessary.

| tissue | field | gene | phenomenon | respond |
| propel | microscopic | inorganic | blood | constituent |

(1) The IT technology will _____ us toward a cashless, mobile payment-driven world.

(2) The height of a person is mainly determined by _____ .

(3) He is very talented in the _____ of painting.

(4) The _____ of job-hopping is common due to low salary.

(5) The microscopic images of the cancer cells have been captured by the scientists.

(6) The _____ space is filled with body fluid.

(7) The best way to _____ to his rude question is to ignore it.

(8) There are both red cells and white cells in human _____.

(9) Caffeine is the active _____ in coffee.

(10) Oxygen and nitrogen molecules are _____ substances.

Translation

1. Translate the following words into Chinese.

(1) evolution   (2) cell

(3) science   (4) urine

(5) information   (6) muscle

(7) heredity   (8) vitamin

(9) oxygen   (10) blood

2. Translate the following phrases into Chinese.

(1) nucleic acid   (2) cell division

(3) composed of   (4) diabetes mellitus

(5) blood vessel   (6) amino acid

Practice

1. Translate the following sentences into Chinese.

(1) The cell is the fundamental unit of life, and all living things are composed of one or more cells.

(2) Finally, cells contain hereditary information (DNA), which is passed from cell to cell during cell division.

(3) The main chemical elements involved in the study of biochemistry are carbon, hydrogen, nitrogen, oxygen, phosphorus, and sulfur, often referred to as CHNOPS.

(4) It provides clinical data for diagnosis, prognosis and treatment of disease.

(5) When the heart beats too fast, too slow or doesn't beat with a regular rhythm, the condition is called an arrhythmia.

2. Translate the following sentences into English.

(1) 血液中90%以上的成分是水。

(2) 红细胞是血液运送氧气的主要媒介。

(3) 空气中含有大量的氧气和氮气。

(4) 细胞生物学研究细胞的结构和生理特性。

(5) 细胞分裂过程中,DNA被平均分到两个细胞内。

> ### Culture salon
>
> #### Basics of human physiology
>
> Human physiology seeks to understand the mechanisms that work to keep the human body alive and functioning, through scientific enquiry into the nature of mechanical, physical, and biochemical functions of humans, their organs, and the cells of which they are composed. The focus of physiology is at the level of organs and systems within systems. The endocrine and nervous systems play major roles in the reception and transmission of signals that integrate function in animals. Homeostasis is a major aspect with regard to such interactions within plants as well as animals. It is achieved through communication that occurs in a variety of ways, both electrical and chemical.

Chapter 1-Section 2-Text 2-看译文

Text 3 A brief introduction to microorganisms

Lead-in questions:

(1) Are microorganisms visible to the naked eye?

(2) Are bacteria eukaryote or prokaryote?

(3) Are all bacteria pathogens?

(4) What are fungi?

(5) What are the simplest organisms?

Microorganisms are very small and invisible with the naked eye. Typical microorganisms include bacteria, viruses, fungi, protozoa and algae, collectively known as "microbes". These microbes play key roles in nutrient cycling, biodegradation, climate change, food spoilage, the cause and control of disease, and biotechnology. Microbes are tiny, simple structures that are invisible to the naked eye and can only be observed under a microscope, only microbial colonies with a cluster of microbes are visible, as shown in Figure 1-3.

Chapter 1-Section 2-Text 3-听课文

Microorganisms and all other living organisms are classified as prokaryotes or eukaryotes. Prokaryotes and eukaryotes are distinguished on the basis of their cellular characteristics. Eukaryotic cells have both a nucleus and organelles, with genetic information housed inside their nucleus. Prokaryotic cells lack a nucleus and other membrane-bound structures known as organelles. Prokaryotes include several kinds of microorganisms, such as bacteria and cyanobacteria. Eukaryotes include such microorganisms as fungi, protozoa, and simple algae. Viruses are considered neither prokaryotes nor eukaryotes because they lack the characteristics of living things, except the ability to replicate accomplishing only in living cells. In this text, typical microorganisms-bacteria, fungi, viruses will be briefly introduced.

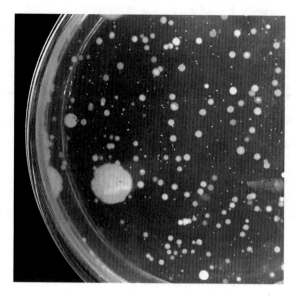

Figure 1-3 Microbial colonies.

(1) Bacteria

Bacteria are a group of very small prokaryote, typically a few micrometers in length. Bacteria have a number of shapes, ranging from spheres to rods and spirals. The smallest bacteria are around 0.35 μm in diameter. The largest common bacteria are around 2 μm long. There are typically 40 million bacterial cells in a gram of soil and a million bacterial cells in a milliliter of fresh water.

Though most of the earliest bacteria discovered were pathogens (disease-causing organisms), the vast majority of bacteria on earth are harmless, or even helpful, to humans. Many animals rely on gut bacteria to provide nutrients from food, which we are unable to synthesize ourselves (for example, E. coli in the gut produce Vitamin K).

Additionally, commensal (non-pathogenic) bacteria fill niches in our body and use resources that would otherwise be available to other pathogenic microorganisms. For instance, our skin and oral cavity are covered with bacteria, most of which will never harm us, as they live in a delicate balance with our immune system. Diseases caused by bacteria include tuberculosis (Mycobacterium tuberculosis), Pseudomonas aeruginosa, and strep throat (Streptococcus pyogenes).

(2) Fungi

Fungi are a group of eukaryotic microorganisms which reproduce by forming spores, including yeasts, molds, as well as the more familiar mushrooms. Most fungi are aerobes, meaning that they require oxygen to support the respiratory actions that release energy from their food.

Yeasts are microscopic, unicellular fungi with a single nucleus and eukaryotic organelles. They reproduce asexually by a process of budding. In this process, a new cell forms at the surface of the original cell, enlarges, and then breaks free to assume an independent existence. A typical yeast cell is 0.003 to 0.004 millimeters (3 to 4 μm) in diameter, which is three times wider than a bacterium and half the size of a red blood cell.

Molds generally have complex life cycles during which they pass through both an asexual and a sexual stage. They exist as multicellular organisms, and are far from being "micro" organisms. Nevertheless, because it is often their spores responsible for diseases, they remain classed as general "microbes."

Fungal spores are important in the identification of a fungus, since the spores are unique in shape, color, and size. A single spore is capable of germinating and reestablishing the entire mycelium. Spores are also the method for spreading fungi in the environment. Finally, the nature of the sexual spores is used for classifying fungi into numerous divisions.

Approximately 100 fungal species(out of ~100, 000 known) are pathogenic for human. Disease due to fungi can be caused by contact with or ingestion of toxins produced by certain species. For example, aflatoxin is a poison produced by the mold Aspergillus flavus. When ingested, symptoms may include vomiting, convulsions or death.

(3) Viruses

From pandemics such as human immunodeficiency virus(HIV), swine flu, and severe acute respiratory syndrome(SARS), we are constantly being bombarded with information about new lethal infections. What are viruses?

Viruses are the simplest organisms we are aware of 100 times smaller than the average bacterium, and they can't be seen with an ordinary microscope. Viruses can only exert influence by invading a cell, because they're not cellular structures. They lack the ability to replicate on their own and have to use a living cell for their replication. Viruses are merely tiny packets of DNA or RNA genes enfolded in a protein coating. The smallest viruses are about 0.02 μm(20 nanometers) in size, while the large viruses are about 0.3 μm(300 nanometers) in size.

On their own, viruses are incapable of any metabolism, including replication. However, viruses contain proteins on their surface, which facilitate binding to and entry into cells of every type imaginable, from bacteria to human cells. Once inside, the viruses use the cell's machinery to churn out more copies of the viral genomes and produce virus-specific proteins. They then assemble into mature virions(virus particles) and leave the cell, either by causing lysis(breaking apart) of the host cell or by "budding" off it, taking portions of the host's cell membrane with it. A single cell may produce anywhere from 10, 000 to 50, 000 new viruses in as little as 48 hours' time. This infection can happen several ways: by air(thanks to coughing and sneezing), via carrier insects like mosquitoes, or by transmission of body fluids such as saliva, blood or semen.

The human body does have some natural defenses against a virus. A cell can initiate RNA interference when it detects viral infection, which works by decreasing the influence of the virus's genetic material in relation to the cell's usual material. The immune system also kicks into gear when it identifies a virus by producing antibodies that bind to the virus and render it unable to replicate. The immune system also releases

T-cells, which work to kill the virus. Antibiotics have no effect on viruses, though vaccinations will provide immunity.

Unfortunately for humans, some viral infections outpace the immune system. Viruses can evolve much more quickly than the immune system can, which gives them a leg up in uninterrupted reproduction. And some viruses, such as HIV, work essentially by tricking the immune system. Viruses cause many diseases, including colds, measles, chicken pox, HPV (human papilloma virus), herpes, rabies, SARS, and the flu. Though they're small, they pack a big punch and they can only sometimes be sent into exile.

Chapter 1-Section 2-Text 3-听词汇

Words and phrases

microorganism	[ˌmaɪkrəʊˈɔːgənɪzəm]	n.	微生物
invisible	[ɪnˈvɪzəbəl]	adj.	看不见的
bacteria	[bækˈtɪərɪə]	n.	细菌(bacterium 的名词复数)
virus	[ˈvaɪrəs]	n.	病毒
fungi	[ˈfʌngiː]	n.	真菌(fungus 的复数)
protozoa	[ˌprəʊtəˈzəʊə]	n.	原生动物
algae	[ˈældʒiː]	n.	藻类
microbe	[ˈmaɪkrəʊb]	n.	微生物,细菌
nutrient	[ˈnjuːtrɪənt]	n.	营养,养分
biodegradation	[ˌbaɪəʊdɪgreɪˈdɪʃən]	n.	生物降解
biotechnology	[ˌbaɪəʊtekˈnɒlədʒi]	n.	生物技术
observe	[əbˈzɜːv]	vt.	观察
tiny	[ˈtaɪni]	adj.	极小的,微小的
microscope	[ˈmaɪkrəskəʊp]	n.	显微镜
prokaryote	[prəʊkærɪɒt]	n.	原核生物
eukaryote	[uːkeərɪˈɒt]	n.	真核生物
nucleus	[ˈnjuːklɪəs]	n.	细胞核
organelle	[ˌɔːgəˈnel]	n.	细胞器
house	[ˌɔːgəˈnel]	vt.	存储,安置
cyanobacteria	[saɪænɒbækˈtɪərɪə]	n.	蓝藻
replicate	[ˈreplɪkeɪt]	vt.	复制
micrometer	[maɪˈkrɒmɪtə(r)]	n.	微米
discover	[dɪˈskʌvə(r)]	vt.	发现
pathogen	[ˈpæθədʒən]	n.	病原体,致病菌
gut	[gʌt]	n.	肠,肠道
commensal	[kəˈmensl]	adj.	共生的

resource	[rɪˈsɔːs]	n.	资源
skin	[skɪn]	n.	皮肤
oral	[ˈɔːrəl]	adj.	口腔的
tuberculosis	[tjuːˌbɜːkjuˈləʊsɪs]	n.	肺结核
reproduce	[ˌriːprəˈdjuːs]	n.	繁殖
spore	[spɔː(r)]	n.	孢子
yeast	[jiːst]	n.	酵母菌
mold	[məʊld]	n.	霉菌
aerobe	[ˈeərəʊb]	n.	需氧菌
asexually	[ˌeɪˈsekʃəlɪ]	adv.	无性地,无性生殖地
bud	[bʌd]	vi.	出芽
enlarge	[ɪnˈlɑːdʒ]	vi.	增大,扩大
germinate	[ˈdʒɜːmɪneɪt]	vi.	出芽,生长,发育
mycelium	[maɪˈsiːlɪəm]	n.	菌丝体
ingestion	[ɪnˈdʒestʃən]	n.	摄入
toxin	[ˈtɒksɪn]	n.	毒素
aflatoxin	[ˌæfləˈtɒksɪn]	n.	黄曲霉毒素
vomit	[ˈvɒmɪt]	vi.	呕吐
convulsion	[kənˈvʌlʃn]	n.	抽搐
pandemic	[pænˈdemɪk]	n.	流行病
flu	[fluː]	n.	流行性感冒,流感
lethal	[ˈliːθl]	adj.	致死的,致命的
infection	[ɪnˈfekʃn]	n.	感染
invade	[ɪnˈveɪd]	vt.	侵入
lysis	[ˈlaɪsɪs]	n.	(细胞)溶解
cold	[kəʊld]	n.	感冒,伤风
measle	[ˈmiːzl]	n.	麻疹
herpes	[ˈhɜːpiːz]	n.	疱疹
rabies	[ˈreɪbiːz]	n.	狂犬病
naked eye			肉眼
nutrient cycling			养分循环
climate change			气候变化
food spoilage			食物腐败
vast majority of			绝大部分
oral cavity			口腔

immune system	免疫系统
Mycobacterium tuberculosis	结核分枝杆菌
Pseudomonas aeruginosa	铜绿假单胞菌,绿脓杆菌
Streptococcus pyogenes	化脓性链球菌
respiratory reaction	呼吸作用
Aspergillus flavus	黄曲霉
human immunodeficiency virus (HIV)	艾滋病病毒
swine flu	猪流感
severe acute respiratory syndrome (SARS)	非典型肺炎
chicken pox	水痘

Key sentences

(1) Microorganisms and all other living organisms are classified as prokaryotes or eukaryotes.

微生物和所有其他生物体都被归类为原核生物或真核生物。

(2) Bacteria are a group of very small prokaryote, typically a few micrometers in length.

细菌是一类非常小的原核生物,长度通常为几微米。

(3) Though most of the earliest bacteria discovered were pathogens (disease-causing organisms), the vast majority of bacteria on earth are harmless, or even helpful, to humans.

虽然大多数最早发现的细菌都是病原体(致病生物),但地球上绝大多数的细菌对人类无害,甚至有益。

(4) Fungi are a group of eukaryotic microorganisms which reproduce by forming spores, including yeasts, molds, as well as the more familiar mushrooms.

真菌是一类真核微生物,通过形成孢子来繁殖,包括酵母菌、霉菌以及更常见的蘑菇。

(5) Viruses are the simplest organisms we are aware of 100 times smaller than the average bacterium, and they can't be seen with an ordinary microscope.

病毒是我们所知道的最简单的有机体,比细菌小100倍,并且不能用普通的显微镜看到。

Learning test

Choose the following words below to fill in the blanks, change the form if necessary.

invisible   virus   nutrient   tiny   observe
micrometer   cold   vomit   lethal   discover

(1) The UV light is _____ to our naked eyes.

(2) This _____ blood sugar monitoring device can be worn on the arm.

(3) The medical students continued to _____ the patient's symptoms after medication.

(4) The _____ seems to have attacked his throat.

(5) He caught a _____ in the cold weather.

(6) The scientists may _____ more secrets of the universe in the near future.

(7) The main _____ in the fruits is vitamin C.

(8) Human hairs are usually 60~90 _____ in diameter.

(9) He _____ up all food he had just eaten.

(10) The chemical is _____ to rats but safe for domestic animals.

Translation

1. Translate the following words into Chinese.

(1) microorganism   (2) virus

(3) pathogen   (4) skin

(5) vomit   (6) flu

(7) infection   (8) microscope

(9) pandemic   (10) bacteria

2. Translate the following phrases into Chinese.

(1) naked eye   (2) oral cavity

(3) immune system   (4) food spoilage

(5) swine flu   (6) vast majority of

Practice

1. Translate the following sentences into Chinese.

(1) Microorganisms are very small and invisible with the naked eye.

(2) Prokaryotic cells lack a nucleus and other membrane-bound structures known as organelles.

(3) The smallest bacteria are around $0.35\mu m$ in diameter.

(4) Most fungi are aerobes, meaning that they require oxygen to support the respiratory actions that release energy from their food.

(5) From pandemics such as HIV, swine flu, and SARS, we are constantly being bombarded with information about new lethal infections.

2. Translate the following sentences into English.

(1) 微生物包括细菌、病毒、真菌以及一些原核生物。

(2) 微生物的生长繁殖速度非常快。

(3) 原核微生物没有细胞器。

(4) 显微镜有两种：光学显微镜和电子显微镜。

(5) 一些细菌对人体是有益的。

Culture salon

HIV and AIDS

The human immunodeficiency virus(HIV) is a lentivirus(a subgroup of retrovirus) that causes HIV infection and over time acquired immunodeficiency syndrome(AIDS). AIDS is a condition in humans in which progressive failure of the immune system allows life-threatening opportunistic infections and cancers to thrive. Without treatment, average survival time after infection with HIV is estimated to be 8 to 10 years, depending on the HIV subtype. Infection with HIV occurs by the transfer of blood, semen, vaginal fluid, pre-ejaculate, or breast milk. Therefore, the route of AIDS transmission includes blood transmission, sexual contact and mother to child transmission.

Chapter 1-Section 2-Text 3-看译文

Chapter 1-Section 2-习题

Chapter 2

Introduction to typical products

Learning objective

In this chapter, typical medical devices, such as consumer products, ECG monitor, Ultrasound imaging devices, and MRI system, will be introduced. Read the following texts, you should meet the following requirements:

(1) Grasp basic concepts, key points, involved professional terms and phrases.

(2) Acquire the working principles of relevant products.

(3) Understand the inherent physics, operation and maintenance method of relevant products.

Section 1 Consumer products

Chapter 2-
Section 1-PPT

Text 1 Contact lenses

Lead-in questions:

(1) Is contact lens a kind of medical device?

(2) What are the best strategies to reduce the risk of eye infection?

(3) How many types of contact lenses?

(4) Can we reuse the contact lens solution?

(5) Can we transfer contact lens solutions into smaller travel size container?

1. Focusing on contact lens safety

Approximately 40 million Americans wear contact lenses, as shown in Figure 2-1. In addition to offering flexibility, convenience, and a "no-glasses" appearance, contact lenses help correct a variety of vision disorders, including nearsightedness, farsightedness, astigmatism, and poor focusing with reading material.

Chapter 2-
Section 1-
Text 1-听课文

But contact lenses also present potential risks. Contact lenses are worn directly on your eyes and can lead to serious eye infections and corneal ulcers if you don't care for the lenses properly—and if you don't wear them properly. Between 40% to 90% of contact lens wearers do not properly follow the care instructions for their contact lenses, increasing their risk of serious injury—and even blindness.

The U. S. Food and Drug Administration regulates contact lenses and certain contact lens care products as medical devices. Best strategies for reducing your risk of infection involve proper hygiene, following recommended wearing schedules, using proper lens care practices for cleaning, disinfecting and storing your lenses(which includes reading and following all product labeling instructions), and having routine eye exams.

Figure 2-1 A pair of contact lenses.

2. Types of contact lenses

(1) **General categories**

• **Soft contact lenses.** These are made of flexible plastics that allow oxygen to pass through to the cornea. Users get used to wearing them within several days. Most soft-contact lenses need to be replaced with new ones after two weeks of use.

• **Rigid gas-permeable(RGP) lenses.** These products are durable, resist deposit buildup, and generally allow for clear, crisp vision. They last longer than soft contacts, and also are easier to handle and less likely to tear.

(2) **Specific types**

• **Extended wear contacts.** These are good for overnight or continuous wear ranging from one to six nights, or up to 30 days. They can be either soft or rigid gas permeable lenses. It's important for your eyes to have a rest without lenses for at least one night following each scheduled removal.

• **Disposable(or "replacement schedule") contacts.** As defined by the FDA, "disposable" means used once and discarded. With a true daily-wear disposable schedule, a brand new pair of lenses is used each day. However, some soft contacts referred to as "disposable" by sellers are actually daily wear lenses used for up to two weeks before being discarded.

• **Lenses designed for "Ortho-K".** Orthokeratology(Ortho-K) is a lens-fitting procedure that uses specially designed RGP contact lenses to change the curvature of the cornea to temporarily improve the eye's ability to focus. It's primarily used for the correction of nearsightedness.

• **Decorative contacts(also called "colored", or "fashion" contacts).** The FDA has often warned people about the serious risks(including eye infection and blindness) associated with wearing these lenses. These lenses don't correct vision and are intended solely to change the appearance of the eye.

3. How to avoid infection or injury

Contact lens users run the risk of infections such as pink eye, corneal abrasions, and eye irritation. A common result of eye infection is corneal ulcers, which are open sores in the outer layer of the cornea. Many of these complications can be avoided through everyday care of the eye and contact lenses.

To reduce your chances of infection:

(1) In general, if you're using multipurpose contact lens solution, replace your contact lens storage case at least every 3 months or as directed by your eye care provider. If you're using contact lens solution that contains hydrogen peroxide, always use the new contact lens case that comes with each box—and follow all directions that are included on or inside the packaging.

(2) Clean and disinfect your lenses properly. When using contact lens solution, read and follow all instructions on the product label to avoid eye injury. This is particularly important if your eye care professional has recommended a solution with hydrogen peroxide, as these solutions require special care.

(3) Always remove contact lenses before swimming.

(4) Never reuse any lens solution. Always discard all of the used solution after each use, and add fresh solution to your lens case.

(5) Do not use any water on your lenses because it can be a source of microorganisms that may cause serious eye infections. (Contact lens solution is sold in "sterile" containers, which means it is free from living germs or microorganisms.)

(6) Never put your lenses in your mouth or put saliva on your lenses. Saliva is not sterile.

(7) Never transfer contact lens solutions into smaller travel size containers. These containers are not sterile, and unsterile solution can damage your eyes.

(8) Do not wear contact lenses overnight unless your eye care provider has prescribed them to be worn that way. Any lenses worn overnight increase your risk of infection. Wearing contact lenses overnight can stress the cornea by reducing the amount of oxygen to the eye. They can also cause microscopic damage to the surface of the cornea, making it more susceptible to infection.

(9) Never ignore symptoms of eye irritation or infection that may be associated with wearing contact lenses. These symptoms include discomfort, excess tearing or other discharge, unusual sensitivity to light, itching, burning, gritty feelings, unusual redness, blurred vision, swelling, or pain. If you experience any of these symptoms, remove your lenses immediately and keep them off. Keep the lenses, because they may help your eye care professional determine the cause of your symptoms.

Words and phrases

disorder	[dɪsˈɔːdə(r)]	n.	障碍，失调
nearsightedness	[ˌnɪəˈsaɪtɪdnəs]	n.	近视
astigmatism	[əˈstɪgmətɪzəm]	n.	散光
infection	[ɪnˈfekʃn]	n.	感染

Chapter 2-
Section 1-
Text 1-听词汇

单词	音标	词性	释义
regulate	[ˈreɡjuleɪt]	vt.	控制,管理
involve	[ɪnˈvɒlv]	vt.	包含,牵涉,围绕
hygiene	[ˈhaɪdʒiːn]	n.	卫生
disinfect	[ˌdɪsɪnˈfekt]	vt.	消毒
label	[ˈleɪbl]	vt. & n.	贴标签,标签,标记
routine	[ruːˈtiːn]	adj.	常规的,例行的
category	[ˈkætəɡəri]	n.	种类,类别
plastic	[ˈplæstɪk]	n.	塑料
cornea	[ˈkɔːniə]	n.	眼角膜
oxygen	[ˈɒksɪdʒən]	n.	氧气
rigid	[ˈrɪdʒɪd]	adj.	硬性,刚硬的
resist	[rɪˈzɪst]	vt.	抗,抵抗
deposit	[dɪˈpɒzɪt]	n.	沉淀
crisp	[krɪsp]	adj.	清晰的
disposable	[dɪˈspəʊzəbl]	adj.	一次性的
define	[dɪˈfaɪn]	vt.	定义
discard	[dɪsˈkɑːd]	vt.	丢弃,抛弃
orthokeratology	[ˌɔːθəʊˌkerəˈtɒlɪdʒɪ]	n.	角膜塑形
curvature	[ˈkɜːvətʃə(r)]	n.	曲率
improve	[ɪmˈpruːv]	vt.	提高
focus	[ˈfəʊkəs]	vi.	聚焦
primarily	[praɪˈmerəli]	adv.	主要地,首要地
decorative	[ˈdekərətɪv]	adj.	装饰的,装饰性
fashion	[ˈfæʃn]	n.	时尚
avoid	[əˈvɔɪd]	vt.	避免,预防
irritation	[ˌɪrɪˈteɪʃn]	n.	刺激
complication	[ˌkɒmplɪˈkeɪʃn]	n.	并发症
solution	[səˈluːʃn]	n.	溶液
professional	[prəˈfeʃnl]	n.	专业人士
remove	[rɪˈmuːv]	vt.	去除,脱掉,拿下
microorganism	[ˌmaɪkrəʊˈɔːɡənɪzəm]	n.	微生物
sterile	[ˈsteraɪl]	adj.	无菌的
germ	[dʒɜːm]	n.	菌,病菌,细菌
saliva	[səˈlaɪvə]	n.	唾液
prescribe	[prɪˈskraɪb]	vt.	指示,规定

susceptible	[sə'septəbl]	*adj.*	易感的,敏感的,易受感染的
symptom	['sɪmptəm]	*n.*	症状
itch	[ɪtʃ]	*vi.*	发痒
gritty	['grɪti]	*adj.*	含沙的
blur	[blɜː(r)]	*vt.*	(使)变模糊
swell	[swel]	*v.*	肿胀
gas-permeable			透气的
contact lens			隐形眼镜,角膜接触镜
focus on			聚焦于
in addition to			除……之外
lead to			导致
free from			无……
hydrogen peroxide			过氧化氢
corneal abrasion			角膜擦伤
corneal ulcer			角膜溃疡
lens-fitting			眼镜验配
associated with			与……相关的

Key sentences

(1) In addition to offering flexibility, convenience, and a "no-glasses" appearance, contacts help correct a variety of vision disorders, including nearsightedness, farsightedness, astigmatism, and poor focusing with reading material.

隐形眼镜除了提供灵活性、便利性和"无眼镜"的外观外,还可以帮助纠正各种视觉障碍,包括近视、远视、散光和对阅读材料的聚焦不良。

(2) Contact lenses are worn directly on your eyes and can lead to serious eye infections and corneal ulcers if you don't care for the lenses properly—and if you don't wear them properly.

隐形眼镜直接戴在你的眼睛上,如果你没有正确护理和佩戴,会导致严重的眼睛感染和角膜溃疡。

(3) Best strategies for reducing your risk of infection involve proper hygiene, following recommended wearing schedules, using proper lens care practices for cleaning, disinfecting and storing your lenses (which includes reading and following all product labeling instructions), and having routine eye exams.

降低感染风险的最佳策略包括恰当的卫生习惯,遵守推荐的佩戴时间表,采用恰当的护理方法进行镜片的清洁、消毒和存放(包括阅读和遵循所有的产品标签说明),并进行常规眼睛检查。

(4) Many of these complications can be avoided through everyday care of the eye and contact lenses.

许多并发症可以通过对眼睛和隐形眼镜的日常护理来避免。

(5) They can also cause microscopic damage to the surface of the cornea, making it more susceptible to

infection.

它们也能造成角膜表面的微小损伤,使其更容易感染。

Learning test

Choose the following words below to fill in the blanks, change the form if necessary.

involve   germ   saliva   deposit   disorder

primarily   oxygen   irritation   regulate   remove

(1) Dirty hands can be full of millions of invisible _____ .

(2) _____ drips from the baby's mouth.

(3) Smoking sets up _____ in the throat and causes cough.

(4) There is some _____ in the bottom of the cup.

(5) The technician has to be able to carry out _____ maintenance of the machine.

(6) Human body is made up _____ of bone, muscle, and fat.

(7) The brain requires a lot of _____ to support its activities.

(8) Don't _____ yourself in some unnecessary arguments.

(9) He is suffering from a severe mental _____ .

(10) _____ the meat from my plate, I don't want to eat meat.

Translation

1. Translate the following words into Chinese.

(1) susceptible   (2) infection

(3) hygiene   (4) gas-permeable

(5) define   (6) prescribe

(7) disposable   (8) curvature

(9) improve   (10) professional

2. Translate the following phrases into Chinese.

(1) contact lens   (2) lead to

(3) free from   (4) in addition to

(5) lens-fitting   (6) corneal ulcer

Practice

1. Translate the following sentences into Chinese.

(1) The U.S. Food and Drug Administration regulates contact lenses and certain contact lens care products as medical devices.

(2) These are made of flexible plastics that allow oxygen to pass through to the cornea.

(3) These lenses don't correct vision and are intended solely to change the appearance of the eye.

(4) In general, if you're using multipurpose contact lens solution, replace your contact lens storage case at least every 3 months or as directed by your eye care provider.

(5) Never put your lenses in your mouth or put saliva on your lenses.

2. Translate the following sentences into English.

(1) 中国的很多年轻女性喜欢佩戴装饰性隐形眼镜。

(2) 这些柔性隐形眼镜是用柔性塑料做的。

(3) 务必正确清洗和消毒隐形眼镜。

(4) 游泳前一定要将隐形眼镜摘下。

(5) 隐形眼镜过夜佩戴会增加眼睛感染风险。

Culture salon

Why choose contact lenses?

A contact lens(CL) is a thin lens placed directly on the surface of the eye. CLs are regulated as medical devices and can be worn to correct vision, or for cosmetic or therapeutic purposes. People choose to wear CLs for many reasons. Aesthetics and cosmetics are the main motivating factors for people who want to avoid wearing glasses or to change the appearance of their eyes. Others wear CLs for functional reasons. Comparing with spectacles, CLs typically provide better peripheral vision, and do not collect moisture(from rain, snow, condensation etc.), this makes them ideal for sports and other outdoor activities. CL wearers can also wear sunglasses, goggles, or other eye products.

Chapter 2-Section 1-Text 1-看译文

Chapter 2-Section 1-Text 1-拓展阅读

Text 2 Blood pressure meters

Lead-in questions:

(1) What equipment do we need to measure the blood pressure?

(2) How many types of blood pressure meter are there?

(3) Which type of blood pressure meter need a rubber bulb?

(4) What's the price range for the aneroid monitors?

(5) What's the price range for the digital monitors?

Chapter 2-Section 1-Text 2-听课文

1. The importance of measuring blood pressure

Measuring your blood pressure at home and keeping a record of the measurements will show you and your doctor how much your blood pressure changes during the day. Your doctor can use your record of measurements to see how well your medicine is working to control your high blood pressure. Also, measuring your own blood pressure is a good way to take part in managing your own health and recognizing changes. A home-use blood pressure meter used to measure blood pressure is shown below in Figure 2-2.

2. Equipment of measuring blood pressure

To measure your blood pressure at home, you can use either an aneroid monitor or a digital monitor. Choose the type of monitor that best suits your needs.

The aneroid monitor has a gauge that is read by looking at a pointer on a dial. The cuff is placed around your upper arm and inflated by hand, by squeezing a rubber bulb.

Digital monitors have either manual or automatic cuffs. The blood pressure reading flashes on a small screen.

3. What are the pros and cons of the aneroid monitor?

One advantage of the aneroid monitor is that it can easily be carried from one place to another. Also, the cuff for the device has a built-in stethoscope, so you don't need

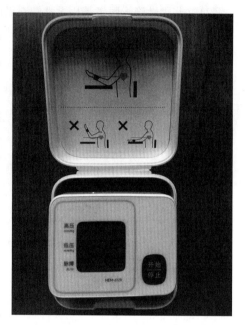

Figure 2-2　A blood pressure meter.

to buy a separate stethoscope. It's also easier to manage this way. The unit may have a special feature that makes it easy to put the cuff on with one hand. In addition, the aneroid monitor usually costs less than digital monitors. Aneroid monitors range in price from about ＄20 to ＄40.

The aneroid monitor also has some disadvantages. First, it is a complicated device that can easily be damaged and become less accurate. The device is also difficult to use if it doesn't have the special feature—a metal ring—that makes it easier to put the cuff on. In addition, the rubber bulb that inflates the cuff may be difficult to squeeze. This type of monitor may not be appropriate for hearing-impaired people, because of the need to listen to heart sounds through the stethoscope.

4. What are the pros and cons of the digital monitor?

Because the digital monitor is automatic, it is the more popular blood-pressure measuring device. The blood pressure measurement is easy to read, because the numbers are shown on a screen. Some electronic monitors also have a paper printout that gives you a record of the blood pressure reading.

The digital monitor is easier to use than the aneroid unit. It has a gauge and stethoscope in one unit, and the numbers are easy to read. It also has an error indicator, and deflation is automatic. Inflation of the cuff is either automatic or manual, depending on the model. This blood pressure monitoring device is good for hearing-impaired patients, since there is no need to listen to heart sounds through the stethoscope.

A disadvantage of the digital monitor is that the accuracy is changed by body movements or an irregular heart rate. In addition, the monitor requires batteries. Some models are designed for use with the left arm only. This may make them hard for some patients to use. Finally, some digital monitors are expensive. They range in price from about $30 to more than $100.

5. How do I know if my monitor is accurate or if I am using it correctly?

Once you buy your monitor, take it to your doctor's office to be checked for accuracy. You should have your monitor checked once a year. Proper care and storage are also necessary. Make sure the tubing is not twisted when the monitor is stored, and keep it away from heat. Periodically check the tubing for cracks and leaks.

Ask your doctor or nurse to teach you how to use your blood pressure monitor correctly. Proper use of it will help you and your doctor achieve good results in controlling your blood pressure.

There are some other types of blood pressure meters such as manual sphygmomanometers, which are used by trained practitioners, and mercury sphygmomanometers, which are considered the gold standard. They show blood pressure by affecting the height of a column of mercury, which does not require recalibration. Because of their accuracy, they are often used in clinical trials of drugs and in clinical evaluations of high-risk patients, including pregnant women.

6. How do I use a manual or aneroid monitor?

(1) Put the stethoscope ear pieces into your ears, with the ear pieces facing forward.

(2) Place the stethoscope disk on the inner side of the crease of your elbow.

(3) Rapidly inflate the cuff by squeezing the rubber bulb to 30 to 40 points higher than your last systolic reading. Inflate the cuff rapidly, not just a little at a time. Inflating the cuff too slowly will cause a false reading.

(4) Slightly loosen the valve and slowly let some air out of the cuff. Deflate the cuff by 2 to 3 millimeters per second. If you loosen the valve too much, you won't be able to determine your blood pressure.

(5) As you let the air out of the cuff, you will begin to hear your heartbeat. Listen carefully for the first sound. Check the blood pressure reading by looking at the pointer on the dial. This number will be your systolic pressure.

(6) Continue to deflate the cuff. Listen to your heartbeat. You will hear your heartbeat stop at some point. Check the reading on the dial. This number is your diastolic pressure.

(7) Write down your blood pressure, with the systolic pressure before the diastolic pressure (for example, 120/80).

(8) If you want to repeat the measurement, wait 2 to 3 minutes before reinflating the cuff.

7. How do I use a digital monitor?

(1) Put the cuff around the arm. Turn the power on, and start the machine.

(2) The cuff will inflate by itself with a push of a button on the automatic models. On the semiautomatic models, the cuff is inflated by squeezing the rubber bulb. After the cuff is inflated, the automatic mechanism will slowly reduce the cuff pressure.

(3) Look at the display window to see your blood pressure reading. The machine will show your systolic and diastolic blood pressures on the screen. Write down your blood pressure, with the systolic pressure before the diastolic pressure (for example, 120/80).

(4) Press the exhaust button to release all of the air from the cuff.

(5) If you want to repeat the measurement, wait 2 to 3 minutes before reinflating the cuff.

Chapter 2-Section 1-Text 2-听词汇

Words and phrases

meter	[ˈmiːtə(r)]	n.	(测量)计,仪,表
measure	[ˈmeʒəmənts]	vt.	测量
aneroid	[ˈænərɔɪd]	adj.	无液的
monitor	[ˈmɒnɪtə(r)]	n.	监护仪
gauge	[ɡeɪdʒ]	n.	仪器,仪表
pointer	[ˈpɔɪntə(r)]	n.	指针(测量仪器上的)
dial	[ˈdaɪəl]	n.	标度盘,表盘
cuff	[kʌf]	n.	袖带,臂带
inflate	[ɪnˈfleɪt]	vt.	(使)充气
squeeze	[skwiːz]	vt.	挤压,捏
rubber	[ˈrʌbə(r)]	n.	橡胶
bulb	[bʌlb]	n.	球
manual	[ˈmænjuəl]	adj.	手动的
automatic	[ˌɔːtəˈmætɪk]	adj.	自动的
flash	[flæʃ]	vi.	闪烁
stethoscope	[ˈsteθəskəʊp]	n.	听诊器
complicated	[ˈkɒmplɪkeɪtɪd]	adj.	复杂的
damage	[ˈdæmɪdʒd]	vt.	损坏
accurate	[ˈækjərət]	adj.	准确的,精确的
appropriate	[əˈprəʊpriət]	adj.	合适的,适当的
printout	[ˈprɪntaʊt]	n.	打印输出
deflation	[ˌdiːˈfleɪʃn]	n.	放气
tubing	[ˈtjuːbɪŋ]	n.	管,管子
twist	[twɪst]	vt.	卷曲,扭曲,交结
periodically	[ˌpɪəriˈɒdɪkli]	adv.	周期性地,定期地
practitioner	[prækˈtɪʃənə]	n.	从业者(尤指医师)
sphygmomanometer	[ˌsfɪɡməʊməˈnɒmɪtə]	n.	血压计
affect	[əˈfekt]	vt.	影响
recalibration	[rɪˌkælɪˈbreɪʃn]	n.	重新校准

trial	[ˈtraɪəl]	n.	试验,测试
valve	[vælv]	n.	阀,阀门
millimeter	[ˈmilimiːtə]	n.	毫米
semiautomatic	[ˌsemɪɔːtəˈmætɪk]	adj.	半自动的
mechanism	[ˈmekənɪzəm]	n.	结构,机械装置
systolic	[ˌsɪsˈtɒlɪk]	adj.	心脏收缩的
diastolic	[ˌdaɪəˈstɒlɪk]	adj.	心脏舒张的
heartbeat	[ˈhɑːtbiːt]	n.	心跳,心搏
exhaust	[ɪɡˈzɔːst]	vt.	排气
blood pressure			血压
upper arm			上臂
pros and cons			利弊,优缺点
hearing-impaired			听觉受损的,听障的
heart rate			心率
mercury sphygmomanometer			汞柱式血压计
systolic pressure			收缩压
diastolic pressure			舒张压
clinical evaluation			临床评价
high-risk			高风险
built-in			嵌入的,内置的

Key sentences

(1) Measuring your blood pressure at home and keeping a record of the measurements will show you and your doctor how much your blood pressure changes during the day.

在家里测量血压并记录测量结果可以给你和你的医生看你白天的血压变化。

(2) To measure your blood pressure at home, you can use either an aneroid monitor or a digital monitor. Choose the type of monitor that best suits your needs.

在家里测量血压,你可以使用无液式血压计或数字血压计。请选择最适合你需要的血压计类型。

(3) One advantage of the aneroid monitor is that it can easily be carried from one place to another. Also, the cuff for the device has a built-in stethoscope, so you don't need to buy a separate stethoscope.

无液式血压计的一个优点是它可以轻松地从一个地方携带到另一个地方。此外,血压计袖带还有一个内置的听诊器,不需要购买单独的。

(4) The digital monitor is easier to use than the aneroid unit. It has a gauge andstethoscope in one unit, and the numbers are easy to read. It also has an error indicator, and deflation is automatic. Inflation of the cuff is either automatic or manual, depending on the model.

数字血压计比无液式血压计更容易使用。一套仪器中有一个仪表和一个听诊器,数字结果很容易读取。它还有一个误差指示器,并且放气是自动的。袖带的充气可以是自动的或手动的,这取决于型号。

(5) There are some other types of blood pressure meters such as manual sphygmomanometers, which are used by trained practitioners, and mercury sphygmomanometers, which are considered the gold standard.

也有一些其他类型的血压计,如由受过训练的专业人员使用的手动血压计,和被认为是金标准的汞柱式血压计。

Learning test

Choose the following words below to fill in the blanks, change the form if necessary.

| periodically | heartbeat | damage | valve | diastolic |
| accurate | meter | appropriate | measure | automatic |

(1) She was taken to hospital because of irregular _____.

(2) He _____ a car with a baseball bat.

(3) Your WeChat password should be updated _____.

(4) The voltage value across the resistor can be measured with the volt _____.

(5) What you said on the occasion was not _____.

(6) This washing machine is _____, and the others are semiautomatic.

(7) Sometimes the weather forecast is _____, sometimes not.

(8) He closed the water _____ to stop the water flow.

(9) _____ the length and width of the gap first.

(10) You should record both systolic and _____ pressures with the blood pressure meter.

Translation

1. Translate the following words into Chinese.

(1) gauge   (2) rubber

(3) deflation   (4) stethoscope

(5) systolic   (6) millimeter

(7) heartbeat   (8) monitor

(9) valve   (10) trial

2. Translate the following phrases into Chinese.

(1) blood pressure   (2) heart rate

(3) high-risk   (4) clinical evaluation

(5) pros and cons   (6) systolic pressure

Practice

1. Translate the following sentences into Chinese.

(1) The aneroid monitor has a gauge that is read by looking at a pointer on a dial.

(2) In addition, the aneroid monitor usually costs less than digital monitors.

(3) As you let the air out of the cuff, you will begin to hear your heartbeat.

(4) Once you buy your monitor, take it to your doctor's office to be checked for accuracy.

(5) If you loosen the valve too much, you won't be able to determine your blood pressure.

2. Translate the following sentences into English.

(1)饮食和运动都对血压有影响。

(2)高血压通常没有明显的症状。

(3)台式血压计价格相对便宜些,电子血压计价格相对高些。

(4)运动是最好的血压调节方法。

(5)降低血压可延长您的寿命。

Culture salon

Choosing the best home blood pressure monitor for you

If you do decide to measure your blood pressure at home, you will need to get a home blood pressure monitor. There is a wide range of home blood pressure monitors available, but it is important to be sure that the blood pressure monitor you choose is accurate and the right one for you.

There are many different kinds of home blood pressure monitor, but it is easiest to use a monitor that is fully automatic (digital). Choose one that measures your blood pressure at your upper arm, rather than at your wrist or finger. Upper-arm blood pressure monitors usually give the most accurate and consistent results.

Chapter 2-Section 1-Text 2-看译文

Chapter 2-Section 1-Text 2-拓展阅读

Text 3 Glucose meters

Lead-in questions:

(1) What's the use of a glucose meter?

(2) What's the use of glucose for our body?

(3) What's the normal level of the blood glucose?

(4) What's the construction of glucose meters?

(5) How can we get information about the meter and test strips?

Chapter 2-Section 1-Text 3-听课文

1. Brief introduction to glucose meters

Glucose meters test and record how much glucose sugar is in your blood (Figure 2-3). They help you track your blood sugar level at different times during the day and night.

Glucose is a sugar that your body uses as a source of energy. Unless you have diabetes, your body regulates the amount of glucose in your blood. People with diabetes may need special diets and medications to control blood glucose.

Meters can help you know how well your diabetes medicines are working. They can also help you learn how the food you eat and your physical activity can change your blood sugar level.

You should take glucose test if you have diabetes and you need to monitor your blood sugar levels. You and your doctor can use the results to:

(1) Determine your daily adjustments in treatment;

(2) Know if you have dangerously high or low levels of glucose;

(3) Understand how your diet and exercise change your glucose levels.

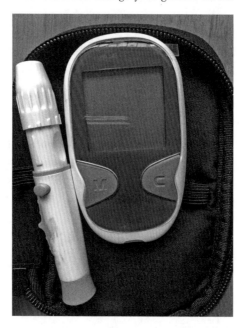

Figure 2-3　A glucose meter.

According to the American Diabetes Association, the blood glucose levels for an adult without diabetes are below 100 mg/dL before meals and fasting, and are less than 140 mg/dL two hours after meals.

2. Construction of glucose meters

People with diabetes should consult their doctor or health care provider to set appropriate blood glucose goals. Or else, you may get complication. You should treat your low or high blood glucose as recommended by your health care provider.

Most meters come with three parts:

(1) Lancet— A needle that is used to get a drop of blood from your finger or another part of your body.

(2) Test strip— The strip where you put the blood you are testing.

(3) Control solutions— Liquid used to make sure your meter is working properly.

Meters come in different sizes and features. Some meters let you track and print out your test results. Others have audio and larger screens to help people who have visual problems. The meter you choose should fit your lifestyle and your needs.

3. How do you choose a glucose meter?

There are many different types of meters available for purchase that differ in several ways, including:

- accuracy
- amount of blood needed for each test
- how easy it is to use
- pain associated with using the product
- testing speed
- overall size
- ability to store test results in memory
- likelihood of interferences
- ability to transmit data to a computer
- cost of the meter
- cost of the test strips used
- doctor's recommendation
- technical support provided by the manufacturer
- special features such as automatic timing, error codes, large display screen, or spoken instructions or results.

Talk to your health care provider about the right glucose meter for you, and how to use it.

4. How can you check your meter's performance?

There are three ways to make sure your meter works properly.

(1) Use liquid control solutions:

- Every time you open a new container of test strips;
- Occasionally as you use the container of test strips.

To test a liquid control solution, you test a drop of these solutions just like you test a drop of your blood. The value you get should match the value written on the test strip vial label.

(2) Use electronic checks.

Every time you turn on your meter, it does an electronic check. If it detects a problem it will give you an error code. Look in your meter's manual to see what the error codes mean and how to fix the problem. If you are unsure if your meter is working properly, call the toll-free number in your meter's manual, or contact your health care provider.

(3) Compare your meter with a blood glucose test performed in a laboratory.

Take your meter with you to your next appointment with your health care provider. Ask your provider to watch your testing technique to make sure you are using the meter correctly. Ask your health care provider to have your blood tested with a laboratory method. If the values you obtain on your glucose meter match the laboratory values, then your meter is working well and you are using good technique.

5. How do you take this test?

Before you test your blood glucose, you must read and understand the instructions for your meter. In general, you prick your finger with a lancet to get a drop of blood. Then you place the blood on a disposable "test strip" that is inserted in your meter. The test strip contains chemicals that react with glucose. Some meters measure the amount of electricity that passes through the test strip. Others measure how much light reflects from it. In the U.S., meters report results in milligrams of glucose per deciliter of blood, or mg/dL.

You can get information about your meter and test strips from several different sources, including the toll-free number in the manual that comes with your meter or on the manufacturer's web site. If you have an urgent problem, always contact your health care provider or a local emergency room for advice.

Words and phrases

glucose	[ˈgluːkəʊs]	n.	葡萄糖
meter	[ˈmiːtə(r)]	n.	（测量）计，仪，表
track	[træk]	vt.	跟踪，监测
diabetes	[ˌdaɪəˈbiːtiːs]	n.	糖尿病
amount	[əˈmaʊnt]	n.	量，数量
diet	[ˈdaɪət]	n.	饮食
medication	[ˌmedɪˈkeɪʃn]	n.	药物治疗，药物
monitor	[ˈmɒnɪtə(r)]	vt.	监测
determine	[dɪˈtɜːmɪn]	vt.	决定，确定
adjustment	[əˈdʒʌstmənts]	n.	调整，调节
treatment	[ˈtriːtmənt]	n.	治疗
complication	[ˌkɒmplɪˈkeɪʃən]	n.	并发症
fasting	[ˈfɑːstɪŋ]	n.	空腹
construction	[kənˈstrʌkʃn]	n.	结构，构造
appropriate	[əˈprəʊpriət]	adj.	适当的，恰当的，合适的
recommend	[ˌrekəˈmend]	vt.	推荐
lancet	[ˈlaːnsɪt]	n.	采血针
needle	[ˈniːdl]	n.	针
finger	[ˈfɪŋgə(r)]	n.	手指
solution	[səˈluːʃn]	n.	溶液
liquid	[ˈlɪkwɪd]	n.	液体

purchase	[ˈpɜːtʃəs]	n.	购买，采购
accuracy	[ˈækjərəsi]	n.	精度，准确性
likelihood	[ˈlaɪklihʊd]	n.	可能性
interference	[ɪntəˈfɪərəns]	n.	干扰
transmit	[trænsˈmɪt]	vt.	传输，传送，传递
recommendation	[rekəmenˈdeɪʃn]	n.	推荐，建议
manufacturer	[ˌmænjuˈfæktʃərə(r)]	n.	制造商，厂商
instruction	[ɪnˈstrʌkʃənz]	vt.	说明，指令
check	[tʃek]	vt.	检查
performance	[pəˈfɔːməns]	n.	性能
vial	[ˈvaɪəl]	n.	小瓶
label	[ˈleɪbl]	n.	标签，标记
manual	[ˈmænjuəl]	n.	手册
compare	[kəmˈpeə(r)]	vt.	比较
laboratory	[ləˈbɒrətri]	n.	实验室
value	[ˈvæljuː]	n.	(测量)值，数值
match	[mætʃ]	vt.	符合，匹配，与……相当
technique	[tekˈniːk]	n.	技术
prick	[prɪk]	n.	刺，扎
disposable	[dɪˈspəʊzəbl]	adj.	一次性的
insert	[ɪnˈsɜːt]	vt.	插入
reflect	[rɪˈflekt]	vi.	反射
milligram	[ˈmɪlɪɡræm]	n.	毫克
deciliter	[ˈdesɪˌliːtə]	n.	分升，1/10 公升
urgent	[ˈɜːdʒənt]	adj.	急迫的，紧急的
glucose meter			血糖仪
emergency room			急诊室
health care			医疗保健，医护
American Diabetes Association			美国糖尿病协会
a drop of			一滴
test strip			试纸
control solution			对照液，校验液
error code			错误代码
toll-free number			免费电话号码

Key sentences

(1) Glucose meters test and record how much glucose sugar is in your blood (Figure 2-3). They help you track your blood sugar level at different times during the day and night.

血糖仪测试并记录你血液中葡萄糖的含量(图2-3)。它们可以帮助你追踪在白天和黑夜的不同时间段的血糖水平。

(2) You should take glucose test if you have diabetes and you need to monitor your blood sugar levels.

如果你患有糖尿病，你应该进行血糖测试，你需要监测血糖水平。

(3) According to the American Diabetes Association, the blood glucose levels for an adult without diabetes are below 100 mg/dL before meals and fasting, and are less than 140 mg/dL two hours after meals.

根据美国糖尿病协会(的标准)，未患糖尿病的成年人饭前和空腹的血糖水平低于100mg/dL，饭后两小时的血糖水平低于140mg/dL。

(4) Talk to your health care provider about the right glucose meter for you, and how to use it.

向你的医疗护理人员咨询适合你的血糖仪，以及如何使用。

(5) Before you test your blood glucose, you must read and understand the instructions for your meter.

测试血糖之前，您必须阅读并理解仪器的说明书。

Learning test

Choose the following words below to fill in the blanks, change the form if necessary.

diet   diabetes   urgent   recommend   glucose

manual   lancet   solution   complication   check

(1) She didn't know how to _____ the accuracy of the glucose meters.

(2) People with high blood pressure are easily to get _____ .

(3) He died of _____ from diabetes.

(4) Poor _____ and excess smoking will seriously damage your health.

(5) It's an _____ task and must be completed right now.

(6) The solid particles were filtered from the _____ .

(7) Ask your doctor to _____ a suitable temperature meter.

(8) I need a user _____ of the ECG monitor.

(9) The doctor injects _____ solution into the patient's vein.

(10) She had just pricked her finger with the _____ .

Translation

1. Translate the following words into Chinese.

(1) track   (2) monitor

(3) treatment   (4) prick

(5) recommend   (6) accuracy

(7) purchase   (8) interference

(9) likelihood (10) manual

2. Translate the following phrases into Chinese.

(1) a drop of (2) glucose meter

(3) test strip (4) error code

(5) control solution (6) toll-free number

Practice

1. Translate the following sentences into Chinese.

(1) Glucose is a sugar that your body uses as a source of energy.

(2) People with diabetes should consult their doctor or health care provider to set appropriate blood glucose goals.

(3) To test a liquid control solution, you test a drop of these solutions just like you test a drop of your blood.

(4) Ask your provider to watch your testing technique to make sure you are using the meter correctly.

(5) In general, you prick your finger with a lancet to get a drop of blood.

2. Translate the following sentences into English.

(1) 大多数血糖仪的价格在￥300以内。

(2) 血糖仪是一种家用医疗器械。

(3) 这个新型血糖仪的精度很高。

(4) 有些消费者可能无法正确使用血糖仪。

(5) 一滴血足够用来测量血糖水平。

Culture salon

Glucose meters

A glucose meter (or glucometer) is a medical device for determining the approximate concentration of glucose in the blood.

Leland C. Clark Jr. (1918—2005) presented his first paper about the oxygen electrode, later named the Clark electrode, on 15 April 1956, at a meeting of the American society for artificial organs. In 1962, Clark and Ann Lyons from the Cincinnati Children's Hospital developed the first glucose enzyme electrode. Due to this work he is considered the "father of biosensors", especially with respect to the glucose sensing for diabetes patients.

The cost of home blood glucose monitoring can be substantial due to the cost of the test strips. In 2006, the consumer cost of each glucose strip ranged from about $0.35 to $1.00. Manufacturers often provide meters at no cost to induce use of the profitable test strips.

Chapter 2-Section 1-Text 3-看译文

Chapter 2-Section 1-Text 3-拓展阅读

Text 4 Hearing aids

Lead-in questions:

(1) What's a hearing aid?
(2) What's the composition of hearing aids?
(3) What's the function of hearing aids?
(4) How can we get a suitable hearing aid?
(5) How many types of hearing aids mentioned in the text?

Chapter 2-
Section 1-
Text 4-听课文

1. Focusing on function of hearing aids

A hearing aid is a small electronic device that you wear in or behind your ear(Figure 2-4). It makes some sounds louder so that a person with hearing loss can listen, communicate, and participate more fully in daily activities. A hearing aid can help people hear more in both quiet and noisy situations. However, only about one out of five people who would benefit from a hearing aid actually.

2. Composition of hearing aids

A hearing aid has three basic parts: a microphone, amplifier, and speaker. The hearing aid receives sound through a microphone, which converts the sound waves to electrical signals and sends them to an amplifier. The amplifier increases the power of the signals and then sends them to the ear through a speaker.

3. How can hearing aids help?

Hearing aids are primarily useful in improving the hearing and speech comprehension of people who have hearing loss that results from damage to the

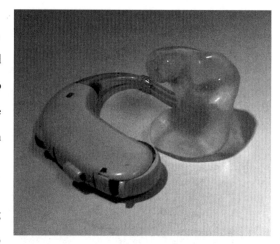

Figure 2-4 A BTE hearing aid with earmold.

small sensory cells in the inner ear, called hair cells. This type of hearing loss is called sensorineural hearing loss. The damage can occur as a result of disease, aging, or injury from noise or certain medicines.

A hearing aid magnifies sound vibrations entering the ear. Surviving hair cells detect the larger vibrations and convert them into neural signals that are passed along to the brain. The greater the damage to a person's hair cells, the more severe the hearing loss, and the greater the hearing aid amplification needed to make up the difference. However, there are practical limits to the amount of amplification a hearing aid can provide. In addition, if the inner ear is too damaged, even large vibrations will not be converted into neural signals. In this situation, a hearing aid would be ineffective.

4. How do I get hearing aids?

Before getting a hearing aid, you should consider having a hearing evaluation to determine the type and amount of your hearing loss. The process can begin with a medical and/or audiological examination.

(1) Medical examination

The medical examination may be performed by any licensed physician including your family doctor or pediatrician, but preferably should be done by an ear, nose, and throat specialist(an otolaryngologist). An examination of your ear, nose, and throat and possibly other testing can be done to rule out any medical reason for your hearing loss, such as infection, injury or deformity, ear wax in the ear canal, and, in rare cases, tumors.

(2) Audiological examination

An audiological exam, or audiogram, involves a hearing evaluation by a hearing health professional that specializes in evaluation, non-medical treatment, and rehabilitation of hearing loss(an audiologist) to identify the type and amount of your hearing loss, to determine the need for medical/surgical treatment and/or referral to a licensed physician, and to provide rehabilitation of the hearing loss.

5. Several types of hearing aids.

Behind-the-ear(BTE) hearing aids consist of a hard plastic case worn behind the ear and connected to a plastic ear mold that fits inside the auris externa. The electronic parts are held in the case behind the ear. Sound travels from the hearing aid through the ear mold and into the ear. BTE aids are used by people of all ages for mild to profound hearing loss.

A new kind of BTE aid is an open-fit hearing aid. Small, open-fit aids fit behind the ear completely, with only a narrow tube inserted into the ear canal, enabling the canal to remain open. For this reason, open-fit hearing aids may be a good choice for people who experience a buildup of earwax, since this type of aid is less likely to be damaged by such substances. In addition, some people may prefer the open-fit hearing aid because their perception of their voice does not sound "plugged up."

In-the-ear(ITE) hearing aids fit completely inside the outer ear and are used for mild to severe hearing loss. The case holding the electronic components is made of hard plastic. Some ITE aids may have certain added features installed, such as atelecoil. A telecoil is a small magnetic coil that allows users to receive sound through the circuitry of the hearing aid, rather than through its microphone. This makes it easier to hear conversations over the telephone. A telecoil also helps people hear in public facilities that have installed special sound systems, called induction loop systems. Induction loop systems can be found in many churches, schools, airports, and auditoriums. ITE aids usually are not worn by young children because the casings need to be replaced often as the ear grows.

Canal aids fit into the ear canal and are available in two styles. The in-the-canal(ITC) hearing aid is made to fit the size and shape of a person's ear canal. A completely-in-canal(CIC) hearing aid is nearly hidden in the ear canal. Both types are used for mild to moderately severe hearing loss.

Because they are small, canal aids may be difficult for a person to adjust and remove. In addition, canal aids have less space available for batteries and additional devices, such as atelecoil. This kind of hearing aids usually are not recommended for young children or for people with severe to profound hearing loss because their reduced size limits their power and volume.

Hearing aids take time and patience to use successfully. Wearing your aids regularly will help you adjust to them.

Words and phrases

participate	[pɑːˈtɪsɪpeɪt]	vi.	参与,参加
situation	[ˌsɪtʃuˈeɪʃn]	n.	情况,局面
benefit	[ˈbenɪfɪt]	vi.	受益,获益
amplifier	[ˈæmplɪfaɪə(r)]	n.	放大器,扩音器
speaker	[ˈspiːkə(r)]	n.	扬声器
convert	[kənˈvɜːt]	vt.	转变,转换
signal	[ˈsɪɡnəl]	n.	信号
primarily	[praɪˈmerəli]	adv.	主要,首要
sensory	[ˈsensəri]	adj.	感觉的,感受的
cell	[sel]	n.	细胞
sensorineural	[ˌsensərɪˈnjʊərəl]	adj.	感觉神经的
magnify	[ˈmæɡnɪfaɪ]	vt.	放大
vibration	[ˈvaɪbreɪʃənz]	n.	振动
survive	[səˈvaɪv]	vi.	幸存,存活
neural	[ˈnjʊərəl]	adj.	神经的
practical	[ˈpræktɪkl]	adj.	实际的,实用的
amplification	[ˌæmplɪfɪˈkeɪʃn]	n.	放大,扩大
pediatrician	[ˌpiːdɪəˈtrɪʃn]	n.	儿科医师
throat	[θrəʊt]	n.	喉,咽喉,喉部
otolaryngologist	[ˌəʊtələˈrɪŋɡɒlədʒɪst]	n.	耳鼻喉科医生
infection	[ɪnˈfekʃn]	n	感染,传染
deformity	[dɪˈfɔːməti]	n.	畸形,残疾
wax	[wæks]	n.	耳垢,耳屎
tumor	[ˈtjuːmə]	n.	肿瘤
audiogram	[ˈɔːdɪəʊɡræm]	n.	听力图
specialize	[ˈspeʃəlaɪz]	vi.	专门从事,专攻
evaluation	[ɪˌvæljʊˈeɪʃn]	n.	诊断,评估
rehabilitation	[ˌriːəˌbɪlɪˈteɪʃn]	n.	康复,复原
audiologist	[ˌɔːdɪˈɒlədʒɪst]	n.	听力学家,听觉病矫治专家
plastic	[ˈplæstɪk]	adj.	塑料的

mold	[məuld]	n.	模,模子,模具
mild	[maɪld]	adj.	轻度的,轻微的
profound	[prə'faʊnd]	adj.	重度的,严重的
buildup	['bɪldˌʌp]	n.	累积,积聚
earwax	['ɪəwæks]	n.	耳垢
substance	['sʌbstənsiz]	n.	物质
perception	[pə'sepʃn]	n.	感知,知觉
install	[ɪns'tɔːl]	vt.	安装
circuitry	['sɜːkɪtri]	n.	电路,线路
component	[kəm'pəʊnənt]	n.	元件,零件,组件
hearing aid			助听器
hearing loss			听力损失
ear canal			耳道
ear wax			耳垢
magnetic coil			电磁线圈
open-fit			开放式适配
plugged up			堵塞,堵住
induction loop system			感应线圈系统
licensed physician			执业医师
hair cell			毛细胞
outer ear			外耳
inner ear			内耳
rule out			排除……的可能性
telecoil			拾音线圈

Key sentences

(1) A hearing aid is a small electronic device that you wear in or behind your ear (Figure 2-4). It makes some sounds louder so that a person with hearing loss can listen, communicate, and participate more fully in daily activities.

助听器是一个小的电子仪器,可以戴在耳内或者耳后(图2-4)。它可以使声音更响亮,使有听力障碍的人在日常生活中能够更好的聆听、交流和参与活动。

(2) A hearing aid has three basic parts: a microphone, amplifier, and speaker.

助听器有三个基本部件:麦克风、扩音器和扬声器。

(3) Hearing aids are primarily useful in improving the hearing and speech comprehension of people who have hearing loss that results from damage to the small sensory cells in the inner ear, called hair cells.

助听器主要用于改善听力受损的人的听力和语言理解能力,这些损伤是由于内耳中微小的感觉细胞受损所致,称为毛细胞。

(4) Before getting a hearing aid, you should consider having a hearing evaluation to determine the type and amount of your hearing loss.

在获取助听器之前,你应该考虑做一个听力评估,以确定听力损伤的类型和程度。

(5) Behind-the-ear(BTE) hearing aids consist of a hard plastic case worn behind the ear and connected to a plastic ear mold that fits inside the auris externa.

耳背式(BTE)助听器包括一个硬塑料盒,硬塑料盒戴在耳后并连接到一个适配在外耳内的塑料耳模中。

Learning test

Choose the following words below to fill in the blanks, change the form if necessary.

| throat | magnify | tumor | situation | profound |
| amplifier | plastic | benefit | substance | participate |

(1) In an accident _____, please contact the nearest police.

(2) Telescope can _____ the beautiful scenery.

(3) She cleared her _____, and started singing.

(4) It wasn't cancer, only a benign _____.

(5) China will actively _____ in international affairs.

(6) These _____ bags cannot be recycled.

(7) Cigarettes contain poisonous _____.

(8) Many small companies _____ from tax reduction.

(9) An _____ is a kind of electrical device which can make the sounds louder.

(10) His illness is _____ now, and will lead to the other complication.

Translation

1. Translate the following words into Chinese.

(1) mild   (2) deformity
(3) sensorineural   (4) audiological
(5) infection   (6) otolaryngologist
(7) tumor   (8) rehabilitation
(9) speaker   (10) signal

2. Translate the following phrases into Chinese.

(1) hearing aid   (2) hearing loss
(3) inner ear   (4) hair cell
(5) ear canal   (6) licensed physician

Practice

1. Translate the following sentences into Chinese.

(1) The hearing aid receives sound through a microphone, which converts the sound waves to electrical signals and sends them to an amplifier.

(2) The amplifier increases the power of the signals and then sends them to the ear through a speaker.

(3) The damage can occur as a result of disease, aging, or injury from noise or certain medicines.

(4) Sound travels from the hearing aid through the ear mold and into the ear.

(5) This kind of hearing aids usually are not recommended for young children or for people with severe to profound hearing loss because their reduced size limits their power and volume.

2. Translate the following sentences into English.

(1)声音是由物体振动产生。

(2)助听器一般都有一个小型的内置电池。

(3)这款助听器体积最小,但价格较高。

(4)老年性耳聋患者更适合佩戴全数字助听器。

(5)他耳朵里戴着助听器。

Culture salon

Hearing aids

Having trouble hearing? Over 35 million children and adults in the United States have some degree of hearing loss. Hearing loss can have a negative effect on communication, relationships, school/work performance, and emotional well-being. However, hearing loss doesn't have to restrict your daily activities. Properly fitted hearing aids and aural rehabilitation (techniques used to identify and diagnose hearing loss, and implement therapies for patients who are hard of hearing, including using amplification devices to aid the patient's hearing abilities), can help in many listening situations. Aural rehabilitation helps a person focus on adjusting to their hearing loss and the use of their hearing aids. Most people who are hearing-impaired will need two hearing aids as both ears may be affected by hearing loss, though some people may only need one hearing aid.

Chapter 2-Section 1-Text 4-看译文

Chapter 2-Section 1-Text 4-拓展阅读

Chapter 2-Section 1-习题

Section 2 Electrocardiograph (ECG) monitors

Chapter 2-Section 2-PPT

Text 1 Introduction to electrocardiograph (ECG) monitors

Lead-in questions:

(1) What is the function of ECG monitors?

(2) Can you list at least three types of heart damage?

(3) How many metrics for the heart's electrical impulses used to determine potential heart damage?

(4) Does an ECG machine send any electricity through your body?

(5) Is the ECG test foolproof?

Electrocardiograph monitors, which is abbreviated as either EKG or ECG monitors (Figure 2-5), is a type of medical device that reads the heart impulses and translates them to a graph-style diagram on specially designed paper strips. To put it simply, an EKG is a test of your heartbeat. Every time your heart beats, an electric wave travels through the heart, which causes it to pump blood. By conducting an electrocardiography, a process of recording an electrocardiograph, doctors can use information about your heart's electrical activity to determine the condition of your heart.

Chapter 2-Section 2-Text 1-听课文

By measuring the intervals between the heart's electrical impulses, a doctor can determine if your heartbeat is regular or irregular, and by measuring the strength of each electrical impulse, the doctor can determine if your heart is working too hard or not hard enough. These two metrics can indicate several types of heart damage, including:

- arrhythmia
- congenital heart defects
- heart valve disease
- pericarditis (an inflammation of the sac surrounding the heart)
- hypertrophy of the heart chamber walls or blood vessels
- heart failure

Figure 2-5 An ECG monitor.

Knowing that you're at risk for heart failure could inspire you to eat better, exercise more and address the risk factors you can control. An ECG is the best way to figure out what your case calls for. It's possible your doctor may ask for an ECG when you're completely fine in order to have baseline data to use later on. You may need an ECG to be cleared for some types of surgery. If you've begun taking new medications or had a pacemaker implanted, the doctor may want to see how those things are affecting your heart. Many times, though, an ECG is ordered when a patient complains of chest pain or other symptoms of heart disease, including shortness of breath, dizziness, weakness or palpitations. Oftentimes, people will ignore such symptoms because they're scared of what the doctor will say. Rather than living with the fact that your heart pounds, races or flutters, it's best to submit to the EKG and find out what's really going on.

It may sound risky to have electricity and your heart in such close proximity, but it's not. That's how your heart works, and an ECG machine doesn't send any electricity through your body, it merely measures the electricity already in your heart. The test isn't painful, either. The only risk of physical hurt may be when they remove the stickers holding the electrodes to your chest. All you'll need to do is show up and lie still on an examining table(one exception: If a doctor calls for a stress ECG, you may need to perform a little exercise). The test should be over in five to 10 minutes, and you'll likely have results almost instantaneously.

ECG monitors include a set of lead wires. The majority of machines use 10 or 12 leads. In a conventional 12-lead ECG, 10 electrodes are placed on the patient's limbs and on the surface of the chest. The overall magnitude of the heart's electrical potential is then measured from 12 different angles("leads") and is recorded over a period of time(usually 10 seconds). In this way, the overall magnitude and direction of the heart's electrical depolarization is captured at each moment throughout the cardiac cycle. The graph of voltage versus time produced by this noninvasive medical procedure is referred to as an electrocardiogram.

Sometimes, certain factors can interfere with the reliability of the electrocardiograph test. Stress and anxiety can cause a rapid and irregular heart rate that would not normally be present under normal circumstances. In addition, consumption of certain medications such as cold and allergy medications may have skew results because they stimulate the heart. Generally, people who drink coffee and consume other foods and drinks containing caffeine may have abnormal electrocardiograph results.

Frequently, cardiac medications called beta blockers can mask certain conditions of the heart. Typically beta blockers slow and regulate the heart rate. Many times when the beta blocker is discontinued, the heart rhythm and rate will revert back to abnormal. It is important to tell the health care provider when taking a beta blocker, or any cardiac medication while undergoing an electrocardiograph evaluation. Pacemakers also slow and regulate the heart, which may also show up on the electrocardiograph as an abnormality.

Generally, based on results of the electrocardiograph, other cardiac tests may be recommended. If the test shows an abnormality, an echocardiogram may be suggested. This test uses sound waves that are bounced off the cardiac structures to visualize the heart, valves, and vessels. In addition, a stress test may be needed to determine if a coronary blockage is present. The electrocardiograph can suggest the presence of cardiac ischemia, which may indicate a blockage.

The electrocardiograph test sometimes elicits false positive or false negative results. Because the test is not foolproof, it is important to supplement it with a thorough physical examination and medical history. Sometimes, cardiac enzyme blood tests will be done to rule out the presence of a myocardial infarction, or heart attack. When the cardiac muscle is damaged, cardiac enzymes slough off into the blood stream and are revealed in this blood test. A combination of diagnostic tests are often needed to rule out cardiac events.

Words and phrases

electrocardiograph	[ɪˌlektrəʊˈkɑːdɪəɡrɑːf]	n.	心电图
abbreviate	[əˈbriːvɪeɪt]	vt.	简写,缩略
impulse	[ˈɪmpʌls]	n.	冲动,搏动
diagram	[ˈdaɪəɡræm]	n.	图表,示意图
heartbeat	[ˈhɑːtbiːt]	n.	心跳,心搏
interval	[ˈɪntevl]	n.	间隔
metric	[ˈmetrɪk]	n.	指标,度量标准
arrhythmia	[əˈrɪθmɪə]	n.	心律失常,心律不齐
congenital	[kənˈdʒenɪtl]	adj.	先天性的,先天的,天生的
pericarditis	[ˌperɪkɑːˈdaɪtɪs]	n.	心包炎
inflammation	[ˌɪnfləˈmeɪʃn]	n.	炎症
sac	[sæk]	n.	心包,囊
hypertrophy	[haɪˈpɜːtrəfi]	n.	肥大,肥厚
baseline	[ˈbeɪslaɪn]	n.	基线
pacemaker	[ˈpeɪsmeɪkə(r)]	n.	心脏起搏器
implant	[ɪmˈplɑːnt]	vt.	植入,移植
breath	[breθ]	n.	呼吸
dizziness	[ˈdɪzɪnəs]	n.	头晕
palpitation	[ˌpælpɪˈteɪʃn]	n.	心悸
oftentimes	[ˈɒfntaɪmz]	adv.	通常情况下,时常
pound	[paʊnd]	vi.	(心脏)狂跳
race	[reɪs]	vi.	剧烈跳动
flutter	[ˈflʌtə(r)]	vi.	快速跳动,扑动
risky	[ˈrɪski]	adj.	危险的,冒险的
proximity	[prɒkˈsɪməti]	n.	接近,邻近
instantaneously	[ˌɪnstənˈteɪniəsli]	adv.	即刻,立刻
conventional	[kənˈvenʃənl]	adj.	一般的,常规的
potential	[pəˈtenʃl]	n.	电位,电势
depolarization	[ˌdiːpəʊləraɪˈzeɪʃən]	n.	去极化,除极

voltage	[ˈvəʊltɪdʒ]	n.	电压
noninvasive	[ˌnɒnɪnˈveɪsɪv]	adj.	无创的
circumstance	[ˈsɜːkəmstəns]	n.	状况,环境
consumption	[kənˈsʌmpʃn]	n.	消耗,消费
allergy	[ˈælədʒi]	n.	过敏
stimulate	[ˈstɪmjuleɪt]	vt.	刺激
caffeine	[ˈkæfiːn]	n.	咖啡因
cardiac	[ˈkɑːdiæk]	adj.	心脏(病)的
echocardiogram	[ˈekəʊˈkɑːdɪəgræm]	n.	超声心动图
visualize	[ˈvɪʒuəlaɪz]	vt.	使可见,可视化
elicit	[iˈlisit]	vt.	引起,产生
foolproof	[ˈfuːlpruːf]	adj.	万无一失的
supplement	[ˈsʌplɪmənt]	vt.	增补,补充
enzyme	[ˈenzaɪm]	n.	酶
graph-style			图式
electrical impulse			电脉冲
congenital heart defect			先天性心脏缺陷
heart valve			心脏瓣膜
heart chamber			心室
blood vessel			血管
heart failure			心脏衰竭
figure out			弄清楚,解决
cardiac cycle			心动周期
beta blocker			β-受体阻断药
coronary blockage			冠状动脉堵塞
cardiac ischemia			心肌缺血
myocardial infarction			心肌梗死
cardiac enzyme			心肌酶
slough off			脱落

Key sentences

(1) Electrocardiograph monitors, which is abbreviated as either EKG or ECG monitors(Figure 2-5), is a type of medical device that reads the heart impulses and translates them to a graph-style diagram on specially designed paper strips.

心电监护仪,简称为 EKG 或 ECG 监护仪(图 2-5),是一种医疗器械,这种器械可以读取心电脉冲,并将其转换成在专门设计的纸带上的图表样式的图。

(2) By measuring the intervals between the heart's electrical impulses, a doctor can determine if your heartbeat is regular or irregular, and by measuring the strength of each electrical impulse, the doctor can determine if your heart is working too hard or not hard enough.

通过测量心脏电脉冲之间的间隔,医生可以确定你的心跳是正常还是异常。通过测量每个电脉冲的强度,医生可以确定你的心脏跳动是太强或者不够强。

(3) Knowing that you're at risk for heart failure could inspire you to eat better, exercise more and address the risk factors you can control.

知道你有患心脏衰竭的风险,这能激励你吃得更好,多锻炼,解决你可以控制的风险因素。

(4) ECG monitors include a set of lead wires. The majority of machines use 10 or 12 leads. In a conventional 12-lead ECG, 10 electrodes are placed on the patient's limbs and on the surface of the chest.

心电监护仪包括一组导联线。大多数仪器使用 10 或 12 个导联。在常规的 12 导联心电图中,大约有 10 个电极要放在患者的四肢和胸部表面。

(5) Sometimes, certain factors can interfere with the reliability of the electrocardiograph test.

有时候,某些因素会影响心电图测试结果的准确性。

Learning test

Choose the following words below to fill in the blanks, change the form if necessary.

dizziness　　instantaneously　　allergy　　cardiac　　visualize
consumption　elicit　　congenital　　inflammation　　breath

(1) The excessive loss of blood will cause lack of oxygen almost _____ .

(2) She felt _____ suddenly and had to lie down.

(3) John died of certain _____ heart disease when he was 17 years old.

(4) The drug can cause _____ of the liver.

(5) He took a deep _____ before entering the interview room.

(6) The _____ cycle can be shown on an electrocardiogram.

(7) I tried to _____ a smile from Serena.

(8) I use a graph to _____ the signal change.

(9) _____ to cats is a common cause of asthma.

(10) Gas and oil _____ will lead to worse weather and air condition.

Translation

1. Translate the following words into Chinese.

(1) electrocardiograph　　　　　　(2) arrhythmia

(3) pericarditis　　　　　　　　　(4) impulse

(5) palpitation　　　　　　　　　 (6) dizziness

(7) baseline　　　　　　　　　　 (8) congenital

(9) caffeine　　　　　　　　　　 (10) enzyme

2. Translate the following phrases into Chinese.

(1) cardiac cycle   (2) heart failure

(3) heart valve   (4) heart chamber

(5) cardiac enzyme   (6) cardiac ischemia

Practice

1. Translate the following sentences into Chinese.

(1) Every time your heart beats, an electric wave travels through the heart, which causes it to pump blood.

(2) If you've begun taking new medications or had a pacemaker implanted, the doctor may want to see how those things are affecting your heart.

(3) Generally, based on results of the electrocardiograph, other cardiac tests may be recommended.

(4) It may sound risky to have electricity and your heart in such close proximity, but it's not.

(5) The electrocardiograph test sometimes elicits false positive or false negative results.

2. Translate the following sentences into English.

(1) 心电图机从体表测量和记录心脏的电活动。

(2) 心电图在临床诊断中应用广泛。

(3) 心电图可用于诊断心律失常。

(4) 心电图机并不产生电信号,所以对人体是安全的。

(5) 心脏有2个心室和2个心房。

Culture salon

The human heart and the origin of ECG wave

The human heart consists of two atria and two ventricles that make up four chambers. The heart functions by pumping blood from the outside of the body into the atria, through the ventricles and back into the body. During each heartbeat, the muscles of the heart chambers are first polarized, meaning that they have electrical charges. In order to generate the energy to contract, these muscles must "spend" their electrical charge and depolarize. The chambers must then recharge in preparation for the next contraction, and this is referred to as repolarization. Depolarization and repolarization describe changes in electric potential that can be detected by electrodes placed on the skin and represented by the ECG wave.

Chapter 2-Section 2-Text 1-看译文

Chapter 2-Section 2-Text 1-拓展阅读

Text 2 How an ECG monitor works

Lead-in questions:

(1) How an ECG monitor works?
(2) How many electrodes are there for a standard 12-lead ECG?
(3) What's the use of conductive electrolyte gel?
(4) What is a lead?
(5) What is the function of the augmented limb leads?

Chapter 2-
Section 2-
Text 2-听课文

ECG monitors works by performing electrocardiography, and produces the electrocardiogram. Electrocardiography is the process of recording the electrical activity of the heart over a period of time using electrodes placed on the skin. These electrodes detect the tiny electrical changes on the skin that arise from the heart muscle's electrophysiologic pattern of depolarizing during each heartbeat. The detected heart impulses are then translated to lead signals to form graph-style diagram on specially designed paper strips or display monitors.

The mechanical work of the heart is accompanied by electrical activities, which always slightly precede the muscle contraction. During each heartbeat, a healthy heart has an orderly progression of depolarization that starts with pacemaker cells in the sinoatrial node, spreads out through the atrium, passes through the atrioventricular node down into the bundle of His and into the Purkinje fibers, spreading down and to the left throughout the ventricles. This orderly pattern of depolarization gives rise to the characteristic ECG tracing. The electrical activities can be studied by picking up electrical potentials on the body surface. During each cardiac cycle, the potential distribution changes in a regular, but quite complex way.

1. Electrodes

An electrode is a conductive pad in contact with the body that makes an electrical circuit with the machine. On a standard 12-lead ECG there are only 10 electrodes, which are listed below, and each electrode should be applied to proper place, as shown in Figure 2-6 and Figure 2-7. Two common electrodes used are a flat paper-thin sticker and a self-adhesive circular pad. The former are typically used in a single ECG recording while the latter are for continuous recordings as they stick longer. Each electrode consists of an electrically conductive electrolyte gel and a silver/silver chloride conductor. The gel typically contains potassium chloride-sometimes silver chloride as well-to permit electron conduction from the skin to the wire and to the electrocardiogram.

Electrode name	Electrode placement
RA	On the right arm, avoiding thick muscle
LA	In the same location where RA was placed, but on the left arm
RL	On the right leg, lateral calf muscle

Electrode name	Electrode placement
LL	In the same location where RL was placed, but on the left leg
V1	In the fourth intercostal space (between ribs 4 and 5) just to the right of the sternum (breastbone)
V2	In the fourth intercostal space (between ribs 4 and 5) just to the left of the sternum
V3	Between leads V2 and V4
V4	In the fifth intercostal space (between ribs 5 and 6) in the mid-clavicular line
V5	Horizontally even with V4, in the left anterior axillary line
V6	Horizontally even with V4 and V5 in the midaxillary line

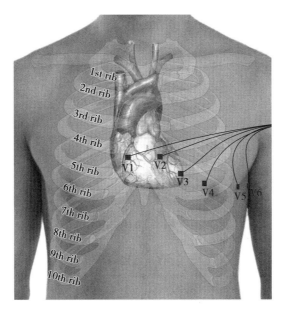

Figure 2-6 Proper placement of the precordial electrodes.

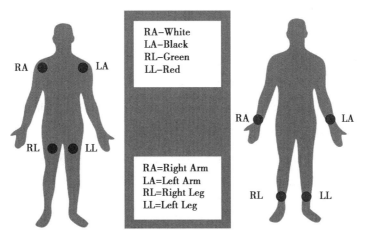

Figure 2-7 Proper placement of the limb electrodes. The limb electrodes can be far down on the limbs or close to the hips/shoulders as long as they are placed symmetrically.

2. Leads

There is a careful distinction between "electrode" and "lead". A lead is the source of measurement of a vector. Most of ECG monitors include 10 or 12 leads. Leads are broken down into three sets: limb, augmented limb, and precordial. The 12-lead ECG has a total of three limb leads and three augmented limb leads arranged like spokes of a wheel in the coronal plane(vertical) and six precordial leads that lie on the perpendicular transverse plane(horizontal).

The common lead, Wilson's central terminal V_w, is produced by averaging the measurements from the electrodes RA, LA and LL to give an average potential across the body:

$$V_w = \frac{1}{3}(RA+LA+LL)$$

In a 12-lead ECG, all leads except the limb leads are unipolar(aVR, aVL, aVF, V1, V2, V3, V4, V5, and V6). The measurement of a voltage requires two contacts and so, electrically, the unipolar leads are measured from the common lead(negative) and the unipolar lead(positive). This averaging for the common lead and the abstract unipolar lead concept makes for a more challenging understanding and is complicated by sloppy usage of "lead" and "electrode".

(1) Limb leads:

Leads Ⅰ, Ⅱ and Ⅲ are called the limb leads. The electrodes that form these signals are located on the limbs—one on each arm and one on the left leg, as shown in Figure 2-8. The limb leads form the points of what is known as Einthoven's triangle.

Lead Ⅰ is the voltage between the(positive) left arm(LA) electrode and right arm(RA) electrode: Ⅰ = LA-RA.

Lead Ⅱ is the voltage between the(positive) left leg(LL) electrode and the right arm(RA) electrode: Ⅱ = LL-RA.

Lead Ⅲ is the voltage between the(positive) left leg(LL) electrode and the left arm(LA) electrode: Ⅲ = LL-LA.

(2) Augmented limb leads

Leads aVR, aVL, and aVF are the augmented limb leads. They are derived from the same three electrodes as leads Ⅰ, Ⅱ, and Ⅲ, but they use Goldberger's central terminal as their negative pole which is a combination of inputs from other two limb electrodes, as shown in Figure 2-8.

- Lead augmented vector right(aVR) has the positive electrode on the right arm. The negative pole is a combination of the left arm electrode and the left leg electrode:

$$aVR = RA - \frac{1}{2}(LA+LL) = \frac{3}{2}(RA-V_w)$$

- Lead augmented vector left(aVL) has the positive electrode on the left arm. The negative pole is a combination of the right arm electrode and the left leg electrode:

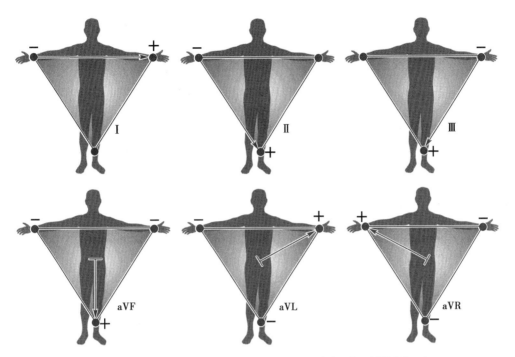

Figure 2-8 The limb leads and augmented limb leads of ECG leads.

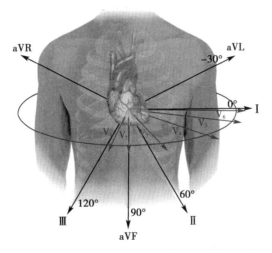

Figure 2-9 The spatial orientation of ECG leads.

$$aVL = LA - \frac{1}{2}(RA+LL) = \frac{3}{2}(LA - V_w)$$

• Lead augmented vector foot (aVF) has the positive electrode on the left leg. The negative pole is a combination of the right arm electrode and the left arm electrode:

$$aVF = LL - \frac{1}{2}(RA+LA) = \frac{3}{2}(LL - V_w)$$

Together with leads I, II, and III, augmented limb leads aVR, aVL, and aVF form the basis of the hexaxial reference system, which is used to calculate the heart's electrical axis in the frontal plane.

(3) Precordial leads

The precordial leads lie in the transverse (horizontal) plane, perpendicular to the other six leads. The six precordial electrodes act as the positive poles for the six corresponding precordial leads: (V1, V2, V3, V4, V5 and V6). Wilson's central terminal is used as the negative pole.

Words and phrases

electrocardiography	[ɪˌlektrəʊkɑːdɪˈɒɡrəfɪ]	n.	心电描迹
electrode	[ɪˈlektrəʊd]	n.	电极
skin	[skɪn]	n.	皮肤
arise	[əˈraɪz]	vi.	产生，出现
muscle	[ˈmʌsl]	n.	肌肉
electrophysiologic	[ɪˌlektrəʊˌfɪzɪˈɒlədʒɪk]	adj.	电生理的
pattern	[ˈpætn]	n.	模式，图案
depolarize	[diːˈpəʊləraɪz]	vt.	去极化
mechanical	[məˈkænɪkl]	adj.	机械的
precede	[prɪˈsiːd]	vt.	先于，在…之前发生或出现
contraction	[kənˈtrækʃn]	n.	收缩
atrium	[ˈeɪtriəm]	n.	心房
fiber	[ˈfaɪbə]	n.	纤维
ventricle	[ˈventrɪkl]	n.	心室
conductive	[kənˈdʌktɪv]	adj.	导电的，传导的
pad	[pæd]	n.	盘，垫，衬垫
circuit	[ˈsɜːkɪt]	n.	电路，线路
sticker	[ˈstɪkə(r)]	n.	贴片
circular	[ˈsɜːkjələ(r)]	adj.	圆形的
electrolyte	[ɪˈlektrəlaɪt]	n.	电解质
gel	[dʒel]	n.	凝胶
silver	[ˈsɪlvə(r)]	n.	银
chloride	[ˈklɔːraɪd]	n.	氯化物
potassium	[pəˈtæsiəm]	n.	钾
intercostal	[ˌɪntəˈkɒstl]	adj.	肋间的
sternum	[ˈstɜːnəm]	n.	胸骨
anterior	[ænˈtɪəriə(r)]	adj.	前面的，前部的
axillary	[ækˈsɪləri]	adj.	腋窝的
transverse	[ˈtrænzvɜːs]	adj.	横向的，横断的
perpendicular	[ˌpɜːpənˈdɪkjələ(r)]	adj.	垂直的，成直角的

precordial	[ˈpriːkɔːdjəl]	adj.	心前(区)的
lead	[liːd]	n.	导联
horizontal	[ˌhɒrɪˈzɒntl]	adj.	水平的
unipolar	[ˈjuːnɪˈpəʊlə]	adj.	单极的
voltage	[ˈvəʊltɪdʒ]	n.	电压
contact	[ˈkɒntækts]	n.	触点
negative	[ˈnegətɪv]	adj.	负的, 负极的
positive	[ˈpɒzətɪv]	adj.	正的, 正极的
augmented	[ɔːgˈmentɪd]	adj.	增强的
sinoatrial node			窦房结
atrioventricular node			房室结
bundle of His			希氏束
Purkinje fibers			浦肯野纤维
self-adhesive			自粘的
cardiac cycle			心动周期
lateral calf			小腿外侧
mid-clavicular			锁骨
midaxillary line			腋中线
coronal plane			冠状平面
Einthoven's triangle			爱因托芬三角
Goldberger's central terminal			戈德伯格中心端
electrical axis			电轴
Wilson's central terminal			威尔逊中心端

Key sentences

(1) Electrocardiography is the process of recording the electrical activity of the heart over a period of time using electrodes placed on the skin.

心电描迹是用放置在皮肤上的电极在一段时间内记录心脏的电活动的过程。

(2) The mechanical work of the heart is accompanied by electrical activities, which always slightly precede the muscle contraction.

心脏的机械运动伴随着电活动,电活动总是稍早于肌肉收缩。

(3) An electrode is a conductive pad in contact with the body that makes an electrical circuit with the machine. On a standard 12-lead ECG there are only 10 electrodes, and each electrod should be applied to proper place, as shown in Figure 2-6 and Figure 2-7.

电极是与身体接触的导电片,与仪器一起构成电路。在一个标准的 12 导联心电图中,有 10 个电极,每个电极应放在适当的位置,如图 2-6、图 2-7 所示。

(4) The 12-lead ECG has a total of three limb leads and three augmented limb leads arranged like spokes of a wheel in the coronal plane (vertical) and six precordial leads that lie on the perpendicular transverse plane (horizontal).

12 导联心电图共有三个肢体导联和三个加压肢体导联,像轮子上的辐条一样在冠状平面上排列,六个胸导联垂直放置在横向平面上。

(5) Leads Ⅰ, Ⅱ and Ⅲ are called the limb leads. The electrodes that form these signals are located on the limbs—one on each arm and one on the left leg, as shown in Figure 2-8.

导联Ⅰ、Ⅱ和Ⅲ称为肢体导联。产生这些信号的电极位于四肢上,每只胳膊上有一个,左腿上有一个,如图 2-8 所示。

Learning test

Choose the following words below to fill in the blanks, change the form if necessary.

horizontal   arise   gel   positive   electrolyte
fiber   electrode   circular   silver   precede

(1) He connected the pair of _____ to form a closed loop circuit.

(2) The conductive _____ can be used to improve the contact effect of the electrodes to the skin.

(3) _____ is a kind of noble metal.

(4) Lightning always _____ thunder.

(5) Electric current always flows from _____ node to the negative node.

(6) The battery consists of an internal _____ solution and two external metal electrodes.

(7) He adjusted the four legs of the instrument so that it is _____ .

(8) The complications may _____ from diabetes.

(9) The full moon is _____ in shape.

(10) In a 12-lead ECG, all leads except the limb leads are _____ .

(11) Eye cornea contains collagen _____ .

Translation

1. Translate the following words into Chinese.

(1) skin   (2) muscle

(3) electrolyte   (4) sternum

(5) conductive   (6) contraction

(7) atrium   (8) voltage

(9) perpendicular   (10) fiber

2. Translate the following phrases into Chinese.

(1) sinoatrial node   (2) cardiac cycle

(3) mid-clavicular   (4) Wilson's central terminal

(5) coronal plane　　　　　　(6) atrioventricular node

Practice

1. Translate the following sentences into Chinese.

(1) The electrical activities can be studied by picking up electrical potentials on the body surface.

(2) Each electrode consists of an electrically conductive electrolyte gel and a silver/silver chloride conductor.

(3) The gel typically contains potassium chloride-sometimes silver chloride as well-to permit electron conduction from the skin to the wire and to the electrocardiogram.

(4) The common lead, Wilson's central terminal V_w, is produced by averaging the measurements from the electrodes RA, LA, and LL to give an average potential across the body.

(5) Together with leads Ⅰ, Ⅱ and Ⅲ, augmented limb leads aVR, aVL, and aVF form the basis of the hexaxial reference system, which is used to calculate the heart's electrical axis in the frontal plane.

2. Translate the following sentences into English.

(1) 两个电极之间组成了双极导联,一个导联为正极,另一个导联为负极。

(2) 威尔森中央电端,是通过一个电阻网络将肢体电极连接而产生的,代表了身体的平均电压。

(3) 心电图的横坐标表示时间,纵坐标表示电压。

(4) 正常心脏的电活动从窦房结开始。

(5) 心房与心室之间有心脏瓣膜,这些瓣膜使血液只能由心房流入心室,而不能倒流。

Culture salon

Understanding ECG waves

Normal rhythm produces four entities — a P wave, a QRS complex, a T wave, and a U wave — that each have a fairly unique pattern (Figure 2-10).

- The P wave represents atrial depolarization.
- The QRS complex represents ventricular depolarization.
- The T wave represents ventricular repolarization.
- The U wave represents papillary muscle repolarization.

However, the U wave is not typically seen and its absence is generally ignored. Changes in the structure of the heart and its surroundings (including blood composition) change the patterns of these four entities.

Chapter 2-Section 2-Text 2-看译文

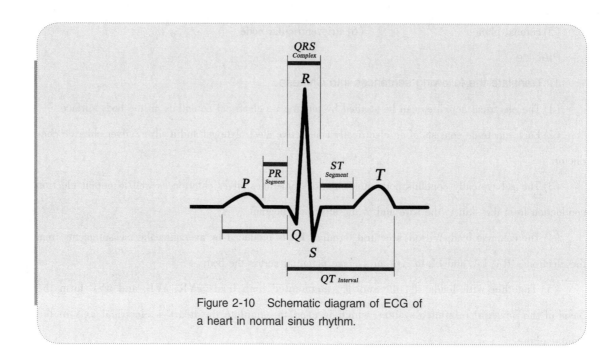

Figure 2-10 Schematic diagram of ECG of a heart in normal sinus rhythm.

Chapter 2-Section 2-Text 2-拓展阅读

Chapter 2-Section 2-习题

Section 3 Ultrasound imaging

Chapter 2-Section 3-PPT

Text 1 Introduction to ultrasound imaging

Lead-in questions:

(1) What is ultrasound imaging?
(2) Why we have to get an ultrasound imaging?
(3) How many procedures are included in common ultrasound imaging?
(4) What are the benefits and risks with ultrasound imaging?
(5) What's the difference between 3D ultrasound and 4D ultrasound?

Chapter 2-Section 3-Text 1-听课文

Ultrasound imaging(sonography) is a diagnostic medical procedure that uses high-frequency sound waves to produce dynamic visual images of organs, tissues or blood flow inside the body. Ultrasound is sound waves with frequencies(typically greater than 20 kHz) higher than the upper audible limit of human hearing.

In an ultrasound exam, a transducer(probe) is placed directly on the body to be visualized. A thin layer of gel is applied to the skin so that the ultrasound waves are transmitted from the transducer through the gel into the body. The sound waves are transmitted to the area to be examined and the returning echoes are captured to provide the physician with a "live" image of the area. Sound energy is attenuated or weakened as it passes through tissue because parts of it are reflected, scattered, absorbed, refracted or diffracted. The strength(amplitude) of the sound signal and the time it takes for the wave to travel through the body provide the information necessary to produce an image.

Ultrasound has been a popular medical imaging technique for many years(Figure 2-11) . There are many reasons to get an ultrasound. Perhaps you're pregnant, and your obstetrician wants you to have an ultrasound to check on the developing baby or determine the due date. Because ultrasound images are captured in real-time, they can also show movement of the body's internal organs as well as blood flowing through the blood vessels.

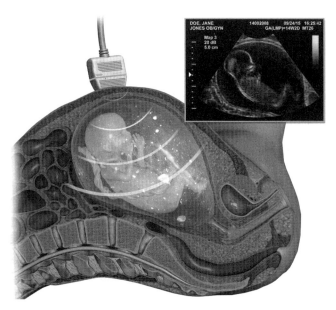

Figure 2-11　A fetal ultrasound exam.

1. Uses

Ultrasound imaging is a medical tool that can help a physician evaluate, diagnose and treat medical conditions. Common ultrasound imaging procedures include:

- Abdominal ultrasound(to visualize abdominal tissues and organs)
- Bone sonometry(to assess bone fragility)
- Breast ultrasound(to visualize breast tissue)
- Doppler fetal heart rate monitors(to listen to the fetal heart beat)
- Doppler ultrasound(to visualize blood flow through a blood vessel, organs, or other structures)
- Echocardiogram(to view the heart)

- Fetal ultrasound(to view the fetus in pregnancy)
- Ultrasound-guided biopsies(to collect a sample of tissue)
- Ophthalmic ultrasound(to visualize ocular structures)
- Ultrasound-guided needle placement(in blood vessels or other tissues of interest)

2. Benefits and risks

Ultrasound imaging has been used for over 20 years and has an excellent safety record. Following aspects are the benefits:

- Most ultrasound scanning is noninvasive(no needles or injections).
- Occasionally, an ultrasound exam may be temporarily uncomfortable, but it should not be painful.
- Ultrasound is widely available, easy-to-use and less expensive than other imaging methods.
- Ultrasound imaging is extremely safe and does not use any ionizing radiation.
- Ultrasound scanning gives a clear picture of soft tissues that do not show up well on X-ray images.
- Ultrasound is the preferred imaging modality for the diagnosis and monitoring of pregnant women and their unborn babies.
- Ultrasound provides real-time imaging, making it a good tool for guiding minimally invasive procedures such as needle biopsies and fluid aspiration.

Although ultrasound imaging is generally considered safe when used prudently by appropriately trained health care providers, ultrasound energy has the potential to produce biological effects on the body. Ultrasound waves can heat the tissues slightly. In some cases, it can also produce small pockets of gas in body fluids or tissues(cavitation). The long-term consequences of these effects are still unknown. Because of the particular concern for effects on the fetus, organizations such as the American Institute of Ultrasound in Medicine have advocated prudent use of ultrasound imaging in pregnancy. Furthermore, the use of ultrasound solely for non-medical purposes such as obtaining fetal "keepsake" videos has been discouraged. Keepsake images or videos are reasonable if they are produced during a medically-indicated exam, and if no additional exposure is required.

3. Information for patients including expectant mothers

For all medical imaging procedures, the FDA recommends that patients talk to their health care provider to understand the reason for the examination, the medical information that will be obtained, the potential risks, and how the results will be used to manage the medical condition or pregnancy. Because ultrasound is not based on ionizing radiation, it is particularly useful for women of child-bearing age when CT or other imaging methods would otherwise result in exposure to radiation.

Ultrasound is the most widely used medical imaging method for viewing the fetus during pregnancy. Routine examinations are performed to assess and monitor the health status of the fetus and mother. Ultrasound examinations provide parents with a valuable opportunity to view the heartbeat of the fetus, bond with the unborn baby, and capture images to share with family and friends.

In fetal ultrasound, three-dimensional(3D) ultrasound allows the visualization of some facial features and possibly other parts such as fingers and toes of the fetus. Four-dimensional(4D) ultrasound is 3D ultrasound in motion. While ultrasound is generally considered to be safe with very low risks, the risks may increase with unnecessary prolonged exposure to ultrasound energy, or when untrained users operate the device.

Words and phrases

Word	Pronunciation	POS	Meaning
sonography	[səˈnɒgrəfɪ]	n.	超声成像
dynamic	[daɪˈnæmɪk]	adj.	动态的
audible	[ˈɔːdəbl]	adj.	听得见的
probe	[prəʊb]	n.	探针
visualize	[ˈvɪʒuəlaɪz]	vt.	可视化
layer	[ˈleɪə(r)]	n.	层
capture	[ˈkæptʃə(r)]	vt.	捕捉,捕获
attenuate	[əˈtenjuet]	vt.	衰减,减弱
scatter	[ˈskætə]	vi.	散射
refract	[rɪˈfrækt]	vt.	使折射
obstetrician	[ˌɒbstəˈtrɪʃn]	n.	产科医师
abdominal	[æbˈdɒmɪnl]	adj.	腹部的
fragility	[frəˈdʒɪləti]	n.	脆性
echocardiogram	[ˈekəʊˈkɑːdɪəgræm]	n.	超声心动图
ophthalmic	[ɒfˈθælmɪk]	adj.	眼的,眼科的
ocular	[ˈɒkjələ(r)]	adj.	眼的,眼球的
noninvasive	[ˌnɒnɪnˈveɪsɪv]	adj.	无创的
injection	[ɪnˈdʒekʃn]	n.	注射,注入
needle	[ˈniːdl]	n.	针,针管
ionize	[ˈaɪənaɪz]	vi.	电离
modality	[məʊˈdæləti]	n.	方式,方法
fluid	[ˈfluːɪd]	n.	流体
aspiration	[ˌæspəˈreɪʃn]	n.	抽吸
prudently	[ˈpruːdntlɪ]	adv.	谨慎地,慎重地
biological	[ˌbaɪəˈlɒdʒɪkl]	adj.	生物的,生物学的
cavitation	[ˌkævɪˈteɪʃən]	n.	气穴(现象)
consequence	[ˈkɒnsɪkwəns]	n.	结果,后果
fetus	[ˈfiːtəs]	n.	胎儿
organization	[ˌɔːgənaɪˈzeɪʃn]	n.	组织,学会,机构
prudent	[ˈpruːdnt]	adj.	谨慎的

Chapter 2-Section 3-Text 1-听词汇

fetal	[ˈfiːtl]	adj.	胎儿的
keepsake	[ˈkiːpseɪk]	n.	纪念品
video	[ˈvɪdiəʊ]	n.	磁带录像,录像
exposure	[ɪkˈspəʊʒə(r)]	n.	曝光
recommend	[ˌrekəˈmend]	vt.	推荐
assess	[əˈses]	vt.	评定,评估
visualization	[ˌvɪʒʊəlaɪˈzeɪʃn]	n.	可视化、形象化
facial	[ˈfeɪʃl]	adj.	面部的,脸的
feature	[ˈfiːtʃə(r)]	n.	特征,容貌
motion	[ˈməʊʃn]	n.	运动,移动
prolong	[prəˈlɒŋ]	vt.	延长
untrained	[ˌʌnˈtreɪnd]	adj.	未受训练的
childbearing	[ˈtʃaɪldbeərɪŋ]	n.	分娩
ultrasound imaging			超声成像
blood flow			血流
abdominal ultrasound			腹部超声
due date			预产期
blood vessel			血管
easy-to-use			易用的
real-time			实时
routine examination			常规检查
expectant mother			孕妇
fetal ultrasound			胎儿超声
child-bearing age			育龄
three-dimensional			三维的

Key sentences

(1) Ultrasound imaging(sonography) is a diagnostic medical procedure that uses high-frequency sound waves to produce dynamic visual images of organs, tissues orblood flow inside the body.

超声成像(超声)是一种医学诊断方法,它使用高频声波生成体内器官、组织或血流的动态可视图像。

(2) The sound waves are transmitted to the area to be examined and the returning echoes are captured to provide the physician with a "live" image of the area.

声波发送到检查部位,捕获回波,给医生提供该区域的"实时"图像。

(3) Sound energy is attenuated or weakened as it passes through tissue because parts of it are reflected, scattered, absorbed, refracted or diffracted. The strength(amplitude) of the sound signal and the time it takes for the wave to travel through the body provide the information necessary to produce an image.

声能通过组织时,因为一部分被反射,散射,吸收,折射或衍射,能量会变得衰减或减弱。声波信号的强度(振幅)及其穿过身体所需的时间提供了产生图像所需的信息。

(4) Because of the particular concern for effects on the fetus, organizations such as theAmerican Institute of Ultrasound in Medicine have advocated prudent use of ultrasound imaging in pregnancy.

由于对胎儿影响的特别关注,美国超声医学协会等组织提倡在怀孕期间谨慎使用超声成像。

(5) In fetal ultrasound, three-dimensional (3D) ultrasound allows the visualization of some facial features and possibly other parts such as fingers and toes of the fetus. Four-dimensional(4D) ultrasound is 3D ultrasound in motion.

在胎儿超声中,三维超声可以看见一些胎儿的面部特征和其他部分,如手指和脚趾。四维超声是动态三维超声。

Learning test

Choose the following words below to fill in the blanks, change the form if necessary.

dynamic   audible   visualize   capture   attenuate

biological   consequence   prudently   exposure   untrained

(1) The _____ of ocean waves are complex.

(2) The noise was _____ even above the roar of the engines.

(3) You can _____ data using annotated diagrams.

(4) The company is out to _____ the European market.

(5) In a forest, wet wood and needles _____ the signals.

(6) In _____ terminology, life is divided into two groups: plants and animals.

(7) Nobody can tell what the _____ may be.

(8) He _____ pursued his plan.

(9) Scientific findings receive regular _____ in the media.

(10) To the _____ eye, the two products look remarkably similar.

Translation

1. Translate the following words into Chinese.

(1) dynamic   (2) fetus

(3) audible   (4) exposure

(5) capture   (6) assess

(7) attenuate   (8) facial

(9) injection   (10) feature

2. Translate the following phases into Chinese.

(1) blood vessel   (2) real-time

(3) due date   (4) ultrasound imaging

(5) blood flow   (6) expectant mother

Practice

1. Translate the following sentences into Chinese.

(1) In an ultrasound exam, a transducer(probe) is placed directly on the body to be visualized.

(2) The sound waves are transmitted to the area to be examined and the returning echoes are captured to provide the physician with a "live" image of the area.

(3) Ultrasound imaging has been used for over 20 years and has an excellent safety record.

(4) Furthermore, the use of ultrasound solely for non-medical purposes such as obtaining fetal "keepsake" videos has been discouraged.

(5) Ultrasound is the most widely used medical imaging method for viewing the fetus during pregnancy.

2. Translate the following sentences into English.

(1)如果你怀孕了,医生会通过超声检查确定你的预产期。

(2)超声换能器的探头能产生和接收超声波,利用的是压电效应。

(3)腹部检查用超声机有多种不同频率的探头。

(4)探头的形状决定了视野的大小,而发射声波的频率则决定了图像的分辨率。

(5)超声波传播时能量会随着传播距离的增加而逐渐减小。

Culture salon

Transducer

Inside the core of the transducer are a number of peizo-electric crystals that have the ability to vibrate and produce sound of a particular frequency when electricity is passed through them. This is how ultrasound waves are formed. These transducers also act as receivers for the reflected echoes and generate a small electric signal when a sound wave is incident upon it.

Chapter 2-
Section 3-
Text 1-看译文

Text 2 Fundamentals of ultrasound imaging

Lead-in questions:

(1) Why ultrasound becomes one of the most widely used imaging technologies in medicine?

(2) What property makes ultrasound useful for medical testing and medical measure?

(3) How many steps are included in the creation of an image from sound?

(4) What's brightness-mode(B-mode) ?

(5) What's Doppler mode?

Chapter 2-
Section 3-
Text 2-听课文

Ultrasound has been used to image the human body for over half a century. Dr. Karl Theo Dussik, an Austrian neurologist, was the first to apply ultrasound as a medical diagnostic tool to image the brain. Today, ultrasound is one of the most widely used imaging technologies in medicine. It is portable, free of radiation risk, and relatively inexpensive when compared with other imaging modalities, such as magnetic resonance and computed tomography. Furthermore, ultrasound images are tomographic, i. e., offering a "cross-sectional" view of anatomical structures. The images can be acquired in "real time", thus providing instantaneous visual guidance for many interventional procedures including those for regional anesthesia and pain management. In this article, we describe some of the relevant fundamental principles.

Ultrasound behaves in a similar manner to audible sound, it has a much shorter wavelength. This means it can be reflected off very small surfaces such as defects inside materials. It is this property that makes ultrasound useful for medical testing and medical measure. Ultrasonic vibrations travel in the form of a wave, similar to the way light travels, as shown in Figure 2-12. However, unlike light waves which can travel in a vacuum(empty space), ultrasound requires an elastic medium such as a liquid or a solid. Ultrasonic nondestructive testing introduces high frequency sound waves into a test object to obtain information about the object without altering or damaging it in any way. Two basic quantities are measured in ultrasonic testing; they are the time of flight or the amount of time for the sound to travel through the sample and the amplitude of received signal. Based on velocity and round trip time of flight through the material, the material thickness can be calculated.

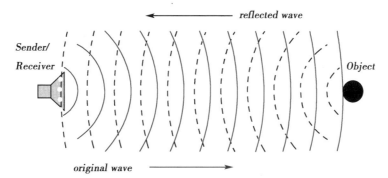

Figure 2-12 Principle of an active sonar.

1. From sound to image

The creation of an image from sound is done in three steps -producing a sound wave, receiving echoes, and image formation.

(1) Producing a sound wave

A sound wave is typically produced by a piezoelectric transducer encased in a plastic housing. Strong, short electrical pulses from the ultrasound machine drive the transducer at the desired frequency. The frequencies can be anywhere between 1 and 18 MHz. Materials on the face of the transducer enable the sound to be transmitted efficiently into the body(often a rubbery coating, a form of impedance matching). That is the reason why a water-based gel is often placed between the patient's skin and the probe.

(2) Receiving the echoes

The return of the sound wave to the transducer results in the same process as sending the sound wave, except in reverse. The returned sound wave vibrates the transducer and the transducer turns the vibrations into electrical pulses that travel to the ultrasonic scanner where they are processed and transformed into a digital image.

(3) Forming the image

To make an image, the ultrasound scanner must determine two things from each received echo:

- How long it took the echo to be received from when the sound was transmitted.
- How strong the echo was.

Once the ultrasonic scanner determines these two things, it can locate which pixel in the image to light up and to what intensity.

2. Modes of sonography

Several modes of ultrasound are used in medical imaging. B-mode and Doppler mode will be introduced in this article.

(1) B-mode or 2D mode

Modern medical ultrasound is performed primarily using a pulse-echo approach with a brightness-mode(B-mode) display. This involves transmitting small pulses of ultrasound echo from a transducer into the body. As the ultrasound waves penetrate body tissues of different acoustic impedances along the path of transmission, some are reflected back to the transducer(echo signals) and some continue to penetrate deeper. The echo signals returned from many sequential coplanar pulses are processed and combined to generate an image. Thus, an ultrasound transducer works both as a speaker(generating sound waves) and a microphone(receiving sound waves) . The ultrasound pulse is in fact quite short, but since it traverses in a straight path, it is often referred to as an ultrasound beam. The direction of ultrasound propagation along the beam line is called the axial direction, and the direction in the image plane perpendicular to axial is called the lateral direction. Usually only a small fraction of the ultrasound pulse returns as a reflected echo after reaching a body tissue interface, while the remainder of the pulse continues along the beam line to greater tissue depths.

On a grey scale, high reflectivity(bone) is white; low reflectivity(muscle) is grey and no reflection (water) is black. Deeper structures are displayed on the lower part of the screen and superficial structures on the upper part.

(2) Doppler mode

This mode makes use of the Dopplereffect in measuring and visualizing blood flow. In this technique, an ultrasound transducer is used to beam sound into the area of interest, and it reads the returning sound. When the sound bounces off a moving target like a blood vessel, the pitch changes as a result of the Doppler effect. The transducer can detect very subtle pitch changes and record them visually, creating an image which shows where blood is flowing, and in which direction.

One obvious application of color Doppler ultrasound is in examination of a patient with a suspected aneurysm or occlusion. The ultrasound will reveal areas where the velocity of the blood flow is changing, acting like a red flag to point out a problem. This technique can also be used to find blood clots, which will also be clearly visible within the color display.

In examination of tumors and venous malformations, doctors can use color Doppler ultrasound to map out the blood supply and learn how far the growth has spread. This can have an impact on which treatments the doctor recommends, and how the doctor wants to approach surgery and other measures.

Words and phrases

ultrasound	[ˈʌltrəsaʊnd]	n.	超声,超声波
neurologist	[ˌnjʊəˈrɒlədʒɪst]	n.	神经学家
diagnostic	[ˌdaɪəgˈnɒstɪk]	adj.	诊断的
audible	[ˈɔːdəbl]	adj.	听得见的
vacuum	[ˈvækjʊəm]	n.	真空
nondestructive	[ˌnɒndɪsˈtrʌktɪv]	adj.	无损的,非破坏性的
tomography	[təˈmɒgrəfi]	n.	层析成像,断层扫描成像
anesthesia	[ˌænəsˈθiːziə]	n.	麻醉
portable	[ˈpɔːtəbl]	adj.	便携的
echo	[ˈekəʊ]	n.	回声,回波
piezoelectric	[piːˌeizəʊiˈlektrik]	adj.	压电的
transducer	[trænzˈdjuːsə(r)]	n.	传感器,换能器
encase	[ɪnˈkeɪs]	vt.	封装,包装
gel	[dʒel]	n.	凝胶,胶体
vibrate	[vaɪˈbreɪt]	v.	振动,振荡
transform	[trænsˈfɔːm]	vt.	变换,转变
transmit	[trænsˈmɪt]	vt. & vi.	发射,播送
pixel	[ˈpɪksl]	n.	像素
intensity	[ɪnˈtensəti]	n.	强度
sonography	[səˈnɒgrəfi]	n.	超声成像
penetrate	[ˈpenɪtreɪt]	vt.	穿入,穿过,透过
tissue	[ˈtɪʃuː]	n.	组织
acoustic	[əˈkuːstɪk]	adj.	声(音)的,声波的
impedance	[ɪmˈpiːdns]	n.	阻抗
sequential	[sɪˈkwenʃl]	adj.	序列的,时序的
coplanar	[kəʊˈpleɪnə]	adj.	共面的

traverse	[trəˈvɜːs]	vt.	穿过,穿越
propagation	[ˌprɒpəˈgeɪʃn]	n.	传播
perpendicular	[ˌpɜːpənˈdɪkjələ(r)]	adj.	垂直的,成直角的
lateral	[ˈlætərəl]	adj.	横向的
reflectivity	[ˌriflekˈtiviti]	n.	反射率
pitch	[pitʃ]	n.	音调,音高
aneurysm	[ˈænjərizəm]	n.	动脉瘤
occlusion	[əˈkluːʒn]	n.	闭塞
tumor	[ˈtjuːmə]	n.	肿瘤
venous	[ˈviːnəs]	adj.	静脉的
malformation	[ˌmælfɔːˈmeɪʃn]	n.	畸形
spread	[spred]	vi.	扩散,传播
treatment	[ˈtriːtmənt]	n.	治疗,疗法
approach	[əˈprəʊtʃ]	vt.	着手,接近
surgery	[ˈsɜːdʒəri]	n.	外科手术
sound wave			声波
cross-sectional			截面的
free of			没有
blood clot			血块,凝血块
rubbery coating			橡胶涂层
impedance matching			阻抗匹配
brightness-mode			亮度模式
Doppler effect			多普勒效应

Key sentences

(1) It is portable, free of radiation risk, and relatively inexpensive when compared with other imaging modalities, such as magnetic resonance and computed tomography.

它是便携式的,无辐射风险,与其他成像方式如磁共振和计算机断层扫描相比,相对便宜一些。

(2) The images can be acquired in "real time", thus providing instantaneous visual guidance for many interventional procedures including those for regional anesthesia and pain management.

图像可以获得"实时"获取,从而为许多介入手术,包括局部麻醉和疼痛控制,提供即时的视觉引导。

(3) Ultrasonic nondestructive testing introduces high frequency sound waves into a test object to obtain information about the object without altering or damaging it in any way.

超声波无损检测引入高频声波到测试对象,以获取有关对象的信息,而不以任何方式改变或破坏它。

(4) The returned sound wave vibrates the transducer and the transducer turns the vibrations into electrical pulses that travel to the ultrasonic scanner where they are processed and transformed into a digital image.

返回的声波使换能器振动,换能器将振动转变成电脉冲,发送到超声波扫描仪,经超声波扫描仪处理后转化成数字图像。

(5) As the ultrasound waves penetrate body tissues of different acoustic impedances along the path of transmission, some are reflected back to the transducer(echo signals) and some continue to penetrate deeper.

由于超声波沿传输路径穿过人体组织的声波阻抗不同,一些超声波被反射回换能器(回波信号),一些超声波继续穿透得更深。

Learning test

Choose the following words below to fill in the blanks, change the form if necessary.

diagnostic   portable   vibrate   transform   transmit
penetrate   pitch   spread   intensity   approach

(1) A stethoscope is a common _____ tool for doctors.

(2) I always carry a _____ notebook with me.

(3) The bus _____ when the driver started the engine.

(4) Our body can _____ food into energy through metabolism.

(5) The tree roots _____ moisture and nutrient to the trunk and branches.

(6) X-rays can _____ many objects.

(7) He raised his voice to an even higher _____ .

(8) The _____ of the tumor can be monitored.

(9) When a surface is perpendicular to the rays, their _____ is maximum.

(10) He was _____ retirement.

Translation

1. Translate the following words into Chinese.

(1) diagnostic   (2) pixel

(3) piezoelectric   (4) intensity

(5) transducer   (6) pitch

(7) vibrate   (8) gel

(9) transmit   (10) malformation

2. Translate the following phases into Chinese.

(1) sound wave   (2) Doppler effect

(3) impedance matching (4) brightness-mode

(5) blood clot   (6) free of

Practice

1. Translate the following sentences into Chinese.

(1) Ultrasound is one of the most widely used imaging technologies in medicine.

(2) Ultrasound behaves in a similar manner to audible sound, it has a much shorter wavelength.

(3) Once the ultrasonic scanner determines these two things, it can locate which pixel in the image to light up and to what intensity.

(4) The echo signals returned from many sequential coplanar pulses are processed and combined to generate an image.

(5) The ultrasound pulse is in fact quite short, but since it traverses in a straight path, it is often referred to as an ultrasound beam.

2. Translate the following sentences into English.

(1) 保持超声设备定期维护保养的记录是一个良好的工作习惯。

(2) 探头采集反射波并把它传递到超声机。

(3) 超声机在屏幕上显示回波的距离和强度,形成一个二维图像。

(4) 在做超声检查的时候医生可能要求患者调整位置以便更好地看清楚。

(5) 当怀疑患者有动脉瘤或心肌梗死时,医生建议做彩色多普勒超声检查。

Culture salon

Frequency and wavelength of sound waves

Frequency refers to the number of cycles of compressions and rarefactions in a sound wave per second, with one cycle per second being 1 hertz. Diagnostic ultrasound uses frequencies in the range of 1~10 million (mega) hertz.

The wavelength is the distance traveled by sound in one cycle, or the distance between two identical points in the wave cycle, i.e. the distance from a point of peak compression to the next point of peak compression. It is inversely proportional to the frequency. Wavelength is one of the main factors affecting axial resolution of an ultrasound image. The smaller the wavelength (and therefore higher the frequency), the higher the resolution, but lesser penetration. Therefore, higher frequency probes (5 to 10 MHz) provide better resolution but can be applied only for superficial structures and in children. Lower frequency probes (2 to 5MHz) provide better penetration albeit lower resolution and can be used to image deeper structures.

Chapter 2-Section 3-Text 2-看译文

Chapter 2-Section 3-习题

Section 4 X-ray imaging

Text 1 Introduction to medical X-ray imaging

Chapter 2-
Section 4-PPT

Lead-in questions:

(1) What is the same basic principle of CT, radiography, and fluoroscopy?

(2) What are the benefits of X-ray imaging exams?

(3) What are the risks of X-ray imaging exams?

(4) What does the risk of developing cancer from medical imaging radiation exposure depend on?

(5) How to get a balance between benefits and risks for X-ray imaging?

Chapter 2-
Section 4-
Text 1-听课文

Medical imaging is the technique and process of creating visual images of the interior of a body for clinical analysis and medical intervention using X-rays. X-rays are basically the same thing as visible light rays. Both are wavelike forms of electromagnetic energy. Our eyes are sensitive to the particular wavelength of visible light, but not to the shorter wavelength of higher energy X-ray waves or the longer wavelength of the lower energy radio waves.

The most common methods of X-ray in medical imaging are X-ray radiography("conventional X-ray" including mammography), computed tomography(CT), and fluoroscopy. CT, radiography, and fluoroscopy all work on the same basic principle: an X-ray beam is passed through the body where a portion of the X-rays are either absorbed or scattered by the internal structures, and the remaining X-ray pattern is transmitted to a detector(e. g. , film or a computer screen) for recording or further processing by a computer. These exams differ in their purpose:

(1) Radiography. A single image is recorded for later evaluation. Mammography is a special type of radiography to image the internal structures of breasts and is commonly used in screening for breast cancer. Today, digital X-ray techniques are replacing film.

(2) Fluoroscopy. A continuous X-ray image is displayed on a monitor, allowing for real-time monitoring of a procedure or passage of a contrast agent("dye") through the body. Fluoroscopy can result in relatively high radiation doses, especially for complex interventional procedures(such as placing stents or other devices inside the body) which require fluoroscopy be administered for a long period of time.

(3) CT. Many X-ray images are recorded as the detector moves around the patient's body (Figure 2-13). A computer reconstructs all the individual images into cross-sectional images or "slices" of internal organs and tissues. A CT exam involves a higher radiation dose than conventional radiography because the CT image is reconstructed from many individual X-ray projections.

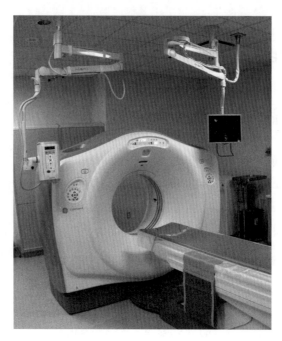

Figure 2-13   A CT scanner.

1. Benefits

The discovery of X-rays and the invention of CT represented major advances in medicine. X-ray imaging exams are recognized as a valuable medical tool for a wide variety of examinations and procedures. They are used to:

(1) Noninvasively and painlessly help to diagnosis disease and monitor therapy;

(2) Support medical and surgical treatment planning;

(3) Guide medical personnel as they insert catheters, stents, or other devices inside the body, treat tumors, or remove blood clots or other blockages.

2. Risks

As in many aspects of medicine, there are risks associated with the use of X-ray imaging, which uses ionizing radiation to generate images of the body. Ionizing radiation is a form of radiation that has enough energy to potentially cause damage to DNA and may elevate a person's lifetime risk of developing cancer. Risks from exposure to ionizing radiation include:

(1) A small increase in the possibility that a person exposed to X-rays will develop cancer later in life.

(2) Tissue effects such as cataracts, skin reddening, and hair loss, which occur at relatively high levels of radiation exposure and are rare for many types of imaging exams. For example, the typical use of a CT scanner or conventional radiography equipment should not result in tissue effects, but the dose to the skin from some long, complex interventional fluoroscopy procedures might, in some circumstances, be high enough to result in such effects.

(3) Another risk of X-ray imaging is possible reactions associated with an intravenously injected contrast agent, or "dye", that is sometimes used to improve visualization.

The risk of developing cancer from medical imaging radiation exposure is generally very small, and it depends on:

(1) Radiation dose-The lifetime risk of cancer increases the larger the dose and the more X-ray exams a patient undergoes.

(2) Patient's age-The lifetime risk of cancer is larger for a patient who receives X-rays at a younger age than for one who receives them at an older age.

(3) Patient's sex-Women are at a somewhat higher lifetime risk than men for developing radiation-associated cancer after receiving the same exposures at the same ages.

(4) Body region-Different organs and tissues have a different sensitivity to radiation. This is why the actual risk to the body from X-ray procedures varies depending on the part of the body being X-rayed.

3. Balancing benefits and risks

While the benefit of a clinically appropriate X-ray imaging exam generally far outweighs the risk, efforts should be made to minimize this risk by reducing unnecessary exposure to ionizing radiation. To help reduce risk to the patient, all exams using ionizing radiation should be performed only when necessary to answer a medical question, treat a disease, or guide a procedure. If there is a medical need for a particular imaging procedure and other exams using no or less radiation are less appropriate, then the benefits exceed the risks, and radiation risk considerations should not influence the physician's decision to perform the study or the patient's decision to have the procedure. However, the "As Low As Reasonably Achievable" (ALARA) principle should always be followed when choosing equipment settings to minimize radiation exposure to the patient.

Words and phrases

Chapter 2-
Section 4-
Text 1-听词汇

interior	[ɪnˈtɪəriə(r)]	n.	内部
intervention	[ˌɪntəˈvenʃn]	n.	干预,介入
wavelike	[ˈweɪvlaɪk]	adj.	波状的
sensitive	[ˈsensətɪv]	adj.	敏感的,灵敏的
electromagnetic	[ɪˌlektrəʊmæɡˈnetɪk]	adj.	电磁的
wavelength	[ˈweɪvleŋθ]	n.	波长
pattern	[ˈpætən]	n.	图案
mammography	[mæˈmɒɡrəfi]	n.	乳房X线造影
fluoroscopy	[ˌfluəˈrɒskəpɪ]	n.	X线透视,透射
radiography	[ˌreɪdiˈɒɡrəfi]	n.	X线成像
evaluation	[ɪˌvæljuˈeɪʃn]	n.	评估
breast	[brest]	n.	乳房,胸部
interventional	[ˌɪntəˈvenʃənəl]	adj.	干预的,介入的
stent	[stent]	n.	支架

reconstruct	[ˌrikən'strʌkt]	vt.	重建,重构
slice	[slaɪs]	n.	薄片
tissue	['tɪʃuː]	n.	组织
projection	[prə'dʒekʃən]	n.	投影,投影图
represent	[ˌreprɪ'zent]	n.	代表,象征
noninvasively	[ˌnɒnɪn'veɪsɪvli]	adv.	无创地
painlessly	['peɪnləslɪ]	adv.	无痛地
therapy	['θerəpɪ]	n.	治疗,疗法
catheter	['kæθɪtə(r)]	n.	导管,导尿管,尿液管
tumor	['tjuːmə]	n.	肿瘤
blockage	['blɒkɪdʒ]	n.	堵塞物
potentially	[pə'tenʃəlɪ]	adv.	潜在地,可能地
elevate	['elɪveɪt]	vt.	提升
cataract	['kætərækt]	n.	白内障
redden	['redn]	vt.	(使)变红
intravenously	[ˌɪntrə'viːnəslɪ]	adv.	静脉地,静脉内地
visualization	[ˌvɪʒʊəlaɪ'zeɪʃn]	n.	可视化
undergo	[ˌʌndə'gəʊ]	vt.	经历,经受
benefit	['benɪfɪt]	n.	益处,好处
inject	[ɪn'dʒekt]	vt.	注射,注入
appropriate	[ə'prəʊprɪət]	adj.	合适的,适当的
outweigh	[ˌaʊt'weɪ]	vt.	超过,大于,胜过
unnecessary	[ʌn'nesəsəri]	adj.	不必要的
physician	[fɪ'zɪʃn]	n.	医生,内科医生
exceed	[ɪk'siːd]	vt.	超过,超越
reasonably	['riːznəblɪ]	adv.	合理地,理性地
achievable	[ə'tʃiːvəbl]	adj.	可完成的,可达到的
radiation	[ˌreɪdi'eɪʃn]	n.	辐射(能),放射
principle	['prɪnsəpl]	n.	原则,原理
minimize	['mɪnɪmaɪz]	vt.	使最小化
X-ray imaging			X射线成像
medical imaging			医学影像
sensitive to			对…敏感
radio wave			无线电波
computed tomography(CT)			计算机断层扫描

contrast agent	造影剂
associate with	与……有关
a form of	一种形式
ionizing radiation	电离辐射
differ in	在……方面不同
real-time	实时

Key sentences

(1) Our eyes are sensitive to the particular wavelength of visible light, but not to the shorter wavelength of higher energy X-ray waves or the longer wavelength of the lower energy radio waves.

我们的眼睛对可见光的特定波长很敏感,但对高能量 X 射线的较短波长或低能量无线电波的较长波长不敏感。

(2) An X-ray beam is passed through the body where a portion of the X-rays are either absorbed or scattered by the internal structures, and the remaining X-ray pattern is transmitted to a detector(e. g. , film or a computer screen) for recording or further processing by a computer.

X 射线波束穿过人体,一部分 X 射线被身体内部结构吸收或散射,其余的 X 射线被传送到检测器(例如胶片或计算机屏幕)以供记录或计算机进一步处理。

(3) As in many aspects of medicine, there are risks associated with the use of X-ray imaging, which usesionizing radiation to generate images of the body.

与医学的许多方面一样,X 射线成像使用电离辐射产生身体图像,会带来相关的风险。

(4) While the benefit of a clinically appropriate X-ray imaging exam generally far outweighs the risk, efforts should be made to minimize this risk by reducing unnecessary exposure to ionizing radiation.

虽然临床合适的 X 线影像检查的益处通常远远超过风险,但应尽量减少不必要的电离辐射暴露,以减少这种风险。

(5) If there is a medical need for a particular imaging procedure and other exams using no or less radiation are less appropriate, then the benefits exceed the risks, and radiation risk considerations should not influence the physician's decision to perform the study or the patient's decision to have the procedure.

如果对特定成像技术有医疗需求,并且使用没有或较低辐射的其他检查不合适的话,那么益处大于风险,并且辐射风险不应该影响医生进行诊断或患者做检查的决定。

Learning test

Choose the following words below to fill in the blanks, change the form if necessary.

intervention	sensitive	wavelength	principle	slice
appropriate	undergo	inject	outweigh	radiation

(1) The government's _____ in this dispute will not help.

(2) Our teeth are _____ to both heat and cold.

(3) Visible light has longer _____ than UV light.

(4) I don't understand the working _____ of the machine.

(5) Try to eat at least four _____ of bread a day.

(6) Casual dress is not _____ for an interview.

(7) The workers in the company _____ much suffering because of heavy workload.

(8) The lab assistant _____ the rat with the new drug.

(9) The advantages greatly _____ the disadvantages.

(10) The radioactive material is stored in a special _____ proof container.

Translation

1. Translate the following words into Chinese.

(1) intervention   (2) sensitive

(3) interior   (4) mammography

(5) wavelength   (6) fluoroscopy

(7) electromagnetic   (8) catheter

(9) wavelike   (10) tumor

2. Translate the following phases into Chinese.

(1) radio wave   (2) a form of

(3) sensitive to   (4) medical imaging

(5) contrast agent   (6) visible light

Practice

1. Translate the following sentences into Chinese.

(1) Mammography is a special type of radiography to image the internal structures of breasts and is commonly used in screening for breast cancer.

(2) A CT exam involves a higher radiation dose than conventional radiography because the CT image is reconstructed from many individual X-ray projections.

(3) As in many aspects of medicine, there are risks associated with the use of X-ray imaging, which uses ionizing radiation to generate images of the body.

(4) This is why the actual risk to the body from X-ray procedures varies depending on the part of the body being X-rayed.

(5) To help reduce risk to the patient, all exams using ionizing radiation should be performed only when necessary to answer a medical question, treat a disease, or guide a procedure.

2. Translate the following sentences into English.

(1) 固态探测器取代了传统射线照相系统上使用的胶片暗盒。

(2) 紧急按钮位于诊疗床的左边和右边，危急的时候用来切断诊疗床的电源。

(3) 医学影像辐射照射引起罹患癌症的风险一般很小。

(4) 应尽量减少不必要的电离辐射照射，以降低患癌风险。

(5) 1972 年第一台 CT 诞生,仅用于颅脑检查。

Culture salon

Who discovered the X-ray?

Have you ever had an X-ray taken? X-rays are used to analyze problems with bones, teeth and organs in the human body; to detect cracks in metal in industry; and even at airports for luggage inspection. The scientific and medical community will forever be indebted to an accidental discovery made by German physicist Wilhelm Conrad Röntgen in 1895. While experimenting with electrical currents through glass cathode-ray tubes, Röntgen discovered that a piece of barium platinocyanide glowed even though the tube was encased in thick black cardboard and was across the room. He theorized that some kind of radiation must be traveling in the space. Röntgen didn't fully understand his discovery so he dubbed it X-radiation for its unexplained nature.

Chapter 2-
Section 4-
Text 1-看译文

Text 2 Radiography (Plain X-rays)

Lead-in questions:

(1) Why can you see different distinct structures inside the body by X-ray imaging?

(2) What's the difference between CR and conventional radiography?

(3) What are the advantages of digital radiography (DR)?

(4) What are the uses of radiography?

(5) What are the major risks associated with radiography?

Chapter 2-
Section 4-
Text 2-听课文

Despite the development of newer technologies such as computed tomography (CT), ultrasound imaging and magnetic resonance imaging (MRI), plain film X-rays remain an important tool for the diagnosis of many disorders. An X-ray machine is essentially a camera. Instead of visible light, however, it uses X-rays. During a radiographic procedure, an X-ray beam is passed through the part of the body to be scanned (Figure 2-14). A portion of the X-rays are absorbed or scattered by the internal structure and the remaining X-ray pattern is transmitted to an X-ray sensitive film or a photonic detector. When the X-rays hit the film, they expose it, and a radiograph appears on the film (Figure 2-15).

The collimator restricts the beam of X-rays so as to irradiate only the region of interest. The antiscatter grid increases tissue contrast by reducing the number of detected X-rays that have been scattered by tissue.

Since bone, fat, muscle, tumors, and other masses all absorb X-rays at different levels, the image on the film lets you see different (distinct) structures inside the body because of the different levels of exposure on the film. For example, bone absorbs X-rays particularly well, and soft tissue such as muscle fiber, which has

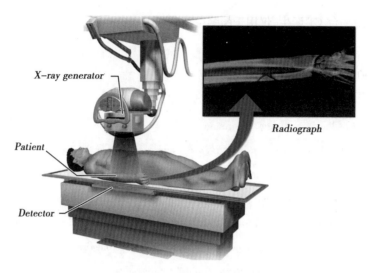

Figure 2-14 The basic setup of radiography.

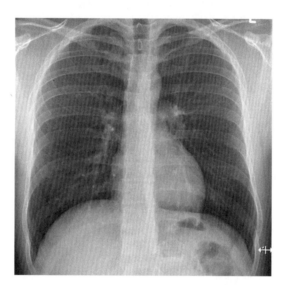

Figure 2-15 A typical X-ray radiograph of the chest,
in which the regions of bone appear white.

a lower density than bone, absorbs fewer X-rays. This results in the familiar contrast seen in X-ray images, with bones shown as clearly defined white areas and darker areas of tissue.

Plain radiography was the only imaging modality available during the first 50 years of radiology. Due to its availability, speed, and lower costs compared to other modalities, radiography is often the first-line test of choice in radiologic diagnosis. Today, film-screen radiography is being replaced by computed radiography (CR) but more recently by digital radiography(DR) and the EOS imaging.

Computed radiography(CR) uses very similar equipment to conventional radiography except that in place of a film to create the image, an imaging plate(IP) made of photo-stimulable phosphor is used. The imaging plate is housed in a special cassette and placed under the body part or object to be examined and the X-ray exposure is made. Hence, instead of taking an exposed film into a darkroom for developing in chemical tanks or an automatic film processor, the imaging plate is run through a special laser scanner, or CR reader, that reads and digitizes the image. The digital image can then be viewed and enhanced using software that has functions very similar to other conventional digital image-processing software, such as contrast, brightness, filtration and zoom.

Instead of X-ray film, digital radiography (DR) uses a digital image capture device. This gives advantages of immediate image preview and availability; elimination of costly film processing steps; a wider dynamic range, which makes it more forgiving for over-exposure and under-exposure; as well as the ability to apply special image processing techniques that enhance overall display quality of the image. EOS is a medical imaging system whose aim is to provide frontal and lateral radiography pictures, while limiting the X-ray dose absorbed by the patient in vertical pose.

X-ray imaging provides fast, high-resolution images and is relatively inexpensive. The average examination for most plain film examinations takes no more than 10~15 minutes and requires no special preparation of the patient. The operator, usually the radiographer(also known as a radiologic technologist), selects the amount and type of X-rays to be used according to the patient's size, the tissue or part of the body being imaged and the amount of image contrast required. Because movement, e. g. of the lungs and diaphragm, blurs the image, patients are usually asked to hold their breath during the exposure. The X-ray picture is stored on a piece of film called a radiograph. These are interpreted by a physician specially trained to interpret them, known as a radiologist.

1. Uses

Radiography is used in many types of examinations and procedures where a record of a static image is desired. It's very suitable for scans of bones and tissue dense in calcium such as in dental images and detection of bone fractures. Other uses of radiography include:

- dental examination;
- the study of the organs in the abdomen, such as the liver and bladder;
- chest radiography for diseases of the lung, such as pneumonia or lung cancer;
- mammography to screen for breast cancer;
- orthopedic evaluations;
- verification of correct placement of surgical markers prior to invasive procedures;
- spot film or static recording during fluoroscopy Chiropractic examinations

2. Risks / benefits

Radiography is a type of X-ray procedure, and it carries the same types of risks as other X-ray procedures. The radiation dose the patient receives varies depending on the individual procedure, but is generally less than that received during fluoroscopy and computed tomography procedures.

The major risks associated with radiography are the small possibilities of developing a radiation-induced cancer or cataracts some time later in life, and causing a disturbance in the growth or development of an embryo or fetus(teratogenic defect) when performed on a pregnant patient or one of childbearing age. Since the developing embryo is much more sensitive to the effects of ionizing radiation than adult patients, X-rays of any part of the body are not recommended for pregnant women.

When an individual has a medical need, the benefit of radiography far exceeds the small cancer risk associated with the procedure. Even when radiography is medically necessary, it should use the lowest possible exposure and the minimum number of images. In most cases many of the possible risks can be reduced or eliminated with proper shielding.

Chapter 2-
Section 4-
Text 2-听词汇

Words and phrases

disorder	[dɪsˈɔːdə]	n.	疾病,失常
radiographic	[ˌreɪdɪəʊˈɡræfɪk]	adj.	射线(影像)的,X射线照相的
portion	[ˈpɔːʃn]	n.	一部分,部分
scatter	[ˈskætə]	vt.	散射
structure	[ˈstrʌktʃə]	n.	结构,构造
pattern	[ˈpætən]	n.	图案
sensitive	[ˈsensɪtɪv]	adj.	敏感的,灵敏的
photonic	[fəʊˈtɒnɪk]	adj.	光子(的)
detector	[dɪˈtektə]	n.	检测器
expose	[ɪkˈspəʊz]	vt.	曝光
setup	[ˈsetʌp]	n.	结构,构成,设置
collimator	[ˈkɒlɪmeɪtə]	n.	准直器
restrict	[rɪˈstrɪkt]	vt.	限制,限定
irradiate	[ɪˈreɪdɪeɪt]	vt.	(用X射线等)照射,辐照
grid	[ɡrɪd]	n.	栅格,格子
distinct	[dɪˈstɪŋkt]	adj.	不同的,独特的
contrast	[ˈkɒntrɑːst]	n.	对比度
particularly	[pəˈtɪkjələli]	adv.	特别,尤其
density	[ˈdensəti]	n.	密度
modality	[məʊˈdæləti]	n.	方式,方法
availability	[əˌveɪləˈbɪləti]	n.	可利用性,可得性
equipment	[ɪˈkwɪpmənt]	n.	配件,装备,设备
phosphor	[ˈfɒsfə]	n.	磷光剂,磷光物质
cassette	[kəˈset]	n.	暗盒
tank	[tæŋk]	n.	(水,油)箱,罐,大容器
digitize	[ˈdɪdʒɪtaɪz]	v.	数字化
conventional	[kənˈvenʃnl]	adj.	常规的,一般的
filtration	[fɪlˈtreɪʃn]	n.	过滤,筛选
zoom	[zuːm]	n.	变焦,放大(小)

preview	[ˈpriːvjuː]	n.	预览
elimination	[ɪˌlɪmɪˈneɪʃn]	n.	去除,淘汰,排除
frontal	[ˈfrʌntl]	adj.	前面的,正面的
lateral	[ˈlætərəl]	adj.	侧面的,横向的
radiographer	[ˌreɪdiˈɒgrəfə(r)]	n.	放射技师
lung	[lʌŋ]	n.	肺
diaphragm	[ˈdaɪəfræm]	n.	横膈肌
abdomen	[ˈæbdəmən]	n.	腹部,下腹,腹腔
bladder	[ˈblædə]	n.	膀胱
pneumonia	[njuːˈməʊniə]	n.	肺炎
mammography	[mæˈmɒgrəfi]	n.	乳房 X 线造影
orthopedic	[ˌɔːθəʊˈpiːdɪk]	adj.	整形的,矫形的
evaluation	[ɪˌvæljʊˈeɪʃn]	n.	评估,评价
fluoroscopy	[ˌflʊəˈrɒskəpɪ]	n.	X 线透视,透射
cataract	[ˈkætərækt]	n.	白内障
embryo	[ˈembriəʊ]	n.	胚胎
fetus	[ˈfiːtəs]	n.	胎儿
teratogenic	[ˌterətəˈdʒenɪk]	adj.	致畸的
shield	[ʃiːld]	vi.	防护,屏蔽
plain X-rays			平板 X 线成像
magnetic resonance imaging			磁共振成像
ultrasound imaging			超声成像
computed tomography			计算机断层扫描
visible light			可见光
a portion of			一部分
computed radiograph(CR)			计算机 X 线成像
digital radiograph(DR)			数字 X 线成像,数字射线照相
in place of			代替
photo-stimulable			光可激发的
high-resolution			高分辨率
over-exposure			曝光过度
under-exposure			曝光不足

Key sentences

(1) Despite the development of newer technologies such as computed tomography(CT), ultrasound imaging and magnetic resonance imaging(MRI), plain film X-rays remain an important tool for the diagnosis of

many disorders.

尽管开发了诸如计算机断层扫描(CT),超声成像和磁共振成像(MRI)等较新的技术,X 射线平片仍然是许多疾病诊断的一个重要工具。

(2) Since bone, fat, muscle, tumors and other masses all absorb X-rays at different levels, the image on the film lets you see different(distinct) structures inside the body because of the different levels of exposure on the film.

由于骨骼、脂肪、肌肉、肿瘤和其他物质对 X 射线的吸收程度不同,在胶片上的曝光程度不同,胶片上的图像可以让你看到体内的不同结构。

(3) Computed radiography(CR) uses very similar equipment to conventional radiography except that in place of a film to create the image, an imaging plate(IP) made of photo-stimulable phosphor is used.

计算机放射成像使用与常规成像非常类似的部件,除了用光激发磷光物质制成的成像板代替胶片产生图像。

(4) Instead of X-ray film, digital radiography(DR) uses a digital image capture device.

数字射线照相(DR)使用数字图像采集设备代替 X 射线胶片。

(5) X-ray imaging provides fast, high-resolution images and is relatively inexpensive. The average examination for most plain film examinations takes no more than 10~15 minutes and requires no special preparation of the patient.

X 射线成像提供了快速、高分辨率的图像,而且相对便宜。大多数平片检查的平均检查时间不超过 10~15 分钟,并且不需要患者进行特别的准备。

Learning test

Choose the following words below to fill in the blanks, change the form if necessary.

disorder　　scatter　　sensitive　　restrict　　density
availability　　structure　　preview　　evaluation　　shield

(1) Mental _____ is usually difficult to cure.

(2) The dusts in the air can _____ sunlight.

(3) This instrument is highly _____ .

(4) The hospital may _____ bookings to local people only.

(5) The region has a very high population _____ .

(6) The _____ of cheap long-term credit would help small business.

(7) Doctors study the _____ of the human body.

(8) You can _____ the photos before saving it.

(9) After analysis and _____ , his article is published.

(10) _____ one's eyes against the glare of the sun.

Translation

1. Translate the following words into Chinese.

(1) disorder　　　　　　　　　　(2) zoom

(3) structure   (4) abdomen

(5) detector   (6) lung

(7) expose   (8) shield

(9) density   (10) contrast

2. Translate the following phases into Chinese.

(1) high-resolution   (2) over-exposure

(3) computed tomography(CT)   (4) visible light

(5) digital radiography(DR)   (6) in place of

Practice

1. Translate the following sentences into Chinese.

(1) During a radiographic procedure, an X-ray beam is passed through the part of the body to be scanned.

(2) The collimator restricts the beam of X-rays so as to irradiate only the region of interest.

(3) This results in the familiar contrast seen in X-ray images, with bones shown as clearly defined white areas and darker areas of tissue.

(4) Radiography is used in many types of examinations and procedures where a record of a static image is desired.

(5) Since the developing embryo is much more sensitive to the effects of ionizing radiation than adult patients, X-rays of any part of the body are not recommended for pregnant women.

2. Translate the following sentences into English.

(1) 为保证 X 射线设备的正常运行,必须建立定期的维护检查。

(2) 应根据每日平均 8 小时的使用时间来确定检查的时间间隔。

(3) CT 在使用前应开机预热半小时以上。

(4) 清洁 X 射线设备时,不要使用任何的清洁剂或者溶剂。

(5) 许多潜在的辐射风险都可以通过适当的屏蔽来减少或消除。

Culture salon

Molecular Imaging

Molecular imaging is a relatively new discipline that allows the biological processes taking place in the body to be viewed at a cellular and molecular level. This breakthrough enables doctors to identify disease in its earliest stages, often well before they would be seen on CT and MR images and would otherwise require invasive surgery or biopsy- the removal of tissue for examination under the microscope. Molecular imaging procedures are used to diagnose and manage the treatment of brain and bone disorders, cancer, gastrointestinal disorders, heart and kidney diseases, lung and thyroid disorders.

Chapter 2-
Section 4-
Text 2-看译文

Text 3　Computed tomography (CT)

Lead-in questions:

(1) How a CT system works?

(2) What can CT help a physician to do?

(3) What are the risks of CT scan?

(4) What are the benefits of CT scan?

(5) What does FDA advice the patients and parents to do a physician recommends a CT scan for you?

Computed tomography (CT), sometimes called "computerized tomography" or "computed axial tomography" (CAT), uses X-rays in conjunction with computing algorithms to image the body. Computed tomography (CT) scanners have been available since the mid 1970s and have revolutionized medical imaging. Today, millions of scans are performed worldwide every year for different clinical questions in a variety of clinical fields.

Chapter 2-Section 4-Text 3-听课文

The most prominent part of a CT scanner is the gantry-a circular, rotating frame with an X-ray tube mounted on one side and a detector on the opposite side(Figure 2-16). A fan-shaped beam of X-rays is created as the rotating frame spins the X-ray tube and detector around the patient. As the scanner rotates, several thousand sectional views of the patient's body are generated in one rotation, which result in reconstructed cross-sectional images of the body. Using on these data, it is possible to create a 3D visualization views from different angles(Figure 2-17).

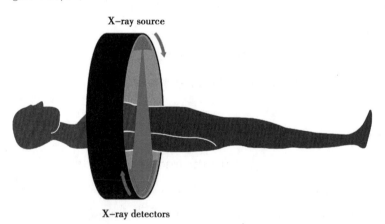

Figure 2-16　Drawing of a CT imaging system.

1. How does a CT system work?

Generally, there are 4 steps for a CT scanning procedure.

(1) A motorized table moves the patient through a circular opening in the CT imaging system.

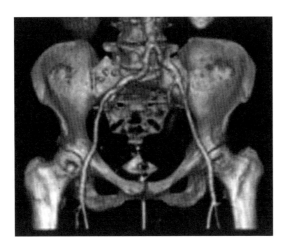

Figure 2-17 A CT image.

(2) While the patient is inside the opening, an X-ray source and a detector assembly within the system rotate around the patient. A single rotation typically takes a second or less. During rotation the X-ray source produces a narrow, fan-shaped beam of X-rays that passes through a section of the patient's body.

(3) Detectors in rows opposite the X-ray source register the X-rays that pass through the patient's body as a snapshot in the process of creating an image. Many different "snapshots" (at many angles through the patient) are collected during one complete rotation.

(4) For each rotation of the X-ray source and detector assembly, the image data are sent to a computer to reconstruct all of the individual "snapshots" into one or multiple cross-sectional images (slices) of the internal organs and tissues.

CT scans provide far more detailed images than conventional X-ray imaging, especially in the case of blood vessels and soft tissue such as internal organs and muscles. With modern CT scans, you can even apply colorcoding to the dataset as seen. Scans can be completed in seconds or even fractions of a second. For some CT scans, a special contrast agent is injected into a vein before the scan as this allows further assessment of the organs and vessels. These preparations for the examination may require additional time.

Modern CT scans provide very detailed images, for example of blood vessels and internal organs, by using relatively low radiation doses. This CT scan of the whole thorax and abdomen was performed in less than 1 sec with only 20ml of contrast media, and a radiation dose of 2.32 mSv.

2. Uses

CT images of internal organs, bones, soft tissue, and blood vessels provide greater clarity and more details than conventional X-ray images, such as a chest X-Ray.

CT is a valuable medical tool that can help a physician:

• Diagnose disease, trauma or abnormality

• Plan and guide interventional or therapeutic procedures

- Monitor the effectiveness of therapy (e. g. , cancer treatment)

CT scans are also invaluable in assessing skeletal injuries, as even very tiny bones are shown clearly.

3. Benefits/risks

CT scans are valuable in emergencies because they are able to provide information very quickly. This is important, for example when assessing strokes, brain injuries, heart disease, and internal injuries. In addition, the short duration of the scanning process benefits patients who are not easily able to keep still, such as children. CT imaging is a very important tool to diagnose cancer and to obtain additional information for different clinical questions. Additionally, the detailed images provided by CT scans may eliminate the need for exploratory surgery.

Concerns about CT scans include the risks from exposure to ionizing radiation and possible reactions to the intravenous contrast agent, or dye, which may be used to improve visualization. A CT scan usually requires a higher radiation exposure dose than a conventional radiography examination. Exposure to ionizing radiation is of particular concern in pediatric patients because the cancer risk per unit dose of ionizing radiation is higher for younger patients than adults, and younger patients have a longer lifetime for the effects of radiation exposure to manifest as cancer. As with conventional X-ray imaging, CT scans are not recommended for pregnant women unless the examination is absolutely necessary. However, when used appropriately, the benefits of a CT scan far exceed the risks.

4. Information for patients and parents

If a physician recommends a CT scan for you or your child, the FDA encourages you to discuss the benefits and risks of the CT scan, as well as any past X-ray procedures you or your child have had, with your physician. A CT scan should always be performed if it is medically necessary and other exams using no or less radiation are unsuitable. At this time, the FDA does not see a benefit to whole-body scanning of individuals without symptoms.

Words and phrases

Chapter 2-Section 4-Text 3-听词汇

algorithm	[ˈælgəˈrɪðəm]	n.	算法
revolutionize	[ˈrevəˈluʃənaɪz]	vt.	使……变革
field	[fild]	n.	领域
prominent	[ˈprɒmɪnənt]	adj.	重要的,显著的
gantry	[ˈɡæntri]	n.	扫描架
circular	[ˈsɜːkjələ(r)]	adj.	圆形的
opposite	[ˈɒpəzɪt]	adj.	相反的,对面的
rotate	[rəʊˈteɪt]	vi.	旋转
reconstruct	[ˌriːkənˈstrʌkt]	vt.	重建
visualization	[ˌvɪʒuəlaɪˈzeɪʃn]	n.	可视化
motorized	[ˈməʊtəˌraɪzd]	adj.	电动的

assembly	[əˈsembli]	n.	组件
register	[ˈredʒɪstə]	vt.	记录
snapshot	[ˈsnæpʃɒt]	n.	快照
vessel	[ˈvesl]	n.	血管
muscle	[ˈmʌsl]	n.	肌肉
dataset	[ˈdeɪtəset]	n.	数据组
fraction	[ˈfrækʃən]	n.	部分,分数
inject	[ɪnˈdʒekt]	vt.	注射,注入
vein	[veɪn]	n.	静脉
assessment	[əˈsesmənt]	n.	评估,评定
preparation	[ˌprepəˈreɪʃn]	n.	准备,准备工作
radiation	[ˌreɪdiˈeɪʃn]	n.	辐射(能),放射
dose	[dəʊs]	n.	剂量
thorax	[ˈθɔːræks]	n.	胸,胸部,胸腔
abdomen	[ˈæbdəmən]	n.	腹部,腹腔,下腹
clarity	[ˈklærəti]	n.	清晰度
chest	[tʃest]	n.	胸部,胸腔
trauma	[ˈtrɔːmə]	n.	外伤,创伤
abnormality	[ˌæbnɔːˈmæləti]	n.	异常,反常
therapeutic	[ˌθerəˈpjuːtɪk]	adj.	治疗的
therapy	[ˈθerəpi]	n.	治疗,疗法
invaluable	[ɪnˈvæljuəbl]	adj.	无价的,非常宝贵的
skeletal	[ˈskelətl]	adj.	骨骼的
stroke	[strəʊk]	n.	中风
eliminate	[ɪˈlɪmɪnet]	vt.	除去,去掉,消除
exploratory	[ɪkˈsplɒrətri]	adj.	探查的,探究的
surgery	[ˈsɜːdʒəri]	n.	手术
intravenous	[ˌɪntrəˈviːnəs]	adj.	静脉的,静脉内的
pediatric	[ˌpiːdɪˈætrɪk]	adj.	儿科的
manifest	[ˈmænɪfest]	vt.	表现,发作
recommend	[ˈrekəˈmend]	vt.	推荐
pregnant	[ˈpregnənt]	adj.	怀孕的,怀胎的
encourage	[ɪnˈkʌrɪdʒ]	vt.	鼓励
perform	[pəˈfɔːm]	vt.	执行,实施,完成
individual	[ˌɪndɪˈvɪdʒuəl]	n.	个人

symptom	[ˈsɪmptəm]	n.	症状,征兆
computed tomography			计算机断层扫描
in conjunction with			结合
cross-sectional			横截面的
in the case of			在…情况下
blood vessel			血管
soft tissue			软组织
radiation dose			辐射剂量
contrast media			造影剂
ionizing radiation			电离辐射
whole-body			全身

Key sentences

(1) Computed tomography (CT), sometimes called "computerized tomography" or "computed axial tomography" (CAT), uses X-rays in conjunction with computing algorithms to image the body.

计算机断层扫描(CT),有时被称为"计算机断层扫描"或"计算机轴向断层扫描"(CAT),用 X 射线结合计算算法对人体进行成像。

(2) As the scanner rotates, several thousand sectional views of the patient's body are generated in one rotation, which result in reconstructed cross-sectional images of the body.

当扫描仪旋转时,旋转一次可生成数千个患者身体的剖视图,从而重建身体的截面图像。

(3) CT scans provide far more detailed images than conventional X-ray imaging, especially in the case of blood vessels and soft tissue such as internal organs and muscles.

CT 扫描成像提供了比传统的 X 射线成像详细得多的图像,特别是在血管和内脏器官和肌肉等软组织的情况下。

(4) Exposure to ionizing radiation is of particular concern in pediatric patients because the cancer risk per unit dose of ionizing radiation is higher for younger patients than adults, and younger patients have a longer lifetime for the effects of radiation exposure to manifest as cancer.

电离辐射暴露是儿科患者特别关心的,因为年幼患者每单位剂量电离辐射的癌症风险高于成年人,而年幼患者有更长的寿命让辐射暴露效应表现为癌症。

(5) If a physician recommends a CT scan for you or your child, the FDA encourages you to discuss the benefits and risks of the CT scan, as well as any past X-ray procedures you or your child have had, with your physician.

如果医生建议为您或您的孩子进行 CT 扫描,FDA 鼓励你与你的医生讨论 CT 扫描的好处和风险,以及你或你的孩子曾经做过的任何 X 射线检查。

Learning test

Choose the following words below to fill in the blanks, change the form if necessary.

prominent   rotation   reconstruct   motorize   assembly

register   fraction   preparation   therapy   manifest

(1) One _____ symptom of the disease is progressive loss of memory.

(2) The _____ of the earth round the sun takes one year.

(3) He had a surgery to help _____ his badly damaged face.

(4) A young man rides a _____ scooter on a narrow road in a rural village.

(5) For the rest of the day, he worked on the _____ of an explosive device.

(6) The instruments will _____ every change of direction or height.

(7) The living cost is only a _____ of his salary.

(8) I don't have enough _____ for the job interview.

(9) He feels better, so the doctor stops drug _____.

(10) Her actions _____ a complete disregard for personal safety.

Translation

1. Translate the following words into Chinese.

(1) rotation   (2) stroke

(3) snapshot   (4) surgery

(5) vessel   (6) skeletal

(7) inject   (8) pediatric

(9) chest   (10) symptom

2. Translate the following phases into Chinese.

(1) blood vessel   (2) in conjunction with

(3) soft tissue   (4) computed tomography

(5) whole-body   (6) cross-sectional

Practice

1. Translate the following sentences into Chinese.

(1) Millions of scans are performed worldwide every year for different clinical questions in a variety of clinical fields.

(2) The most prominent part of a CT scanner is the gantry-a circular, rotating frame with an X-ray tube mounted on one side and a detector on the opposite side.

(3) A fan-shaped beam of X-rays is created as the rotating frame spins the X-ray tube and detector around the patient.

(4) While the patient is inside the opening, an X-ray source and a detector assembly within the system rotate around the patient.

(5) CT scans provide far more detailed images than conventional X-ray imaging, especially in the case of blood vessels and soft tissue such as internal organs and muscles.

2. Translate the following sentences into English.

(1) CT 是用 X 射线束对人体某部位一定厚度的层面进行扫描。

(2) CT 设备主要由三部分组成:扫描部分、计算机系统和图像显示存储系统。

(3) CT 特别适用于心血管的造影检查。

(4) CT 的扫描方式有平扫、造影增强扫描和造影扫描三种。

(5) CT 检查对脑损伤、脑梗死等脑部疾病的诊断效果好。

Culture salon

Types of electromagnetic spectrums

The electromagnetic spectrum is the term for all known frequencies and their linked wavelengths of the known photons. Visible light is electromagnetic radiation within a certain portion of the electromagnetic spectrum, which is visible to the human eye and is responsible for the sense of sight. Visible light is usually defined as having wavelengths in the range of 400~700 nanometres(nm). The types of electromagnetic waves are broadly classified into the following classes:

1. Gamma radiation
2. X-ray radiation
3. Ultraviolet radiation
4. Visible radiation
5. Infrared radiation
6. Terahertz radiation
7. Microwave radiation
8. Radio waves

This classification goes in the increasing order of wavelength, which is characteristic of the type of radiation.

Chapter 2-Section 4-习题

Section 5　Magnetic resonance imaging（MRI）

Text 1　Introduction to magnetic resonance imaging（MRI）

Lead-in questions:

(1) What's the full name of MRI?

(2) What are the advantages of MRI?

(3) What's the clinical use of MRI?

(4) How many steps are there for an MRI exam?

(5) Are there any dangers involved with MRI?

An X-ray is very effective for showing doctors a broken bone, but if they want a look at a patient's soft tissue, then they will likely want a magnetic resonance imaging（MRI）. An MRI scanner can be used to take images of any part of the body（e. g. , head, joints, abdomen, legs, etc. ）, in any imaging direction（Figure 2-18）. MRI provides better soft tissue contrast than CT and can differentiate better between fat, water, muscle, and other soft tissue than CT. These images provide information to physicians and can be useful in diagnosing a wide variety of diseases and conditions（Figure 2-19）.

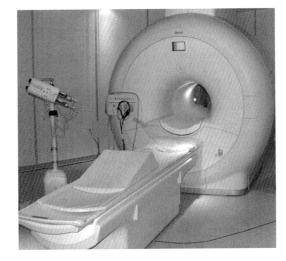

Figure 2-18　A commercial MRI machine.

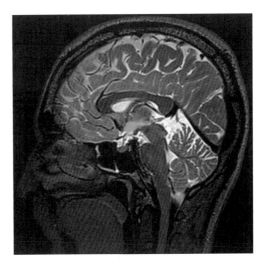

Figure 2-19　An MR image.

MRI is widely used in hospitals to help identify diseases. In 2014, there were approximately 36, 000 MRI scanners in use worldwide. At present, about 2, 500 MR imaging units are sold worldwide every year. The biggest markets are still the United States of America with around 45% of worldwide sales, followed by Europe which buys approximately one quarter of all units. The Chinese market is huge and moving, but even approximate data are difficult to get.

One major advantage of MRI is its ability to image in any plane. Computerized tomography(CT) , for example, is limited to one plane, the axial plane(in the loaf-of-bread analogy, the axial plane would be how a loaf of bread is normally sliced). An MRI system can create axial images as well as sagittal(slicing the bread side-to-side lengthwise) and coronal(think of the layers in a layer cake) images, or any degree in between, without the patient ever moving.

The use of MRI as a diagnostic technique continues to grow, allowing the study of more and more body parts. Initially, MR was mainly used to image the brain and spinal column, and each exam could last up to an hour.

MRI can be used for a huge range of clinical applications in the fields of clinical neurology, cardiology, cancer, and soft tissue damage. It gives health care providers useful information about a variety of conditions and diagnostic procedures including:

- abnormalities of the brain and spinal cord
- abnormalities in various parts of the body such as breast, prostate, and liver
- injuries or abnormalities of the joints(such as the shoulder, knee, elbow, wrist)
- the structure and function of the heart(cardiac imaging)
- areas of activation within the brain(functional MRI or fMRI)
- blood flow through blood vessels and arteries(angiography)
- the chemical composition of tissues(spectroscopy)

In the field of oncology, MRI can help visualize both the presence and the extent of underlying pathology with outstanding anatomical details. The boundaries of the tumor can be clearly defined and target tissues can be clearly delineated from healthy surrounding tissue. MRI examinations have also become much faster, in some cases rivaling the speed of spiral CT. Skilled MR operators with the latest equipment can now do complete routine studies(e. g. brain or knee) in as little as ten minutes. In some cases, MR systems equipped with echo planar imaging(EPI) packages can use special emergency or fast protocols to do a basic head study in as little as five to twenty seconds.

Image contrast may be weighted to demonstrate different anatomical structures or pathology. For some MRI exams, intravenous(IV) drugs, such as gadolinium-based contrast agents(GBCAs) are used to change the contrast of the MR image. Gadolinium-based contrast agents are rare earth metals that are usually given through an IV in the arm.

MR creates images from the manipulation of hydrogen atoms of human body in magnetic fields, based on the principles of nuclear magnetic resonance(NMR) . The MR images are obtained by placing the patient or area of interest within a powerful, highly uniform, static magnetic field. Magnetized protons(hydrogen nuclei) within the patient align like small magnets in this field. Radiofrequency pulses are then utilized to create an oscillating magnetic field perpendicular to the main field, from which the nuclei absorb energy and move out of alignment with the static field, in a state of excitation. As the nuclei return from excitation to the equilibrium state, a signal induced in the receiver coil of the instrument by the nuclear magnetization can then be transformed by a series of algorithms into diagnostic images. Images based on different tissue characteristics can be obtained by varying the number and sequence of pulsed radiofrequency fields in order to take advantage of magnetic relaxation properties of the tissues.

An MRI scanning mainly involve following steps:

(1) Magnetize your body by using a really big, strong magnet.

(2) Make your net magnetization spin using a radio-frequency pulse.

(3) This spinning magnetization generates an electric current in the radio receiver in the MRI machine.

(4) Tell where in your body this signal came from by its frequency.

(5) Putting all the signals from all over your body together to make an image.

MRI is generally safe. Risks are primarily related to the static and oscillating magnetic fields used in MRI. These fields are capable of producing adverse biologic effects at a sufficiently high exposure, but effects have not been observed at the levels currently employed in clinical practice. The most important known risk is the projectile effect, which involves the forceful attraction of ferromagnetic objects to the magnet. MRI should not be performed on patients when there are ferromagnetic objects embedded in the patient, such as shrapnel, or implants such as pacemaker wires.

Words and phrases

Chapter 2-Section 5-Text 1-听词汇

tissue	[ˈtɪʃuː]	n.	组织
scanner	[ˈskænə(r)]	n.	扫描机,扫描仪
magnetic	[mæɡˈnetɪk]	adj.	磁的,有磁性的
resonance	[ˈrezənəns]	n.	共振
joint	[dʒɔɪnt]	n.	关节
abdomen	[ˈæbdəmən]	n.	腹部
differentiate	[ˌdɪfəˈrenʃieɪt]	vt.	区分,辨别
fat	[fæt]	n.	脂肪
physician	[fɪˈzɪʃn]	n.	内科医生,医生
diagnose	[ˈdaɪəɡnəʊz]	vt.	诊断
identify	[aɪˈdentɪfaɪ]	vt.	确定,识别
approximately	[əˈprɒksɪmətli]	adv.	大约,近似地
plane	[pleɪn]	n.	平面
analogy	[əˈnælədʒi]	n.	类比,比拟
slice	[slaɪs]	vt.	切片,切下
sagittal	[ˈsædʒətəl]	adj.	矢状的
coronal	[kəˈrəʊnəl]	adj.	冠状的

spinal	[ˈspaɪnl]	adj.	脊柱的
neurology	[njʊəˈrɒlədʒi]	n.	神经学
cardiology	[ˌkɑːdiˈɒlədʒi]	n.	心脏学
abnormality	[ˌæbnɔːˈmæləti]	n.	畸形
prostate	[ˈprɒsteɪt]	n.	前列腺
liver	[ˈlɪvə(r)]	n.	肝脏
cardiac	[ˈkɑːdiæk]	adj.	心脏的
artery	[ˈɑːtəri]	n.	动脉
oncology	[ɒŋˈkɒlədʒi]	n.	肿瘤学
pathology	[pəˈθɒlədʒi]	n.	病理(学)
outstanding	[aʊtˈstændɪŋ]	adj.	杰出的,显著的
anatomical	[ˌænəˈtɒmɪkl]	adj.	解剖(学)的
boundary	[ˈbaʊndri]	n.	边界
tumor	[ˈtjuːmə]	n.	肿瘤
delineate	[dɪˈlɪnieɪt]	vt.	划分,勾画
rival	[ˈraɪvl]	vt.	媲美,比得上
spiral	[ˈspaɪrəl]	adj.	螺旋形的
intravenous	[ˌɪntrəˈviːnəs]	adj.	静脉的
gadolinium	[ˌgædəˈlɪniəm]	n.	钆
manipulation	[məˌnɪpjuˈleɪʃn]	n.	控制,操纵
hydrogen	[ˈhaɪdrədʒən]	n.	氢
proton	[ˈprəʊtɒn]	n.	质子
align	[əˈlaɪn]	vt.	对齐,排列
utilize	[ˈjuːtəlaɪz]	vt.	利用
oscillating	[ˈɒsɪleɪt]	adj.	震荡的
perpendicular	[ˌpɜːpənˈdɪkjələ]	adj.	垂直的
nuclei	[ˈnjuːklɪaɪ]	n.	原子核
static	[ˈstætɪk]	adj.	静态的
excitation	[ˌeksaɪˈteɪʃən]	n.	激发
equilibrium	[ˌiːkwɪˈlɪbriəm]	n.	平衡
induce	[ɪnˈdjuːs]	vt.	感应
coil	[kɔɪl]	n.	线圈
transform	[trænsˈfɔːm]	vt.	转换,转变
algorithm	[ˈælgərɪðəm]	n.	算法
generate	[ˈdʒenəreɪt]	vt.	产生

projectile	[prəˈdʒektaɪl]	adj.	投射的,弹射的
ferromagnetic	[ˌferəʊmæɡˈnetɪk]	adj.	铁磁的
shrapnel	[ˈʃræpnəl]	n.	弹片
pacemaker	[ˈpeɪsmeɪkə(r)]	n.	心脏起搏器
magnetic resonance imaging(MRI)			磁共振成像
nuclear magnetic resonance(NMR)			核磁共振
soft tissue			软组织
spinal column			脊柱
spinal cord			脊髓
axial plane			轴面
radio frequency pulse			无线射频脉冲
in the field of			在……领域
take advantage of			利用
be related to			与……有关
contrast agent			造影剂
echo planar imaging(EPI)			回波平面成像
functional MRI(fMRI)			功能性磁共振成像

Key sentences

(1) MRI provides better soft tissue contrast than CT and can differentiate better between fat, water, muscle, and other soft tissue than CT.

MRI 可提供比 CT 更好的软组织对比度,也能更好地区分脂肪、水、肌肉及其他软组织。

(2) MRI can be used for a huge range of clinical applications in the fields of clinical neurology, cardiology, cancer and soft tissue damage.

MRI 在临床神经学、心脏病学、癌症和软组织损伤等领域有广泛的临床应用。

(3) MRI creates images from the manipulation of hydrogen atoms of human body in magnetic fields, based on the principles of nuclear magnetic resonance(NMR).

MRI 是基于核磁共振原理,通过在磁场中操纵人体氢原子来产生图像。

(4) Images based on different tissue characteristics can be obtained by varying the number and sequence of pulsed radio frequency fields in order to take advantage of magnetic relaxation properties of the tissues.

基于不同组织特征的图像可以通过改变脉冲射频场的数量和序列来获得,以便利用组织的磁弛豫特性。

(5) The most important known risk is the projectile effect, which involves the forceful attraction of ferromagnetic objects to the magnet.

已知的最大风险是投射效应,其涉及铁磁体对磁铁的强吸引力。

Learning test

Choose the following words below to fill in the blanks, change the form if necessary.

tissue   scanner   intravenous   diagnose   spinal

magnetic   utilize   identify   abnormality   rival

(1) The doctor has to insert an _____ catheter for parenteral nutrition, as she cannot eat by herself.

(2) A portable ultrasound is a convenient _____ tool.

(3) The new diagnostic tool has been used to _____ the sex of fetus.

(4) Athletes have hardly any fatty _____ .

(5) The earth is a giant object with strong _____ field.

(6) The central nervous system of body includes the brain and the _____ cord.

(7) The MRI _____ is too expensive for the small hospital.

(8) The performance of this machine almost _____ the new model, but it's much cheaper.

(9) We _____ waterfalls for producing electric power.

(10) Further scans are required to confirm the _____ of her brain.

Translation

1. Translate the following words into Chinese.

(1) magnetic   (2) hydrogen

(3) align   (4) transform

(5) nuclei   (6) liver

(7) oncology   (8) artery

(9) intravenous   (10) generate

2. Translate the following phrases into Chinese.

(1) magnetic resonance imaging(MRI)   (2) take advantage of

(3) spinal cord   (4) spinal column

(5) radio frequency pulse   (6) echo planar imaging(EPI)

Practice

1. Translate the following sentences into Chinese.

(1) An X-ray is very effective for showing doctors a broken bone, but if they want a look at a patient's soft tissue, then they will likely want a magnetic resonance imaging(MRI).

(2) One major advantage of MRI is its ability to image in any plane.

(3) In the field of oncology, MRI can help visualize both the presence and the extent of underlying pathology with outstanding anatomical details.

(4) Image contrast may be weighted to demonstrate different anatomical structures or pathology.

(5) MRI should not be performed on patients when there are ferromagnetic objects embedded in the pa-

tient, such as shrapnel, or implants such as pacemaker wires.

2. Translate the following sentences into English.

（1）MRI 在医疗诊断中有广泛的应用。

（2）MRI 利用磁共振现象从人体中获得电磁信号，并重建出人体信息。

（3）人体中存在大量的氢原子。

（4）MRI 对人体没有电离辐射损伤。

（5）回波平面成像是目前为止最快速的磁共振成像方法。

> **Culture salon**
>
> ### The brief history of MRI
>
> In 1977, RaymondDamadian demonstrated MRI called field-focusing nuclear magnetic resonance. Edelstein and coworkers demonstrated imaging of the body using Ernst's technique in 1980. A single image could be acquired in approximately five minutes by this technique. In 1987 echo-planar imaging was used to perform real-time movie imaging of a single cardiac cycle. In this same year Charles Dumoulin was perfecting magnetic resonance angiography(MRA), which allowed imaging of flowing blood without the use of contrast agents.
>
> In 1991, Richard Ernst was rewarded for his achievements in pulsed Fourier Transform NMR and MRI with the Nobel Prize in Chemistry. In 1992 functional MRI(fMRI) was developed. This technique allows the mapping of the function of the various regions of the human brain.
>
> In 2003, Paul C. Lauterbur and Sir Peter were awarded the Nobel Prize in Medicine for their discoveries concerning magnetic resonance imaging.

Chapter 2-Section 5-Text 1-看译文

Text 2 Physics of magnetic resonance imaging (MRI)

Lead-in questions:

(1) What are the working principles of MRI?

(2) What are the major advantages of MRI over CT?

(3) What is the most important component of MRI?

(4) Is MRI a safe procedure?

(5) Do you know the contraindications for MRI examination?

Magnetic resonance imaging(MRI) is a medical imaging procedure to produce detailed images of the body's organs and structures. MRI scanners use strong magnetic fields, radio waves, and field gradients to generate images of the inside of the body. It is based on the principles of nuclear magnetic resonance(NMR) , a technique used by scientist to obtain microscopic chemical and physical information about molecules. Nuclear magnetic resonance

Chapter 2-Section 5-Text 2-听课文

(NMR) is a physical phenomenon in which nuclei in a magnetic field absorb and re-emit electromagnetic radiation. This radiation is at a specific resonance frequency which depends on the strength of the magnetic field and the magnetic properties of the isotope of the atoms. The signal in an MR image comes mainly from the protons in fat and water molecules in the body.

During an MRI exam, an electric current is passed through coiled wires to create a temporary magnetic field in a patient's body. Radio waves are emitted and received by a transmitter/receiver in the machine, and these signals are used to make digital images of the scanned area of the body. A typical MRI scan last from 20 to 90 minutes, depending on the part of the body being imaged.

The biggest and most important component in an MRI system is the magnet. The magnet in an MRI system is rated using a unit of measure known as a tesla. Another unit of measure commonly used with magnets is the gauss(1 tesla = 10,000 gauss). The magnets in use today in MRI systems create a magnetic field of 0.5 tesla to 2.0 tesla, or 5,000 to 20,000 gauss. When you realize that the earth's magnetic field measures 0.5 gauss, you can see how powerful these magnets are. Another part of the MRI system is a set of coils that transmit radiofrequency waves into the patient's body. There are different coils for different parts of the body: knees, shoulders, wrists, heads, necks and so on. These coils usually conform to the contour of the body part being imaged, or at least reside very close to it during the exam. Other parts of the machine include a very powerful computer system and a patient table, which slides the patient into the bore. Whether the patient goes in head or feet first is determined by what part of the body needs examining. Once the body part to be scanned is in the exact center of the magnetic field, the scan can begin.

When patients slide into an MRI machine, they take with them the billions of atoms that make up the human body. For the purposes of an MRI scan, we are only concerned with the hydrogen atom, which is abundant since the body is mostly made up of water. These atoms are randomly spinning on their axis, like a child's top. All of the atoms are spinning in various directions around their individual magnetic fields(Figure 2-20).

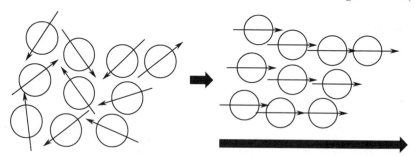

Figure 2-20　Magnetic moments of hydrogen atoms. Usually, all the magnetic moments in your body are jumbled in all directions. That's why you are not magnetic. When this is the case, we say that your net magnetic moment is zero. But if we apply a magnetic field from the outside, we can get the moments in your body to line up with the field.

These hydrogen atoms have a strong magnetic moment, which means that in a magnetic field, the atoms line up in the direction of the field. If the nuclei of hydrogen atoms—single protons, all spinning randomly—are caught suddenly in a strong magnetic field, they tend to line up like so many compass needles. When radio frequency pulse is applied, the protons are then hit with a short, precisely tuned burst of radio waves. They will absorb the energy, flip around, and spin again in a different direction. This is the "resonance" part of MRI. When the radio frequency is turned off, the hydrogen protons slowly return to their natural alignment within the magnetic field and release the energy absorbed from the RF pulses. When they do this, they give off a signal that the coils pick up and send to the computer system. The computer receives the signal from the spinning protons as mathematical data; the data is converted into an image.

Figure 2-20 Magnetic moments of hydrogen atoms. Usually, all the magnetic moments in your body are jumbled in all directions. That's why you are not magnetic. When this is the case, we say that your net magnetic moment is zero. But if we apply a magnetic field from the outside, we can get the moments in your body to line up with the field.

MRI and computed (axial) tomography (CT or CAT) are complementary imaging technologies and each has advantages and limitations for particular applications. MRI is particularly useful for showing soft tissues, such as the brain, muscle, connective tissue and most tumors. Another advantage of MRI is that MRI does not use X-rays, so there is no exposure to ionizing radiation. They can be safely used in people who may be vulnerable to the effects of radiation, such as pregnant women and babies. While the hazards of X-rays are now well-controlled in most medical contexts, MRI can still be seen as superior to CT in this regard.

MRI is in general a very safe procedure. Maybe you are concerned about the long-term impact of having all your atoms mixed up, but once you're out of the magnetic field, your body and its chemistry return to normal. There are no known biological hazards to humans from being exposed to magnetic fields of the strength used in medical imaging today. However, the strong magnet used for scanning can attract any metal object containing iron. People with some medical implants or other non-removable metal inside the body may be unable to safely undergo an MRI examination. Therefore, patients who are having an MRI should be prescreened for the presence of any metal objects such as cochlear implants, cardiac pacemakers, and metallic fragments in the eye. Some patients may also experience symptoms of claustrophobia while in the scanner. Compared with CT, MRI scans typically: take more time, are louder, and usually require that the subject go into a narrow tube.

Words and phrases

gradient	[ˈgreɪdiənt]	n.	梯度
microscopic	[ˌmaɪkrəˈskɒpɪk]	adj.	微观的,显微的
molecule	[ˈmɒlɪkjuːl]	n.	分子
physical	[ˈfɪzɪkl]	adj.	物理的
phenomenon	[fəˈnɒmɪnən]	n.	现象
absorb	[əbˈsɔːb]	vt.	吸收
frequency	[ˈfriːkwənsi]	n.	频率
property	[ˈprɒpəti]	n.	性质
isotope	[ˈaɪsətəʊp]	n.	同位素

Chapter 2-
Section 5-
Text 2-听词汇

atom	[ˈætəm]	n.	原子
current	[ˈkʌrənt]	n.	电流
temporary	[ˈtemprəri]	adj.	暂时的,临时的
transmitter	[trænsˈmɪtə(r)]	n.	发射机
component	[kəmˈpəʊnənt]	n.	部件,零件
magnet	[ˈmæɡnət]	n.	磁体
rate	[reɪt]	v.	计量
tesla	[ˈteslə]	n.	特斯拉
unit	[ˈjuːnɪt]	n.	单位
gauss	[ɡaʊs]	n.	高斯
transmit	[trænsˈmɪt]	vt.	传送,发射
radiofrequency	[reɪdiəfˈriːkwənsɪ]	n.	射频
conform	[kənˈfɔːm]	v.	符合,使一致
contour	[ˈkɒntʊə]	n.	外形,轮廓
bore	[bɔː]	n.	孔
randomly	[ˈrændəmli]	adv.	随机
spin	[spɪn]	vi.	旋转
axis	[ˈæksɪs]	n.	轴
compass	[ˈkʌmpəs]	n.	指南针
mathematical	[ˌmæθəˈmætɪkl]	adj.	数学的
convert	[kənˈvɜːt]	v.	转换,转变
jumbled	[ˈdʒʌmbld]	adj.	混乱的
apply	[əˈplaɪ]	vt.	应用
complementary	[ˌkɒmplɪˈmentri]	adj.	互补的
tumor	[ˈtjuːmə]	n.	肿瘤
ionizing	[ˈaɪənaɪzɪŋ]	adj.	电离的
radiation	[ˌreɪdiˈeɪʃn]	n.	辐射
pregnant	[ˈpreɡnənt]	adj.	怀孕的
hazard	[ˈhæzəd]	n.	危害
superior	[suːˈpɪəriə(r)]	adj.	较好的,优胜的
undergo	[ˌʌndəˈɡəʊ]	vt.	经历,承受
cochlear	[ˈkɒklɪə]	adj.	耳蜗的
metallic	[məˈtælɪk]	adj.	金属的
fragment	[ˈfræɡmənt]	n.	碎片,片段
symptom	[ˈsɪmptəm]	n.	症状

claustrophobia	[ˌklɔːstrəˈfəʊbɪə]	n.	幽闭恐惧症
based on			基于
unit of measure			度量单位，计量单位
magnetic field			磁场
radiofrequency wave			射频波
computed tomography (CT)			计算机断层扫描
ionizing radiation			电离辐射
connective tissue			结缔组织
depend on			取决于
in this regard			在这点上
superior to			优越于
long-term			长期的

Key sentences

(1) It is based on the principles of nuclear magnetic resonance (NMR), a technique used by scientist to obtain microscopic chemical and physical information about molecules.

它（磁共振成像）是基于核磁共振的原理，这是科学家用来获取分子的微观化学和物理信息的技术。

(2) The biggest and most important component in an MRI system is the magnet. The magnet in an MRI system is rated using a unit of measure known as a tesla.

MRI 系统中最大和最重要的部件是磁铁。MRI 系统中的磁体测量单位为特斯拉。

(3) When patients slide into an MRI machine, they take with them the billions of atoms that make up the human body.

当患者滑入 MRI 机器时，他们随身携带了数十亿个构成人体的原子。

(4) MRI and computed (axial) tomography (CT or CAT) are complementary imaging technologies and each has advantages and limitations for particular applications.

MRI 和计算机（轴向）断层扫描（CT 或 CAT）是互补的成像技术，每种都具有优点和局限性，适于特定的应用。

(5) MRI is in general a very safe procedure. Maybe you are concerned about the long-term impact of having all your atoms mixed up, but once you're out of the magnetic field, your body and its chemistry return to normal.

MRI 检测通常是一个非常安全的过程。也许你关心的是让所有原子混合在一起的长期影响，但是一旦你离开磁场，你的身体及化学性质就会恢复正常。

Learning test

Choose the following words below to fill in the blanks, change the form if necessary.

apply	phenomenon	symptom	radiation	complementary
undergo	physical	pregnant	convert	absorb

(1) I just don't know why I have to _____ this disease.

(2) It is a common _____ phenomenon that substances expand when heated.

(3) Nobody on the bus offered a seat for the _____ woman.

(4) Photons can be _____ to electrical signals based on photoelectric effect.

(5) Paper can _____ much water very quickly.

(6) The scientists _____ this new technique to improve the performance of MRI.

(7) It's an interesting scientific _____, but no practical use yet.

(8) _____ of a cold include a headache and sore throat.

(9) He refused to have a CT scan because he fears the _____ hazard.

(10) The sequence of the two DNA strands is always _____ to each other.

Translation

1. Translate the following words into Chinese.

(1) tesla   (2) gauss

(3) cochlear   (4) hazard

(5) axis   (6) mathematical

(7) magnet   (8) molecule

(9) nuclei   (10) radiofrequency

2. Translate the following phrases into Chinese.

(1) unit of measure   (2) magnetic field

(3) long-term   (4) connective tissue

(5) ionizing radiation   (6) superior to

Practice

1. Translate the following sentences into Chinese.

(1) Magnetic resonance imaging(MRI) is a medical imaging procedure to produce detailed images of the body's organs and structures.

(2) MRI scanners use strong magnetic fields, radio waves, and field gradients to generate images of the inside of the body.

(3) During an MRI exam, an electric current is passed through coiled wires to create a temporary magnetic field in a patient's body.

(4) Another advantage of MRI is that MRI does not use X-rays, so there is no exposure to ionizing radiation.

(5) Therefore, patients who are having an MRI should be prescreened for the presence of any metal objects such as cochlear implants, cardiac pacemakers, and metallic fragments in the eye.

2. Translate the following sentences into English.

(1) 磁共振成像是一个无疼痛且安全的诊断过程。

(2) 磁共振成像有别于计算机断层扫描,它不需要用 X 射线。

(3) MRI 对胎儿无明显影响。

(4) MRI 检查已成为美国医疗机构收入的重要来源。

(5) 在法国,MRI 检查的费用约为 150 欧元。

Culture salon

The hydrogen atom basics

Matter consists of atoms(e. g, ^{1}H, ^{12}C, ^{16}O, ^{31}P, etc.). An atom of one element differs from an atom of another element in its internal structure: the number of protons, neutrons, and electrons. Protons and neutrons comprise the nucleus; protons are positively charged, neutrons possess no charge, and the electrons orbiting around the nucleus are negatively charged. Different nuclear compositions and numbers of surrounding electrons are reflected in different physical properties.

The proton(^{1}H) is the most commonly used because the two major components of the human body are water and fat, both of which contain hydrogen. They all have magnetic properties which distinguish them from nonmagnetic isotopes. The hydrogen atom(^{1}H) consists of a single positively charged proton which spins around its axis. Spinning charged particles create an electromagnetic field, analogous to that from a bar magnet.

Chapter 2-
Section 5
Text 2-看译文

Chapter 2-Section 5-习题

Section 6　Clinical laboratory equipment

Chapter 2
Section 6-PPT

Text 1　Hematology analyzer

Lead-in questions:

(1) What's the use of hematology analyzers?

(2) What is the function of blood?

(3) What does blood consist of?

(4) How do automated cell counters work?

(5) How to utilize the results of complete blood count test to diagnose disease?

Hematology analyzers are used to perform complete blood counts(CBC), erythrocyte sedimentation rates(ESRs), or coagulation tests. Abnormally high or low counts of cells may

Chapter 2
Section 6
Text 1-听课文

indicate the presence of many forms of disease, and hence blood counts are amongst the most commonly performed blood tests in medicine.

The average human adult has more than 5 liters(6 quarts) of blood in his or her body. Blood carries oxygen and nutrients to living cells and takes away their waste products. It also delivers immune cells to fight infections and contains platelets that can form a plug in a damaged blood vessel to prevent blood loss.

Through the circulatory system, blood adapts to the body's needs. When you are exercising, your heart pumps harder and faster to provide more blood and hence oxygen to your muscles. During an infection, the blood delivers more immune cells to the site of infection, where they accumulate to ward off harmful invaders.

Blood is composed of plasma and several kinds of cells. All of the cells found in the blood come from bone marrow. They begin their life as stem cells, and they mature into three main types of cells— red blood cells(RBCs), leukocytes(white blood cells, WBCs), and thrombocytes(platelets). In turn, there are three types of WBC—lymphocytes, monocytes, and granulocytes—and three main types of granulocytes (neutrophils, eosinophils, and basophils).

Plasma is mainly water, but it also contains many important substances such as proteins(albumin, clotting factors, antibodies, enzymes, and hormones), sugars(glucose), and fat particles.

Among the many remarkable advances in medical science in the last few decades, hematology analyzers have quietly revolutionized medical practice. Internal medicine, pediatrics andoncologyare among the many specialties of medicine that require frequent, up-to-date and accurate CBC tests to determine if treatments are efficacious. It is now taken for granted that CBC tests can be ordered, the blood drawn and results available within the hour in facilities with an analyzer(Figure 2-21).

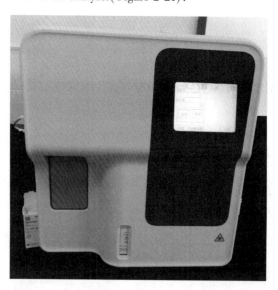

Figure 2-21 A commercial hematology analyzer.

Automated cell counters sample the blood, and quantify, classify, and describe cell populations using both electrical and optical techniques. Electrical analysis involves passing a dilute solution of the blood through an aperture across which an electrical current is flowing. The passage of cells through the current changes the impedance between the terminals(the Coulter principle). A lytic reagent is added to the blood solution to selectively lyse the red cells(RBCs), leaving only white cells(WBCs), and platelets intact. Then the solution is passed through a second detector. This allows the counts of RBCs, WBCs, and platelets to be obtained. The platelet count is easily separated from the WBC count by the smaller impedance spikes they produce in the detector due to their lower cell volumes.

Optical detection may be utilized to gain a differential count of the populations of white cell types. A dilute suspension of cells is passed through a flow cell, which passes cells one at a time through a capillary tube past a laser beam. The reflectance, transmission and scattering of light from each cell are analyzed by sophisticated software giving a numerical representation of the likely overall distribution of cell populations.

A complete blood count will normally include flowing results.

1. Red cells

Total red blood cells—The number of red cells is given as an absolute number per litre.

Hemoglobin—The amount of hemoglobin in the blood, expressed in grams per litre. Low hemoglobin is called anemia.

Hematocrit(HCT) or packed cell volume(PCV)— This is the fraction of whole blood volume that consists of red blood cells.

Mean corpuscular volume (MCV)— the average volume of the red cells, measured in femto litres. Anemia is classified as microcytic or macrocytic based on whether this value is above or below the expected normal range. Other conditions that can affect MCV include thalassemia and reticulocytosis.

Mean corpuscular hemoglobin(MCH)— the average amount of hemoglobin per red blood cell, in picograms.

Mean corpuscular hemoglobin concentration(MCHC)— the average concentration of hemoglobin in the cells.

Red blood cell volume distribution width(RDW)— a measure of the variation of the RBC population.

2. White cells

Each type has a specific purpose, so increased or decreased levels will determine how well a patient is recovering, or what type of foreign object the body is trying to fight. For example, lymphocytes respond to viral infections. If high levels of this type of cell appeared on the test results, the patient could be quickly determined that he or she had a viral disease.

Total white blood cells — All the white cell types are given as a percentage and as an absolute number per litre. A complete blood count with differential will also include:

Neutrophil granulocytes — May indicate bacterial infection. May also be raised in acute viral infections.

Lymphocytes — Higher with some viral infections such as glandular fever and are also raised in chronic lymphocytic leukaemia (CLL).

Monocytes — May be raised in bacterial infections.

Eosinophil granulocytes — Increased in parasitic infections.

Basophil granulocytes — A manual count will also give information about other cells that are not normally present in peripheral blood, but may be released in certain disease processes.

3. Platelets

Platelet numbers are given, as well as information about their size (MPV) and the range of sizes (PDW) in the blood. But in the USA, the PDW and Pct parameters are not for diagnostic use. The value for PDW is used as an internal check on the reported platelet parameters PLT and MPV.

Many disease states are heralded by changes in the blood count: leukocytosis can be a sign of infection; thrombocytopenia can result from drug toxicity; pancytopenia is generally as the result of decreased production from the bone marrow, and is a common complication of cancer chemotherapy.

Words and phrases

Chapter 2-
Section 6
Text 1-听词汇

hematology	[hiːməˈtɒlədʒɪ]	n.	血液学
erythrocyte	[ɪˈrɪθrəsaɪt]	n.	红细胞
coagulation	[kəʊægjuˈleɪʃn]	n.	凝结
indicate	[ˈɪndɪkeɪt]	vt.	显示, 指示
liter	[ˈliːtə(r)]	n.	升
oxygen	[ˈɒksɪdʒən]	n.	氧气
nutrient	[ˈnjuːtriənt]	n.	营养, 营养物质
immune	[ɪˈmjuːn]	adj.	免疫的
platelet	[ˈpleɪtlɪt]	n.	血小板
circulatory	[ˈsɜːkjələˌtɔːriː]	adj.	(血液)循环的
plasma	[ˈplæzmə]	n.	血浆
mature	[məˈtʃʊə]	vi.	成熟, 长成
leukocyte	[ˈljuːkəʊsaɪt]	n.	白细胞
thrombocyte	[ˈθrɒmbəsaɪt]	n.	血小板, 凝血细胞
lymphocyte	[ˈlɪmfəsaɪt]	n.	淋巴细胞
monocyte	[ˈmɒnəsaɪt]	n.	单核细胞
granulocyte	[ˈgrænjʊləsaɪt]	n.	粒细胞
neutrophil	[ˈnjuːtrəfɪl]	n.	嗜中性粒细胞
eosinophil	[ˌiːəˈsɪnəfɪl]	n.	嗜酸性粒细胞
basophil	[ˈbæsəfɪl]	n.	嗜碱性粒细胞
albumin	[ˈælbjʊmɪn]	n.	清蛋白, 白蛋白

antibody	[ˈæntɪbɒdɪ]	n.	抗体
enzyme	[ˈenzaɪm]	n.	酶
hormone	[ˈhɔːməʊn]	n.	荷尔蒙,激素
glucose	[ˈgluːˌkəʊs]	n.	葡萄糖
efficacious	[ˌefɪˈkeɪʃəs]	adj.	有效的
sample	[ˈsɑːmpl]	vt.	取……的样品,取样
quantify	[ˈkwɒntɪfaɪ]	vt.	定量,量化
optical	[ˈɒptɪkl]	adj.	光学的
dilute	[daɪˈluːt]	adj.	稀释的
aperture	[ˈæpətʃə]	n.	孔,隙缝
impedance	[ɪmˈpiːdns]	n.	阻抗
lyse	[laɪz]	vt.	溶解
detector	[dɪˈtektə]	n.	检测器
suspension	[səˈspenʃn]	n.	悬浮液
reflectance	[rɪˈflektns]	n.	反射
transmission	[trænzˈmɪʃn]	n.	透射
scattering	[ˈskætərɪŋ]	n.	散射
analyze	[ˈænəlaɪz]	vt.	分析
hemoglobin	[ˌhiːməʊˈgləʊbɪn]	n.	血红蛋白
anemia	[əˈniːmiə]	n.	贫血
hematocrit	[ˈhemətəʊkrɪt]	n.	红细胞比容
thalassemia	[θæləˈsiːmiə]	n.	地中海贫血
reticulocytosis	[rɪtɪkjʊləʊsaɪˈtəʊsɪs]	n.	网状细胞过多症
variation	[ˌveərɪˈeɪʃn]	n.	变化
bacterial	[bækˈtɪərɪəl]	adj.	细菌的
parameter	[pəˈræmɪtə]	n.	参数
herald	[ˈherəld]	v.	预示,预告
parasitic	[ˌpærəˈsɪtɪk]	adj.	由寄生虫引起的
leukocytosis	[ˌljuːkəʊsaɪˈtəʊsɪs]	n.	白细胞增多
chemotherapy	[ˌkiːməʊˈθerəpɪ]	n.	化学疗法
toxicity	[tɒkˈsɪsəti]	n.	毒性
pancytopenia	[ˈpænsaɪtəˈpiːniə]	n.	全血细胞减少症
complication	[ˌkɒmplɪˈkeɪʃ(ə)n]	n.	并发症
thrombocytopenia	[θrɒmbəʊsaɪtəʊˈpiːniə]	n.	血小板减少(症)
stem cell			干细胞

complete blood count (CBC)	全血细胞计数
erythrocyte sedimentation rate (ESR)	红细胞沉降率
red blood cell (RBC)	红细胞
bone marrow	骨髓
white blood cell (WBC)	白细胞
clotting factor	凝血因子
internal medicine	内科,内科医学
lytic reagent	溶解剂
capillary tube	毛细管
packed cell volume (PCV)	红细胞比积
mean corpuscular volume (MCV)	平均红细胞体积
mean corpuscular hemoglobin (MCH)	平均红细胞血红蛋白含量
mean corpuscular hemoglobin concentration (MCHC)	平均红细胞血红蛋白浓度
red blood cell volume distribution width (RDW)	红细胞体积分布宽度
chronic lymphocytic leukaemia	慢性淋巴细胞性白血病

Key sentences

(1) Hematology analyzers are used to perform complete blood counts (CBC), erythrocyte sedimentation rates (ESRs), or coagulation tests.

血液分析仪用于进行全血细胞计数(CBC),红细胞沉降率(ESR)或凝血试验。

(2) Blood is composed of plasma and several kinds of cells. All of the cells found in the blood come from bone marrow.

血液由血浆和几种细胞组成。血液中发现的所有细胞都来自骨髓。

(3) Automated cell counters sample the blood, and quantify, classify, and describe cell populations using both electrical and optical techniques.

自动细胞计数器可对血液进行取样,并利用电和光学技术量化、分类和描述细胞。

(4) The reflectance, transmission and scattering of light from each cell are analyzed by sophisticated software giving a numerical representation of the likely overall distribution of cell populations.

用复杂的软件来分析来自每个细胞的反射,透射和散射,给出细胞的总体分布数值。

(5) Many disease states are heralded by changes in the blood count: leukocytosis can be a sign of infection; thrombocytopenia can result from drug toxicity; pancytopenia is generally as the result of decreased production from the bone marrow, and is a common complication of cancer chemotherapy.

血细胞计数的变化预示着很多疾病的状态:如白细胞增多可能是感染的征兆,血小板减少可能是药物毒性所致,全血细胞减少可能是因为骨髓的生产能力降低,同时,这也是癌症化疗的常见并发症。

Learning test

Choose the following words below to fill in the blanks, change the form if necessary.

toxicity   variation   hematology   hormone   bacterial

indicate   quantify   classify   analyze   anemia

(1) The chronic _____ of the detergent need to be tested.

(2) The content of glucose in our saliva is difficult to _____ .

(3) Patients are _____ into three categories.

(4) We should develop the students' ability to _____ and solve the problems.

(5) She was diagnosed with _____ caused by malnutrition.

(6) The red light _____ that the instrument is out of work.

(7) The _____ of glucose in urine can be used to monitor diabetes.

(8) This is the website of Shanghai Institute of _____ .

(9) Vitamin C is needed for _____ production.

(10) Cholera is a _____ infection.

Translation

1. Translate the following words into Chinese.

(1) hematology   (2) immune

(3) plasma   (4) leukocyte

(5) platelet   (6) lymphocyte

(7) monocyte   (8) granulocyte

(9) hemoglobin   (10) coagulation

2. Translate the following phrases into Chinese.

(1) complete blood count (CBC)   (2) erythrocyte sedimentation rate

(3) clotting factor   (4) white blood cell

(5) mean corpuscular volume   (6) bone marrow

Practice

1. Translate the following sentences into Chinese.

(1) It is now taken for granted that CBC tests can be ordered, the blood drawn and results available within the hour in facilities with an analyzer.

(2) Automated cell counters sample the blood, and quantify, classify, and describe cell populations using both electrical and optical techniques.

(3) The reflectance, transmission and scattering of light from each cell is analysed by sophisticated software giving a numerical representation of the likely overall distribution of cell populations.

(4) They begin their life as stem cells, and they mature into three main types of cells— red blood cells (RBCs), leukocytes (white blood cells, WBCs), and thrombocytes (platelets).

(5) Each type has a specific purpose, so increased or decreased levels will determine how well a patient is recovering, or what type of foreign object the body is trying to fight.

2. Translate the following sentences into English.

(1)成人的血液约占体重的十三分之一。

(2)葡萄糖为运动提供能量来源。

(3)凝血因子从肝脏被输送到破裂的血管,以确保凝血。

(4)血液循环系统中的细胞一般分为三种:白细胞、红细胞和血小板。

(5)血常规检查主要用血液分析仪。

> **Culture salon**
>
> Hematology analyzer for coagulation tests
>
> Automated coagulation machines orcoagulometers measure the ability of blood to clot by performing any of several types of tests including partial thromboplastin times, prothrombin times (and the calculated INRs commonly used for therapeutic evaluation), lupus anticoagulant screens, D dimer assays, and factor assays.
>
> Coagulometers require blood samples that have been drawn in tubes containing sodium citrate as an anticoagulant. These are used because the mechanism behind the anticoagulant effect of sodium citrate is reversible. Depending on the test, different substances can be added to the blood plasma to trigger a clotting reaction. The progress of clotting may be monitored optically by measuring the absorbance of a particular wavelength of light by the sample and how it changes over time.

Chapter 2-
Section 6
Text 1-看译文

Text 2　Biochemistry analyzer

Lead-in questions:

(1) What is a biochemistry analyzer?

(2) What's the principle involved in the photometric biochemistry analyzer?

(3) What substances can be detected using biochemistry analyzer?

(4) Does the photometric biochemistry analyzer need enough space to operate?

(5) Is the throughput high enough to test 400 samples per hour?

The clinical biochemistry analyzer is an instrument that uses the pale yellow supernatant portion (serum) of centrifuged blood sample or a urine sample, and induces reactions using reagents to measure various components, such as sugar, cholesterol, protein, enzyme, etc. These tests are performed for routine health checks or at hospitals, and the results provide objective data enabling early detection and diagnosis of disease, as well as

Chapter 2-
Section 6
Text 2-听课文

indicating the effects of treatment and patient prognosis. In this article, an automatic photometric biochemistry analyzer will be introduced.

1. Product description

The photometric biochemistry analyzer is a compact version of abenchtop, random access biochemistry analyzer, based on the photometric measurement principle. It offers a nearly unlimited, expandable range of measurement applications. A broad range of interesting substances can be detected including substrates, metabolites and products. Kinetic and end point determination of assays are possible. Due to the stability of the photometric assays, recalibration is necessary only from lot to lot of the selected test method. Barcode detection of reagents and samples ensures a safe and traceable handling of the analysis results. Reagent kits for the detection of ammonia, glucose, glutamine, glutamate, human IgG, inorganic phosphate, Na^+, K^+, and Cl^-, and enzymes, such as alanine aminotransferase(ALT), aspartate aminotransferase(AST), gamma glutamyl transferase(GGT), lactate dehydrogenase(LDH), amylase(AMS), are currently available. However, possible analytes can be expanded to a nearly unlimited range due to the technical measurement principle of the photometric biochemistry analyzer.

2. Benefits

The photometric biochemistry analyzer delivers asemi-automated platform technology that guarantees highly reproducible results due to its membrane-free measurement principle. The design is geared towards future expansion of the assay portfolio. The quality assurance tools(calibrators and controls) allow for system suitability checks at any time. The compact size of the photometric biochemistry analyzer enables the user to operate the system even in laboratories with limited space.

3. Technology

The photometric biochemistry analyzer is equipped with a photometric measurement unit that allows for automatic wavelength selection from 340 to 800 nm. Endpoint and rate analysis are possible from 0.1 to 3.0 absorbance.

The integrated ISE(ion selective electrode) module delivers data on Na^+, K^+, and Cl^-. The temperature-controlled incubation reaction unit guarantees reproducible test results. Long-term stability of the reagents onboard is possible due to the cooled reagent container unit.

The system works with disposable, high precision, resin cuvettes which are automatically loaded and discarded(Figure 2-22). The photometric biochemistry analyzer uses a built-in CPU(central processing unit) with an LCD(liquid crystal display) touch panel(Figure 2-23). In the future, the additional sample port may allow for integration of the photometric biochemistry analyzer into fully automated bioprocesses.

4. Technical Features

(1) Photometric measurement

The membrane-free technology of the photometric biochemistry analyzer enables highly reproducible test results over a long period of time without requiring permanent recalibration cycles.

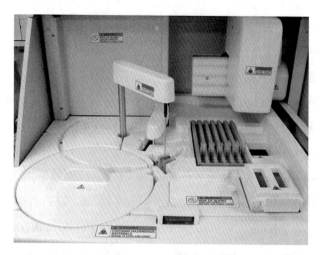

Figure 2-22 Automatically loaded and discarded resin cuvettes with reagent and sample arms.

Figure 2-23 Front view of the photmetric analyzer. It has a built-in CPU and LCD touch panel.

(2) Sample capacity

Up to 90 results per one hour are possible; with ISE unit expandable to up to 360 results(Figure 2-24).

Figure 2-24 The reagent and sample container unit.

(3) Calibrators and controls

Up to three controls can be defined for each assay. Photometric biochemistry analyzer can be run in compliance with Westguard rules. Levey-Jennings plots document the system and assay performance.

(4) Flexible software platform

The photometric biochemistry analyzer can be expanded to a wider assay range without any hardware changes required. The hardware and software architecture allows for easy addition of new analysis methods and it is designed for projects in current GMP(Good Manufacturing Practice) regulated environments.

(5) Technical Specifications

Type	Desktop fully automated random access analyzer
Working principle	General chemistry as photometric or immunoturbidity assay
Throughput	Up to 90 tests/hour(with ISE up to 360 tests/hour)
Actual test scenario (expandable)	Ammonium, glucose, lactate, glutamine (in presence of glutamate), glutamate, human hs IgG, IgG, inorganic phosphate, LDH, Na^+, K^+, Cl^- (to be expanded)
Wavelength range	Automatic wavelength selection by 8-position filter wheel(range between 340 nm and 800 nm), bi-Chrom
Reagent containment	20 bottles(20 ml, or 50 ml volume); cooled by peltier element(+8℃ to +15℃)
Reagent consumption monitoring	Optional external barcode reader for reagents available. Typical reagent consumption ($\sim 180 \mu l$ = 96 Test/20ml)
Sample containment	Tray for 10 samples/controls/calibrators(plus emergency port). $2 \sim 35 \mu l$ sample volume(increment by $0.1 \mu m$). Several sample tubes/containers formats possible. Optional external barcode reader for sample detection available
Measurement	Absorbance measurements in cuvettes
Discrete	Single line random access. Multi-test analysis
Calibrations	Linear. Spline. Logit-Log. Exponential. Point-to-Point. Factor
Analytic modes	End point. 2 point end. Rate. 2 point rate. ISE
Sample pre-dilution	Programmable ratios
Incubation reaction unit(IRU)	24 single-use plastic cuvettes, cuvette temperature adjusted to 37℃ in IRU, monitored by 2 temperature sensors
LIS Capability	Host query mode. Broadcast download mode. Uni-directional mode. Bi-directional mode. Built-in PC with touch panel

(6) Data management system(DMS)

The data management system(DMS) software system of the photometric biochemistry analyzer tracks quality controls and archives sample data and monitors system status. The system is LIS interface capable.

Generally speaking, the photometric biochemistry analyzer has many characteristics, including robustness and stability, quality assurance, high throughput, traceability and what are listed above. A biochemistry analyzer is widely used to determine materials in some kinds of samples, such as blood, urine, and other body fluid quantitatively or qualitatively. And is also one of the most important equipments of laboratory department in a hospital.

Words and phrases

biochemistry	[ˌbaɪəʊˈkemɪstri]	n.	生物化学，生化
pale	[peɪl]	adj.	浅色的
supernatant	[ˌsjuːpəˈneɪtənt]	adj.	上清液
serum	[ˈsɪərəm]	n.	血清
sample	[ˈsɑːmpl]	n.	样品，样本
cholesterol	[kəˈlestərɒl]	n.	胆固醇
prognosis	[prɒgˈnəʊsɪs]	n.	预后
photometric	[ˌfəʊtəˈmetrɪk]	adj.	光度的，光度计的
benchtop	[ˈbentʃtɒp]	adj.	台式的
kinetic	[kɪˈnetɪk]	adj.	动力学的
barcode	[bɑːˈkəʊd]	n.	条码，条形码
traceable	[ˈtreɪsəbl]	adj.	可追踪的，可溯源的
reagent	[riˈeɪdʒənt]	n.	试剂，反应物
glutamine	[ˈgluːtəmiːn]	n.	谷氨酰胺
glutamate	[ˈgluːtəmeɪt]	n.	谷氨酸
amylase	[ˈæmɪleɪz]	n.	淀粉酶
guarantee	[ˌgærənˈtiː]	v.	确保，保证
reproducible	[ˌriːprəˈdjuːsəbl]	adj.	可重复的
endpoint	[ˈendpɔɪt]	n.	终点
absorbance	[əbˈsɔːbəns]	n.	吸收，吸光度
resin	[ˈrezɪn]	n.	树脂
cuvette	[kjuːˈvet]	n.	比色皿
load	[ləʊd]	v.	装载
discard	[dɪsˈkɑːd]	v.	丢弃
permanent	[ˈpɜːmənənt]	adj.	永久的
expandable	[ɪkˈspændəbl]	adj.	可扩展的
flexible	[ˈfleksəbl]	adj.	灵活的
platform	[ˈplætfɔːm]	n.	平台
hardware	[ˈhɑːdweə(r)]	n.	硬件
throughput	[ˈθruːpʊt]	n.	通量
containment	[kənˈteɪnmənt]	n.	容量
increment	[ˈɪŋkrəmənt]	n.	增大，增加
discrete	[dɪˈskriːt]	adj.	分立式
calibration	[ˌkælɪˈbreɪʃn]	n.	校准，校正

pre-dilution	[ˌpriːdaɪˈjuːʃn]	n.	预稀释
interface	[ˈɪntəfeɪs]	n.	界面，接口
quantitatively	[ˈkwɒntɪtətɪvlɪ]	adv.	定量地
qualitatively	[ˈkwɒlɪtətɪvlɪ]	adv.	定性地
reagent kit			试剂盒
alanine aminotransferase (ALT)			丙氨酸氨基转移酶
aspartate aminotransferase (AST)			天门冬氨酸氨基转移酶
gamma glutamyl transferase (GGT)			γ-谷氨酰转移酶
lactate dehydrogenase (LDH)			乳酸脱氢酶
semi-automated			半自动的
ISE (ion selective electrode)			离子选择电极
LCD (liquid crystal display)			液晶显示器
CPU (central processing unit)			中央处理器
temperature-controlled			温度控制的
GMP (Good Manufacturing Practice)			生产质量管理规范
working principle			工作原理
body fluid			体液

Key sentences

(1) The clinical biochemistry analyzer is an instrument that uses the pale yellow supernatant portion (serum) of centrifuged blood sample or a urine sample, and induces reactions using reagents to measure various components, such as sugar, cholesterol, protein, enzyme, etc.

临床生物化学分析仪是一种使用离心血液样品或尿样的淡黄色上清液（血清）的仪器，并使用试剂诱导反应，测量各种成分，如糖、胆固醇、蛋白质、酶等。

(2) The photometric biochemistry analyzer is a compact version of a benchtop, random access biochemistry analyzer, based on the photometric measurement principle.

这台光度生化分析仪是一个台式紧凑型、随机存取的生化分析仪，基于光度测量原理。

(3) The photometric biochemistry analyzer delivers a semi-automated platform technology that guarantees highly reproducible results due to its membrane-free measurement principle.

这台光度生化分析仪采用了一种半自动化平台技术，其无膜测量原理确保了高度可重复的结果。

(4) The photometric biochemistry analyzer is equipped with a photometric measurement unit that allows for automatic wavelength selection from 340~800 nm.

这台光度生化分析仪配备了一个光度测量单元，可从340~800 nm 中自动选择波长。

(5) A biochemistry analyzer is widely used to determine materials in some kinds of samples, such as blood, urine, and other body fluid quantitatively or qualitatively.

生物化学分析仪广泛用于定量或定性地测定血液、尿液等体液样品中的物质。

Learning test

Choose the following words below to fill in the blanks, change the form if necessary.

biochemistry　　　pale　　　expandable　　　absorbance　　　calibration

guarantee　　　permanent　　　discard　　　qualitatively　　　hardware

(1) I _____ that this drug is effective for the disease.

(2) Heavy drinking can cause _____ damage to the brain.

(3) Please read the manual carefully before you _____ it.

(4) The Raman scattering method is used to analyze chemical composition _____ .

(5) This software is not compatible with your _____ .

(6) The _____ course involves a lot of cellular knowledge.

(7) The walls in the child's bedroom are painted with _____ blue.

(8) The microscope is _____ with several external ports.

(9) The organic material has an _____ peak in UV region.

(10) His job includes equipment installation, maintenance, and _____ .

Translation

1. Translate the following words into Chinese.

(1) serum　　　　　　　　(2) sample

(3) biochemistry　　　　　(4) quantitatively

(5) endpoint　　　　　　　(6) containment

(7) interface　　　　　　　(8) calibration

(9) cholesterol　　　　　　(10) reproducible

2. Translate the following phrases into Chinese.

(1) reagent kit　　　　　　(2) LCD (liquid crystal display)

(3) working principle　　　(4) GMP (Good Manufacturing Practice)

(5) semi-automated　　　　(6) body fluid

Practice

1. Translate the following sentences into Chinese

(1) A broad range of interesting substances can be detected including substrates, metabolites and products.

(2) Due to the stability of the photometric assays, recalibration is necessary only from lot to lot of the selected test method.

(3) The compact size of the photometric biochemistry analyzer enables the user to operate the system even in laboratories with limited space.

(4) Long-term stability of the reagents onboard is possible due to the cooled reagent container unit.

(5) The photometric biochemistry analyzer can be expanded to a wider assay range without any hardware changes required.

2. Translate the following sentences into English.

(1) 自动生化分析仪具有快速、简便、灵敏等优点。

(2) 世界上第一台自动生化分析仪是1957年制造的连续流动式自动生化分析仪。

(3) 自动生化分析仪可自动完成生化分析中的取样、加试剂、混合等所有步骤。

(4) 便携式半自动生化分析仪目前仍在广泛使用。

(5) 生化分析仪采用的分析方法有终点法、动力学法等。

> **Culture salon**
>
> **Dialyzer**
>
> In medical testing applications and industrial samples with high concentrations or interfering material, there is often adialyzer module in the instrument in which the analyte permeates through a dialysis membrane into a separate flow path going on to further analysis. The purpose of a dialyzer is to separate the analyte from interfering substances such as protein, whose large molecules do not go through the dialysis membrane but go to a separate waste stream. The reagents, sample and reagent volumes, flow rates, and other aspects of the instrument analysis depend on which analyte is being measured.

Chapter 2-
Section 6
Text 2-看译文

Text 3 Urine analyzer

Lead-in questions:

(1) How many types of urine analyzers are there?

(2) Do you know which is the most cost-effective device used to screen urine?

(3) Can you list some examples of conditions that can be screened using urinalysis?

(4) Do you know how to standardize urine test strip results for the analyzer?

(5) What parameters can be determined by the chemstrip urine analyzer?

Clinical urine tests, one of the most common methods of medical diagnosis, are various tests of urine for diagnostic purpose. The most common urine test is urinalysis(UA). Other tests are urine culture(a microbiological cultureof urine) and urine electrolyte levels.

Urinalysis(UA) simply means analysis of urine. This is a very commonly ordered test which is performed in many clinical settings such as hospitals, clinics, emergency departments, and outpatient laboratories. Urinalysis is often used prior to surgery, as a preemptive screening during a pregnancy checkup, or as part of a routine medical or physical ex-

Chapter 2
Section 6
Text 3 听课文

am. Your doctor may also order urinalysis if they suspect that you have certain conditions, such as: diabetes, kidney disease, liver disease and urinary tract infection, to name just a few. Urinalysis can be classified into three types: macroscopic urinalysis(direct visual observation), urine dipstick chemical analysis, and microscopic urinalysis.

A urine analyzer is a device used in the clinical setting to perform urine testing(Figure 2-25). The units can detect and quantify a number of analytes including bilirubin, protein, glucose and red blood cells. There are various types of urine analyzers, either dipstick or table reagent urinalysis, which can be non-automated, semi-automated, and fully automated. Many models contain urine strip readers, a type of reflectance photometer that can process several hundred strips per hour.

The most cost-effective device used to screen urine is a paper or plastic dipstick(Figure 2-26). Urine dipstick is used to check the invisible components in urine. A dipstick is a paper strip with patches impregnated with chemicals that undergo a color change when certain constituents of the urine are present or in a certain concentration. The strip is dipped into the urine sample, and after the appropriate number of seconds, the color change is compared to a standard chart to determine the findings. Although this method is time-saving and cost-effective, a careless doctor, nurse, or assistant is entirely capable of misreading or misinterpreting the results.

Figure 2-25 A urine analyzer.

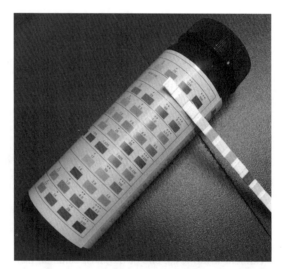

Figure 2-26 A urine test strip.

Chemstrip urine analyzer is a dry chemistry urine analyzer designed to read and evaluate the results of a urine test strip. It can be used to determine specific gravity(SG), power of hydrogen(pH), protein(albumin) (PRO), glucose(GLU), ketone body(KET), bilirubin(BIL), urobilinogen(URO), leukocytes(esterase) (LEU), erythrocytes(hemoglobin) (ERY), ascorbic acid(VitC), and nitrite(NIT).

These urine test strips are multi-parameter strips used for the determination of the chemicals that are stated above. The urine analyzer is a reflectance photometer. It reads the urine test strips under standardized conditions, saves the results to memory and outputs them via its own built-in printer and/or serial interface. The urine analyzer standardizes urine test strip results by eliminating factors known to affect visual evaluation of urine test strips, such as variable lighting conditions at the workplace, individual skill levels at matching test strip pad colors, failure to keep the prescribed testing times, and clerical errors.

The test strip is placed on a sliding tray, and a stepping motor moves it under the reading head, which remains stationary. The analyzer reads the reference pad, followed by each of the test pads on the strip. The reading head contains LED that emits light at various wavelengths. Reading is done electro-optically as follows: The LED emits light of a defined wavelength onto the surface of the test pad at an optimum angle. The light hitting the test zone is reflected proportionally to the color produced on the test pad and is picked up by the detector, a phototransistor positioned directly above the test zone. The phototransistor sends an analog electrical signal to an A/D converter, which changes it to digital form. The microprocessor then converts this digital reading to a relative reflectance value by referring it to a calibration standard. Finally, the system compares the reflectance value with the defined range limits and outputs a semi-quantitative result.

Microscopic urinalysis requires only a relatively inexpensive light microscope-urine sediment analyzer. Red blood cell, white blood cell, epithelial cells, casts can be tested. The methodology is done as below. A sample of well-mixed urine(usually 10~15 ml) is centrifuged in a test tube at relatively low speed (about2~3,000 rpm) for 5~10 minutes until a moderately cohesive sediment is produced at the bottom of the tube. The supernate is decanted and a volume of 0.2 to 0.5 ml is left inside the tube. The sediment is resuspended in the remaining supernate by flicking the bottom of the tube several times. A drop of resuspended sediment is poured onto a glass slide and coverslipped. The sediment is first examined under low power to identify most crystals, casts, squamous cells, and other large objects.

The numbers of casts seen are usually reported as number of each type found per low power field (LPF). Example: 5~10 hyaline casts/LPF. Since the number of elements found in each field may vary considerably from one field to another, several fields are averaged. Next, examination is carried out at high power to identify crystals, cells, and bacteria. The various types of cells are usually described as the number of each type found per average high power field(HPF). Example: 1~5 WBC/HPF.

Words and phrases

urine	[ˈjʊərɪn]	n.	尿
urinalysis	[ˌjʊərɪˈnælɪsɪs]	n.	尿液分析,验尿
kidney	[ˈkɪdni]	n.	肾,肾脏
culture	[ˈkʌltʃə(r)]	n.	培养
electrolyte	[ɪˈlektrəlaɪt]	n.	电解质,电解液
bilirubin	[ˌbɪlɪˈruːbɪn]	n.	胆红素
urobilinogen	[ˌjʊərəbaɪˈlɪnədʒɪn]	n.	尿胆原

Chapter 2 Section 6-Text 3-听词汇

英文	音标	词性	中文
nitrite	[ˈnaɪtraɪt]	n.	亚硝酸盐
photometer	[fəʊˈtɒmɪtə]	n.	光度计
macroscopic	[ˌmækrəˈskɒpɪk]	adj.	宏观的,肉眼可见的
dipstick	[ˈdɪpstɪk]	n.	试纸
patch	[pætʃ]	n.	小片,小块
impregnate	[ˈɪmpregneɪt]	v.	注入,浸渍
constituent	[kənˈstɪtjuənt]	n.	成分
concentration	[ˌkɒnsenˈtreɪʃən]	n.	浓度
dip	[dɪp]	v.	浸,蘸
misinterpret	[ˌmɪsɪnˈtɜːprɪt]	v.	曲解
via	[ˈvaɪə]	prep.	经由,通过
eliminate	[ɪˈlɪmɪneɪt]	v.	除去,消除
clerical	[ˈklerɪkl]	adj.	文书的,抄写的
stationary	[ˈsteɪʃənri]	adj.	不动的,固定的
prescribe	[prɪˈskraɪbd]	vt.	指定,规定
wavelength	[ˈweɪvleŋθ]	n.	波长
emit	[iˈmɪt]	vt.	发出,发射
optimum	[ˈɒptɪməm]	adj.	最佳的,最适宜的
proportionally	[prəˈpɔːʃnli]	adv.	按比例地
reflect	[rɪˈflekt]	vt.	反射
analog	[ˈænəlɔːg]	adj.	模拟的
converter	[kənˈvɜːtə(r)]	n.	转换器
phototransistor	[fəʊtəʊtrænˈsɪstə]	n.	光电晶体管
microprocessor	[ˌmaɪkrəʊˈprəʊsesə(r)]	n.	微处理器
reflectance	[rɪˈflektəns]	n.	反射,反射率
sediment	[ˈsedɪmənt]	n.	沉淀物
cast	[kɑːst]	n.	(尿)管型
methodology	[ˌmeθəˈdɒlədʒi]	n.	方法
centrifuge	[ˈsentrɪfjuːdʒ]	v.	离心
moderately	[ˈmɒdərətli]	adv.	适度地,一般地
cohesive	[kəʊˈhiːsɪv]	adj.	黏性的
supernate	[ˈsjuːpəneɪt]	n.	上清液
decant	[dɪˈkænt]	v.	倒出
flick	[flɪk]	vt.	轻弹,抖动
coverslip	[ˈkʌvəslɪp]	n.	盖玻片

crystal	[ˈkrɪstl]	n.	晶体
hyaline	[ˈhaɪəlɪn]	adj.	玻璃似的，透明的
squamous	[ˈskweɪməs]	adj.	鳞状的，有鳞的
bacteria	[bækˈtɪəriə]	n.	细菌（bacterium 复数）
paper strip			试纸
time-saving			省时的
cost-effective			经济的，划算的
serial interface			串行接口
specific gravity(SG)			尿比重
clerical error			笔误
epithelial cell			上皮细胞
power of hydrogen(pH)			酸碱度
ketone body(KET)			尿酮体
Leukocyte esterase(LEU)			白细胞酯酶
ascorbic acid			抗坏血酸，维生素 C
low power field(LPF)			低倍镜视场
high power field(HPF)			高倍镜视场

Key sentences

(1) Urinalysis can be classified into three types: macroscopic urinalysis (direct visual observation), urine dipstick chemical analysis and microscopic urinalysis.

尿液分析可以分为三种类型：肉眼检查（直观观察），尿试纸化学分析和显微镜尿检。

(2) A dipstick is a paper strip with patches impregnated with chemicals that undergo a color change when certain constituents of the urine are present or in a certain concentration. The strip is dipped into the urine sample, and after the appropriate number of seconds, the color change is compared to a standard chart to determine the findings.

试纸是一种纸条，带有浸渍了化学物质的小条块，当尿液的某些成分存在或到达一定浓度时，小条块会发生颜色变化。将试纸浸入尿样中，在适当的秒数之后，将颜色变化与标准图进行比较以确定结果。

(3) It (Chemstrip urine analyzer) can be used to determine specific gravity(SG), power of hydrogen (pH), protein(albumin)(PRO), glucose(GLU), ketone body(KET), bilirubin(BIL), urobilinogen(URO), leukocytes(esterase)(LEU), erythrocytes(hemoglobin)(ERY), ascorbic acid(VitC), and nitrite(NIT).

干化学尿液分析仪可用于测定尿比重（SG）、酸碱度（pH）、蛋白质（白蛋白）（PRO）、葡萄糖（GLU）、尿酮体（KET）、胆红素（BIL）、尿胆原（URO）、白细胞（酯酶）（LEU）、红细胞（血红蛋白）（ERY）、抗坏血酸（VitC）和亚硝酸盐（NIT）。

(4) The urine analyzer standardizes urine test strip results by eliminating factors known to affect visual

evaluation of urine test strips, such as variable lighting conditions at the workplace, individual skill levels at matching test strip pad colors, failure to keep the prescribed testing times, and clerical errors.

尿液分析仪通过消除已知会影响尿液测试试纸的视觉评估的因素来标准化试纸测试结果,例如,工作场所的照明条件,匹配试纸条块颜色的个人技能水平,未能保持指定的测试次数,和笔误等等。

(5) Microscopic urinalysis requires only a relatively inexpensive light microscope-urine sediment analyzer. Red blood cell, white blood cell, epithelial cells, casts can be tested.

显微镜尿检只需要相对便宜的光学显微镜——尿沉渣分析仪。红细胞、白细胞、上皮细胞、管型都可以测试。

Learning test

Choose the following words below to fill in the blanks, change the form if necessary.

optimum   concentration   eliminate   reflect   proportionally
urinalysis   clerical   emit   culture   bacteria

(1) They collected a urine specimen for _____ .

(2) Please excuse this _____ error in this article.

(3) The LEDs _____ soft yellow light.

(4) The DNA and cell _____ tests currently used are time-consuming and expensive.

(5) If your blood sample shows the presence of _____, you'll be prescribed antibiotics.

(6) The _____ weight for me is 45kg.

(7) The global ozone _____ has dropped several percent over the last decade.

(8) If you are allergic to any food or drink, _____ it from your diet.

(9) A mirror can _____ light, that's why you can see yourself in it.

(10) The concentration of the substance varies _____ with its absorbance.

Translation

1. Translate the following words into Chinese.

(1) kidney   (2) urine
(3) calibration   (4) supernate
(5) concentration   (6) wavelength
(7) nitrite   (8) methodology
(9) photometer   (10) centrifuge

2. Translate the following phrases into Chinese.

(1) time-saving   (2) cost-effective
(3) specific gravity   (4) ascorbic acid
(5) power of hydrogen (pH) (6) clerical error

Practice

1. Translate the following sentences into Chinese.

(1) Clinical urine tests, one of the most common methods of medical diagnosis, are various tests of urine for diagnostic purpose.

(2) Urinalysis is often used prior to surgery, as a preemptive screening during a pregnancy checkup, or as part of a routine medical or physical exam.

(3) There are various types of urine analyzers, either dipstick or table reagent urinalysis, which can be non-automated, semi-automated, and fully automated.

(4) Although this method is time-saving and cost-effective, a careless doctor, nurse, or assistant is entirely capable of misreading or misinterpreting the results.

(5) The phototransistor sends an analog electrical signal to an A/D converter, which changes it to digital form.

2. Translate the following sentences into English.

(1) 尿的 pH 范围一般是 5~9。
(2) 尿检是怀孕最简单的初步筛查方法。
(3) 当状态指示灯为红色时,请远离尿液分析仪。
(4) 标准尿检试纸可以包括多达 10 种不同的化学试剂。
(5) 在许多文化中,尿液曾被视为神秘的液体。

Culture salon

Urine

Urine is a liquid by-product of metabolism in the bodies of many animals, including humans. It is expelled from the kidneys and flows through the ureters to the urinary bladder, from which it is soon excreted from the body through the urethra during urination. About 91%~96% of urine consists of water. Urine also contains an assortment of inorganic salts and organic compounds, including proteins, hormones, and a wide range of metabolites, varying by what is introduced into the body. Producing too much or too little urine needs medical attention. Polyuria is a condition of excessive production of urine(> 2.5 L/day), oliguria when< 400 ml are produced, and anuria one of < 100 ml per day.

Chapter 2-Section 6 Text 3-看译文

Chapter 2-Section 6-习题

Section 7 Biomaterials

Chapter 2-
Section 7-PPT

Text 1 Medical dressing materials

Lead-in questions:

(1) What is a medical dressing?

(2) What are the key medical uses of a medical dressing?

(3) How many types of wound dressings available on the market today?

(4) What materials are used to make non-resorbable gauze dressings?

(5) What are the main four classes of wound dressings?

Whenever you have a wound, whether it's a minor cut or a major incision, it's crucial to care for it properly. Part of the process includes wound care dressings. There are a variety of options when it comes to dressings, and to determine which is the best and most effective depends on what sort of wound you have.

Chapter 2-
Section 7-
Text 1-听课文

A doctor or other medical professional will examine the wound and determine what is necessary to keep it free from complications and to assist with healing. The dressing options will also depend on where the wound is, how large it is and other related factors.

A dressing is a sterile pad compress applied to a wound to promote healing and protect the wound from further harm. A dressing is designed to be in direct contact with the wound, as distinguished from a bandage, which is most often used to hold a dressing in place. Many modern dressings are self-adhesive.

1. Medical uses

A dressing can have a number of purposes, depending on the type, severity and position of the wound, although all purposes are focused towards promoting recovery and protecting from further harm. Key purposes of a dressing are:

(1) stem bleeding-to help to seal the wound to expedite the clotting process;

(2) protection from infection-to defend the wound against germs and mechanical damage;

(3) absorb exudates-to soak up blood, plasma, and other fluids exuded from the wound, containing it/them in one place and preventing maceration;

(4) ease pain-either by a medicated analgesic effect, compression or simply preventing pain from further trauma;

(5) debride the wound-to remove slough and foreign objects from the wound to expedite healing;

(6) reduce psychological stress-to obscure a healing wound from the view of others.

Ultimately, the aim of a dressing is to promote healing of the wound by providing a sterile, breathable and moist environment that facilitates granulation and epithelialization. This will then reduces the risk of infection, help the wound heal more quickly, and reduce scarring.

2. Types

There are more than 3,000 types of wound dressings available on the market today, making it easy to become overwhelmed by the options. The secret to understanding the various types of wound dressings is to learn the basic properties of the main categories of wound dressings. The dressings within each category are not identical; however, they do possess many of the same properties. This section provides detailed information on the properties of the four main classes of wound dressings, as suggested by the Food and Drug Administration(FDA) on November 4, 1999.

(1) Non-resorbable gauze/sponge dressings

Non-resorbable gauze/sponge dressings are made of woven or no-woven cotton-mesh cellulose or cellulose derivatives and can be manufactured in the form of pads or strips. Gauze/sponge dressings are used mainly to cover a wound surface area following application of topical preparation(e. g. : antibacterial creams or advanced spreadable preparations.

Gauze dressings continue to be the most readily available wound dressings in use today (Figure 2-27). Gauze is highly permeable and relatively non-occlusive. Therefore, gauze dressings may promote desiccation in wounds with minimal exudate unless used in combination with another dressing or topical agent. Gauze may be used as a primary or secondary wound dressing. Gauze dressings are inexpensive for one-time or short-term use. Gauze dressings come in many forms: squares, sheets, rolls, and packing strips.

Figure 2-27 Commonly used gauze dressings

(2) Hydrophilic/absorptive dressings

Hydrocolloid dressings are used on burns, light to moderately draining wounds, necrotic wounds, under compression wraps, pressure ulcers and venous ulcers(Figure 2-28). Hydrocolloid wound dressings contain hydrophilic colloidal particles such as gelatin, pectin, and cellulose, and have a very strong film or foam adhesive backing. This class of dressings varies greatly in absorption abilities. Hydrocolloids absorb exudate slowly by swelling into a gel-like mass. Upon removal, a residue commonly remains within the wound bed. Because this residue may have a foul odor, it is often mistaken as a sign of infection. Hydrocolloids come in a variety of sizes and precut shapes. Several hydrocolloids have beveled edges to reduce the tendency for the dressing to roll when placed in high-friction areas. Hydrocolloids provide thermal insulation to the wound and are impermeable to water, oxygen, and bacteria. Wounds dressed with hydrocolloids have lower infection rates than wounds covered with gauze, semipermeable films, sheet hydrogels, or semi-permeable foams.

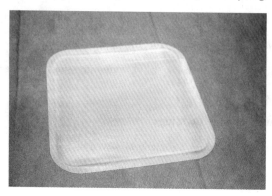

Figure 2-28 Hydrocolloid dressings.

(3) Occlusive dressings

Occlusive dressings, in general, are used mainly to maintain a moist environment within the ulcer area. The significance of a moist environment for all the complex processes of wound healing was noted earlier in this chapter. This approach was confirmed by a variety of research studies, demonstrating the beneficial effect of occlusive dressings on surgical wounds and chronic cutaneous ulcers. In most of these studies, a more efficient healing was manifested by improved granulation tissue formation as well as enhanced epithelialization. However, one should avoid 'over-moisturizing' cutaneous ulcers, since this may lead to maceration, skin breakdown, and infection.

(4) Hydrogel dressings

This type of dressing is for wounds with little to no excess fluid, painful wounds, necrotic wounds, pressure ulcers, donor sites, second degree or higher burns and infected wounds. Hydrogel wound dressings are 80% to 99% water-or-glycerin-based wound dressings that are available in sheets, gels, or impregnated gauzes. Hydrogels can only absorb a minimal amount of fluid. But since they contain high water/glycerin content they are able to donate moisture to dry wounds. When applied to the skin or wound, they feel cool and may decrease wound pain. But, hydrogels are permeable to gas and water, making them less effective bacterial barriers than semipermeable films or hydrocolloids. And, they may dehydrate easily, particularly if water based. Almost all hydrogels are non-adhesive and require a secondary dressing.

3. Usage

Applying a dressing is a first aidskill, although many people undertake the practice with no training-especially on minor wounds. Modern dressings will almost all come in a prepackaged sterile wrapping, date coded to ensure sterility. Sterility is necessary to prevent infection from pathogens resident within the dressing.

Words and phrases

Chapter 2-
Section 7-
Text 1-听词汇

dressing	[ˈdresɪŋ]	n.	敷料
wound	[wuːnd]	n.	伤口
incision	[ɪnˈsɪʒn]	n.	切口
crucial	[ˈkruːʃl]	adj.	至关重要的
complication	[ˌkɒmplɪˈkeɪʃn]	n.	并发症
sterile	[ˈsteraɪl]	adj.	无菌的
severity	[sɪˈverɪtɪ]	n.	严重程度
stem	[stem]	vt.	阻止
bleed	[bliːd]	vi.	出血,流血
infection	[ɪnˈfekʃn]	n.	感染,传染
germ	[dʒɜːm]	n.	病菌,细菌
exudate	[ˈegzjʊdeɪt]	n.	渗出物,渗出液
plasma	[ˈplæzmə]	n.	血浆
ease	[iːz]	vt.	减轻,缓和
analgesic	[ˌænəlˈdʒiːzɪk]	adj.	止痛的
trauma	[ˈtrɔːmə]	n.	外伤,创伤
debride	[drˈbraid]	vt.	清创
slough	[slʌf]	n.	腐肉
expedite	[ˈekspɪdaɪt]	vt.	加快,促进
psychological	[ˌsaɪkəˈlɒdʒɪkl]	adj.	心理的,心理学的,精神上的
obscure	[əbˈskjʊə]	vt.	掩盖,隐藏
epithelialization	[epɪθiːlɪəlaɪˈzeɪʃn]	n.	上皮形成,外皮形成
moist	[mɒɪst]	adj.	潮湿的
gauze	[gɔːz]	n.	纱布
sponge	[spʌndʒ]	n.	海绵,海绵状物
cellulose	[ˈseljʊləʊz]	n.	纤维素
derivative	[dɪˈrɪvətɪv]	n.	衍生物
antibacterial	[ˌæntibækˈtɪərɪəl]	adj.	抗菌的
desiccation	[ˌdesɪˈkeɪʃn]	n.	干燥,脱水
inexpensive	[ˌɪnɪkˈspensɪv]	adj.	便宜的,不贵的
sheet	[ʃiːt]	n.	薄片,片
roll	[rəʊl]	n.	卷,卷形物
hydrophilic	[ˌhaɪdrəˈfɪlɪk]	adj.	亲水的,吸水的
drain	[dreɪn]	v.	引流

necrotic	[neˈkrɔtik]	adj.	坏死的
venous	[ˈviːnəs]	adj.	静脉的
hydrocolloid	[haɪdrəˈkɒlɔɪd]	n.	水状胶质，水状胶体
gelatin	[ˈdʒelətɪn]	n.	明胶
pectin	[ˈpektɪn]	n.	果胶
chronic	[ˈkrɒnɪk]	adj.	慢性的，长期的
cutaneous	[kjuːˈteɪnɪəs]	adj.	皮肤的
ulcer	[ˈʌlsə]	n.	溃疡
impregnate	[ˈɪmpregneɪt]	vt.	灌注，浸渍
maceration	[ˌmæsəˈreɪʃən]	n.	浸渍
glycerin	[ˈglɪsərɪn]	n.	甘油
permeable	[ˈpɜːmɪəbl]	adj.	可渗透的
hydrogel	[ˈhaɪdrədʒel]	n.	水凝胶
dehydrate	[diːhaɪˈdreɪt]	v.	脱水
self-adhesive			自粘的，自动附着的
stembleeding			止血
psychological stress			心理压力
wound dressing			伤口敷料
analgesic effect			镇痛作用
draining wound			引流伤口
cutaneous ulcer			皮肤溃疡
over-moisturizing			过度保湿
non-resorbable			不可吸收的
packing strip			打包带
soak up			吸收

Key sentences

(1) Whenever you have a wound, whether it's a minor cut or a major incision, it's crucial to care for it properly.

每当你有伤口，无论是小伤口还是大切口，妥善护理都是至关重要的。

(2) The dressing options will also depend on where the wound is, how large it is and other related factors.

敷料的选择也取决于伤口在哪里，有多大和其他相关因素。

(3) A dressing can have a number of purposes, depending on the type, severity and position of the wound, although all purposes are focused towards promoting recovery and protecting from further harm.

根据伤口的类型、严重程度以及位置，敷料可以有很多的用途，尽管所有用途的目的都是为了促进伤口的愈合和防止伤口进一步恶化。

(4) Ultimately, the aim of a dressing is to promote healing of the wound by providing a sterile, breathable and moist environment that facilitates granulation and epithelialization.

最终,敷料的作用是为促进伤口愈合提供一个能促进肉芽生长和上皮形成的无菌、透气而且湿润的环境。

(5) Applying a dressing is a first aid skill, although many people undertake the practice with no training-especially on minor wounds. Modern dressings will almost all come in a prepackaged sterile wrapping, date coded to ensure sterility.

使用敷料是一种急救技能,尽管许多人在没有受过训练的情况下进行了这种实践,特别是在小伤口上。现代敷料基本都是无菌包装,并印有灭菌时间。

Learning test

Choose the following words below to fill in the blanks, change the form if necessary.

| desiccation | infection | moist | expedite | ease |
| chronic | venous | wound | plasma | crucial |

(1) She broke into the conversation at a _____ moment.

(2) A cold is an _____ of the upper respiratory tract.

(3) Blood cells like red blood cells float in the _____.

(4) The pain began to _____ up after medication.

(5) Several experts join the group to _____ the product development process.

(6) Hydrocolloid dressings create a _____ environment at the wound site that promotes cell migration.

(7) _____ is the state of extreme dryness, or the process of extreme drying.

(8) The _____ was infected with germs.

(9) They think his illness is acute rather than _____.

(10) The catheter is a double-cavity central _____ catheter.

Translation

1. Translate the following words into Chinese.

(1) crucial   (2) bleed

(3) infection   (4) glycerin

(5) trauma   (6) expedite

(7) psychological   (8) moist

(9) ulcer   (10) hydrogel

2. Translate the following phrases into Chinese.

(1) soak up   (2) wound dressing

(3) non-resorbable   (4) cutaneous ulcer

(5) psychological stress   (6) analgesic effect

Practice

1. Translate the following sentences into Chinese.

(1) The secret to understanding the various types of wound dressings is to learn the basic properties of the main categories of wound dressings.

(2) Non-resorbable gauze/sponge dressings are made of woven or no-woven cotton-mesh cellulose or cellulose derivatives and can be manufactured in the form of pads or strips.

(3) Hydrocolloid dressings are used on burns, light to moderately draining wounds, necrotic wounds, under compression wraps, pressure ulcers and venous ulcers.

(4) Occlusive dressings, in general, are used mainly to maintain a moist environment within the ulcer area.

(5) But, hydrogels are permeable to gas and water, making them less effective bacterial barriers than semipermeable films or hydrocolloids.

2. Translate the following sentences into English.

(1) 随着医用材料的发展,敷料的种类越来越多。

(2) 银离子可以提高敷料的抗菌能力。

(3) 湿润环境比干燥环境更有益于伤口的愈合。

(4) 胶原敷料因其优异的生物相容性和可吸收性而受到广泛的关注。

(5) 海藻酸盐敷料可以有效抑制有害细菌的生长。

Culture salon

Collagen used in wound care

Collagen is a natural product, therefore it is used as a natural wound dressing and has properties that artificial wound dressings do not have. It is resistant against bacteria, which is of vital importance in a wound dressing. It helps to keep the wound sterile, because of its natural ability to fight infection. When collagen is used as a burn dressing, healthy granulation tissue is able to form very quickly over the burn, helping it to heal rapidly.

Chapter 2-Section 7-Text 1-看译文

Text 2 Dental materials

Lead-in questions:

(1) What is the common use of temporary dressing in dentistry?

(2) What are dental impressions?

(3) What are denture adhesives?

(4) What is the medical use of dental amalgam?

(5) What are the chemical compositions of dental amalgam?

Dental materials are specially fabricated materials, designed for use in dentistry. There are many different types of dental material, and their characteristics vary according to their intended purpose. Examples include temporary dressings, dental restorations (fillings, crowns, bridges), endodontic materials (used in root canal therapy), impression materials, prosthetic materials (dentures), dental implants (Figure 2-29), and many others.

Chapter 2-
Section 7-
Text 2-听课文

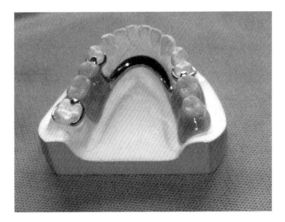

Figure 2-29 Dental implants.

1. Temporary dressings

A temporary dressing is a dental filling which is not intended to last in the long term. They are interim materials that may have therapeutic properties. A common use of temporary dressing occurs if root canal therapy is carried out over more than one appointment. In between each visit, the pulp canal system must be protected from contamination from the oral cavity, and a temporary filling is placed in the access cavity.

2. Impression materials

Dental impressions are negative imprints of teeth and oral soft tissues from which a positive representation can be cast. They are used in prosthodontics (to make dentures), orthodontics, restorative dentistry, dental implantology and oral and maxillofacial surgery.

Impression materials are designed to be liquid or semi-solid when first mixed, then set hard in a few minutes, leaving imprints of oral structures. Common dental impression materials include sodium alginate, polyether, and silicones.

3. Denture adhesives

Denture adhesives are pastes, powders or adhesive pads that may be placed in/on dentures to help them stay in place. Sometimes denture adhesives contain zinc to enhance adhesion. Zinc is a mineral that is an essential ingredient for good health. It is found in protein-rich foods such as shellfish, beef, chicken and nuts, as well as in some dietary supplements. However, an excess of zinc in the body can lead to health problems such as nerve damage, especially in the hands and feet. This damage appears slowly, over an extended period of time. Overuse of zinc-containing denture adhesives, especially when combined with dietary supplements that contain zinc and other sources of zinc, can contribute to an excess of zinc in your body.

In most cases, properly fitted and maintained dentures should not require the use of denture adhesives. Over time, shrinkage in the bone structure in the mouth causes dentures to gradually become loose. When this occurs, the dentures should be relined or new dentures made that fit the mouth properly. Denture adhesives fill gaps caused by shrinking bone and give temporary relief from loosening dentures.

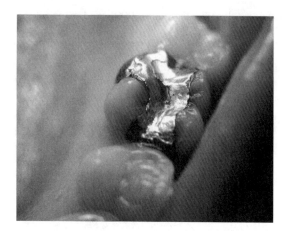

Figure 2-30　Amalgam filling on first molar.

4. Dental amalgam

Dental amalgam is a dental filling material which is used to fill cavities caused by tooth decay(Figure 2-30). Its primary component is elemental mercury. Tooth decay results in a loss of tooth structure.

Dental amalgam is a mixture of metals, consisting of liquid(elemental) mercury and a powdered alloy composed of silver, tin, and copper. Approximately 50% of dental amalgam is elemental mercury by weight. The chemical properties of elemental mercury allow it to react with and bind together the silver/copper/tin alloy particles to form an amalgam. Dental amalgam fillings are also known as "silver fillings" because of their silver-like appearance. Despite the name, "silver fillings" do contain elemental mercury.

Dental amalgam fillings are strong and long-lasting, so they are less likely to break than some other types of fillings. Dental amalgam is the least expensive type of filling material.

You may ask why mercury is used in dental amalgam. Approximately half of a dental amalgam filling is liquid mercury and the other half is a powdered alloy of silver, tin, and copper. Mercury is used to bind the alloy particles together into a strong, durable, and solid filling. Mercury's unique properties(it is a liquid at room temperature and that bonds well with the alloy powder) make it an important component of dental amalgam that contributes to its durability.

Mercury from dental amalgam and other sources(e. g. , fish) is bioaccumulative. Studies of healthy subjects with amalgam fillings have shown that mercury from exposure to mercury vapor bioaccumulates in certain tissues of the body including kidneys and brain. Studies have not shown that bioaccumulation of mercury from dental amalgam results in damage to target organs. For more information about bioaccumulation, please see Related Resources.

The mercury in dental amalgam is not the same as the mercury in some types of fish. There are several different chemical forms of mercury: elemental mercury, inorganic mercury, and methylmercury. The form of mercury associated with dental amalgam is elemental mercury, which releases mercury vapor. The form of mercury found in fish is methylmercury, a type of organic mercury. Mercury vapor is mainly absorbed by the lungs. Methylmercury is mainly absorbed through the digestive tract. The body processes these forms of mercury differently and has different levels of tolerance for mercury vapor and methylmercury.

Other materials can also be used to fill cavities caused by dental decay. Like dental amalgam, these direct filling materials are used to restore the biting surface of a tooth that has been damaged by decay. Your dentist can discuss treatment options based on the location of cavities in your mouth and the amount of tooth decay.

The primary alternatives to dental amalgam are composite resin fillings and glass ionomer cement fillings.

Words and phrases

Chapter 2-Section 7-Text 2-听词汇

dental	[ˈdentl]	adj.	牙科的,牙齿的
fabricate	[ˈfæbrɪkeɪt]	vt.	制造,加工
dentistry	[ˈdentɪstrɪ]	n.	牙科,牙医业
intended	[ɪnˈtendɪd]	adj.	预期的
temporary	[ˈtemprərɪ]	adj.	临时的,暂时的
restoration	[restəˈreɪʃn]	n.	修复,复原
crown	[kraʊn]	n.	牙冠,齿冠
filling	[ˈfɪlɪŋ]	n.	填充物
endodontic	[endəʊˈdɒntɪk]	adj.	牙髓的,根管的
prosthetic	[prɒsˈθetɪk]	adj.	假体的
denture	[ˈdentʃə]	n.	义齿,假牙
interim	[ˈɪnt(ə)rɪm]	adj.	临时的,暂时的
contamination	[kənˌtæmɪˈneɪʃən]	n.	污染,弄脏
oral	[ˈɔːrl]	adj.	口的,口腔的
cavity	[ˈkævɪtɪ]	n.	腔,洞
prosthodontic	[ˌprɒsθəˈdɒntɪk]	n.	口腔修复
orthodontic	[ˌɔːθəˈdɒntɪk]	n.	正畸
implantology	[ɪmplɑːnˈtɒlədʒɪ]	n.	种植(学),移植(学)
maxillofacial	[mækˌsɪləʊˈfeɪʃl]	adj.	上颌面的
sodium	[ˈsəʊdɪəm]	n.	钠
alginate	[ˈældʒɪneɪt]	n.	藻酸盐

polyether	[ˌpɒlɪˈiːθə]	n.	聚醚,多醚
silicone	[ˈsɪlɪkəʊn]	n.	硅橡胶
adhesive	[ədˈhiːsɪv]	n.	粘合剂,胶黏剂
zinc	[zɪŋk]	n.	锌
mineral	[ˈmɪnərəl]	n.	矿物
paste	[peɪst]	n.	糊状物,膏体
ingredient	[ɪnˈgriːdɪənt]	n.	成分,要素
protein	[ˈprəʊtiːn]	n.	蛋白质
shellfish	[ˈʃelfɪʃ]	n.	贝类
shrink	[ʃrɪŋk]	v.	收缩
loosen	[ˈluːsn]	vt.	变松,松开
amalgam	[əˈmælgəm]	n.	汞合金,银汞合金
decay	[dɪˈkeɪ]	n.	腐败,腐烂
mercury	[ˈmɜːkjəri]	n.	汞,水银
powdered	[ˈpaʊdəd]	adj.	粉状的
alloy	[ˈælɔɪ]	n.	合金
silver	[ˈsɪlvə]	n.	银
tin	[tɪn]	n.	锡
copper	[ˈkɒpə]	n.	铜
durable	[ˈdjʊərəbl]	adj.	耐用的,持久的
solid	[ˈsɒlɪd]	adj.	固体的,固态的
bioaccumulative	[ˌbaɪəʊəˈkjuːmjulətɪv]	adj.	生物累积性的,生物积聚的
vapor	[ˈveɪpə(r)]	n.	蒸汽
kidney	[ˈkɪdni]	n.	肾脏
organ	[ˈɔːgən]	n.	器官
methylmercury	[ˌmeθɪlˈmɜːkjʊrɪ]	n.	甲基水银,甲基汞
dental restoration			牙齿修复
endodontic material			牙髓材料
impression material			印模材料
prosthetic material			假体材料
dental implant			牙种植体
root canal			(牙)根管
pulp canal			(牙)根管,牙髓管
negative imprint			阴模
semi-solid			半固体的

set hard	凝固,固化
sodium alginate	海藻酸钠
denture adhesive	义齿黏附剂
dietary supplement	膳食补充剂
tooth decay	蛀牙,龋齿
glass ionomer cement	玻璃离子水门汀
soft tissue	软组织
carry out	执行,实行,贯彻

Key sentences

(1) Dental materials are specially fabricated materials, designed for use in dentistry. There are many different types of dental material, and their characteristics vary according to their intended purpose.

牙科材料是为牙科使用而设计的特殊加工材料。牙科材料有很多种,其特性根据预期目的而不同。

(2) A common use of temporary dressing occurs if root canal therapy is carried out over more than one appointment.

如果根管治疗要做一次以上,则常常使用临时敷料。

(3) Dental impressions are negative imprints of teeth and oral soft tissues from which a positive representation can be cast.

牙齿印模是牙齿和口腔软组织的阴模,用于铸造阳模。

(4) Denture adhesives are pastes, powders or adhesive pads that may be placed in/on dentures to help them stay in place.

义齿粘结剂是可以放置在义齿上的膏体、粉末或粘合片,使其保持原位。

(5) Dental amalgam is a dental filling material which is used to fill cavities caused by tooth decay (Figure 2-30).

银汞合金是一种牙科填充材料,用于填补由龋齿造成的龋洞。

Learning test

Choose the following words below to fill in the blanks, change the form if necessary.

silicone	durable	oral	alloy	contamination
interim	solid	loosen	dental	fabricate

(1) Certain _____ and eye care services are also covered by the medical insurance.

(2) The dental implants can be _____ using ceramics or titanium alloy.

(3) The _____ particles were filtered from the solution.

(4) The _____ of drinking water will directly harm the health of the public.

(5) Stainless steel is a type of _____ and its main ingredient is iron.

(6) I agreed to be the _____ CEO and to stay only six months.

(7) Our _____ cavities should be cleaned twice a day.

(8) You need to take some exercise to _____ up your muscles.

(9) _____ rubber is a commonly used material to fabricate medical devices.

(10) Our product is highly _____ and reliable.

Translation

1. Translate the following words into Chinese.

(1) denture   (2) crown

(3) restoration   (4) oral

(5) amalgam   (6) alloy

(7) mineral   (8) decay

(9) mercury   (10) endodontic

2. Translate the following phrases into Chinese.

(1) dental implant   (2) root canal

(3) negative imprint   (4) set hard

(5) tooth decay   (6) denture adhesive

Practice

1. Translate the following sentences into Chinese.

(1) A temporary dressing is a dental filling which is not intended to last in the long term. They are interim materials that may have therapeutic properties.

(2) Overuse of zinc-containing denture adhesives, especially when combined with dietary supplements that contain zinc and other sources of zinc, can contribute to an excess of zinc in your body.

(3) Dental amalgam is a mixture of metals, consisting of liquid (elemental) mercury and a powdered alloy composed of silver, tin, and copper.

(4) Dental amalgam fillings are strong and long-lasting, so they are less likely to break than some other types of fillings.

(5) There are several different chemical forms of mercury: elemental mercury, inorganic mercury, and methylmercury.

2. Translate the following sentences into English.

(1) 牙科材料也称为口腔材料,是一类生物医用材料。

(2) 牙髓在密封的牙髓腔里,一般不会受到细菌感染。

(3) 大部分金属都溶于汞。

(4) 口腔是牙齿的外环境,与龋齿的发生密切相关。

(5) 黄金也是有效的补牙材料,但价格昂贵。

Culture salon

Too young for implants?

Many parents experience the agony of seeing their children lose permanent teeth in sporting accidents or other injuries. Dental implants might seem like a great option for replacing those teeth, but children and young adults typically are bad candidates for implants because their jaws are still growing. Implants in moving jaws will move too, resulting in unstable and crooked bites over time. Sometimes, though, implants are successful when used in conjunction with orthodontics, so check with your dentist and orthodontist for options.

Chapter 2-Section 7-Text 2-看译文

Text 3   Metallic biomaterials in orthopaedic surgery

Lead-in questions:

(1) What are the two main groups of biomaterials used in orthopaedic surgery?

(2) When did metals start to be used for orthopaedic devices?

(3) What metallic implants are widely used in orthopaedic surgery?

(4) What is stainless steel?

(5) What are the two main types of Co-Cr alloys?

Metallic biomaterials are one of the most commonly used biomaterial groups along with ceramics, synthetic polymers, and naturally derived materials. Firstly, it is used for augmentation of bone such as in the cases of fracture treatment (Figure 2-31). Secondly, they are also in very good use in replacement and reconstruction of musculoskeletal organs like joints, ligaments, intervertebral disc, resected bone due to malignancy, and defective joints due to degenerative changes. The definition of biomaterials itself is any material intended to perform the function of an organ or a tissue partly or completely in living organisms. Those usually used in orthopaedics are commonly called implants.

Chapter 2-Section 7-Text 3-听课文

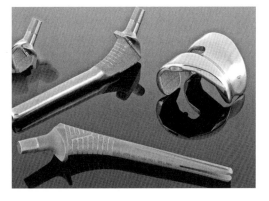

Figure 2-31   Hip joint implants.

Biocompatibility, appropriate design, manufacturability, mechanical reliability, biological stabilities, special properties such as tensile strength, yield strength, elastic modulus, corrosion and fatigue resistance, surface finish, creep, hardness, resistance to implant wear, aseptic loosening, and corrosion of implants are all contributing to characteristic of implants. There are several basic requirements that need to be fulfilled in order for the devices to function properly. The main biomaterials used in orthopaedic surgery are divided into two groups: metals and non-metals.

The use of metals for orthopaedic devices started long time ago. In the 17th century, metals have started been used, and later on, in the 18th century a metal screw implant was used for the first time. Metals have their main applications in load-bearing systems such as hip and knee prostheses and for the fixation of internal and external bone fractures. The metallic implants that are widely used in orthopaedic surgery are: low carbon grade austenitic stainless steels, titanium(Ti) and its alloys, and cobalt-chromium(Co-Cr) alloys.

1. Stainless steels

Like any steel, it is an iron-carbon alloy. Alloying elements were added and contributed to the ability of stainless steel to resist corrosion. The first biomedical 18-8 steel alloy contained vanadium(V) but subsequent studies indicate that it provides the alloy with insufficient corrosion resistance, and therefore the medical use of vanadium steel has been discontinued.

Further study yielded new resistance property in which modern corrosion resistant(stainless) steel is made by adding typically 18 wt% of Cr, which resulting in a thin Cr oxide(Cr_2O_3) film on the surface and 8 wt% of nickel(Ni) which contributes to austenite phase formation in the alloy. This Cr_2O_3 film is characterized by its low chemical reactivity and displays a self-healing capability in contact with oxygen. Based on phase composition, these steels are classified into three categories: austenitic, martensitic or ferritic stainless steels. They considered as the good choice for number of biomedical applications. In practice, stainless steels are used predominantly in temporary implant devices, such as fracture plates, screws, and hip nails, although stainless steel also is used in some Charnley-style femoral components for hip replacement.

The microstructure of austenitic stainless steels is characterized by its gamma(γ) phase with faced-centered cubic structure which contributes excellent workability(both forming and machining) and non-magnetic properties. Those belong to this group are the types 302, 316 and the famous 316L which remains popular as an implant material and a low-cost alternative compared to Ti and Co-Cr alloys. Despite its special attributes, there is some downward of stainless steels such as the reported Ni toxicity to the host organism, and vulnerability of the alloy to pitting and crevice corrosion as well as stress-corrosion cracking. Therefore, the use of stainless steels is generally limited to temporary use as biomedical devices despite its role in modern use as orthopaedics and load-bearing implants.

2. Ti and its alloys

They are commonly used for dental and orthopaedic implants. Different alpha and beta structure leads to three different structural types of Ti alloys: alpha(α), alpha-beta(α-β), and metastable β and β-phase alloys. Their excellent corrosion resistance is a result of inert TiO_2 film formation on their surface. For dental implants, commercially pure (CP) Ti, typically with single-phase alpha microstructure, is commonly used, while Ti-6Al-4V, with a biphasic alpha-beta microstructure, is most commonly used in orthopaedic applications. The presence of aluminum (Al) and V alloying elements not only contributes to the mechanical stability but also improves the microstructure of alpha-beta in relation to CP Ti. These properties are modulated by heat treatment and deformational works.

Despite Ti and its alloys combine a range of excellent properties, i. e. mechanical properties, corrosion resistance, fatigue-corrosion resistance, low density and relatively low modulus, their processing is not easy whether it is machining, forging or heat treating. The β-phase in Ti alloys tends to exhibit a much lower Young's modulus than that of the α-phase, therefore it satisfies most of the necessities or requirements for orthopaedic applications. However, the presence of Al and V in long term use of Ti alloys (especially Ti64) poses potential hazard of toxicity. The release Al and V ions from the alloys has been associated with the potential for Alzheimer and neuropathy disease.

3. Co-Cr alloys

These alloys, which have been used for many decades, can be basically categorized into two types: Co-Cr-Mo alloys [Cr(27%~30%), Mo(5%~7%), Ni(2.5%)] generally used in dentistry and artificial joints; and Co-Ni-Cr-Mo alloys [Cr(19%~21%), Ni(33%~37%), and Mo(9%~11%)] mostly used for making the stems of prostheses of heavily loaded joints, such as knee and hip. Involvement of heat during processing of Co-Cr-Mo alloys modifies the alloy's microstructure and leads to the alteration of electrochemical and mechanical properties of the alloys.

In term of corrosion resistance, Co-Cr alloys are considerably superior to stainless steels as it demonstrates better performance in chloride-rich environment. However, the corrosion products of Co-Cr-Mo, especially the possible release of Cr ions is potentially more toxic than that of stainless steel 316L.

Words and phrases

biomaterial	[baɪəʊˈmætɪərɪəl]	n.	生物材料
orthopaedic	[ˌɔːθəˈpiːdɪk]	adj.	骨科的,矫形外科的
augmentation	[ɔːɡmenˈteɪʃn]	n.	增强,增加
fracture	[ˈfræktʃə]	n.	骨折
reconstruction	[riːkənˈstrʌkʃn]	n.	重建,再建
musculoskeletal	[ˌmʌskjʊləʊˈskelɪtəl]	adj.	肌肉骨骼的
organ	[ˈɔːɡən]	n.	器官
joint	[dʒɔɪnt]	n.	关节
ligament	[ˈlɪɡəmənt]	n.	韧带
intervertebral	[ˌɪntəˈvɜːtɪbrəl]	adj.	椎间的,椎骨间的
resect	[rɪˈsekt]	vt.	切除,割除
malignancy	[məˈlɪɡnənsɪ]	n.	恶性肿瘤

Chapter 2-
Section 7-
Text 3-听词汇

defective	[dɪˈfektɪv]	adj.	有缺陷的
degenerative	[dɪˈdʒenərətɪv]	adj.	退行性的,退化的
reliability	[rɪˌlaɪəˈbɪlətɪ]	n.	可靠性
stability	[stəˈbɪləti]	n.	稳定性
biocompatibility	[ˈbaɪəʊkəmˌpætəˈbɪləti]	n.	生物相容性
corrosion	[kəˈrəʊʒ(ə)n]	n.	耐腐蚀性,腐蚀
creep	[kriːp]	n.	蠕变
hardness	[ˈhɑːdnəs]	n.	硬度
wear	[weə(r)]	n.	磨损
austenitic	[ˌɔːstəˈnɪtɪk]	adj.	奥氏体的
titanium	[taɪˈteɪnɪəm]	n.	钛
cobalt	[ˈkəʊbɔːlt]	n.	钴
chromium	[ˈkrəʊmɪəm]	n.	铬
resist	[rɪˈzɪst]	vt.	抵抗,抗
vanadium	[vəˈneɪdiəm]	n.	钒
nickel	[ˈnɪkl]	n.	镍
martensitic	[mɑːtɪnˈzɪtɪk]	adj.	马氏体的
ferritic	[fəˈritik]	adj.	铁素体的
screw	[skruː]	n.	螺丝钉
femoral	[ˈfemərəl]	adj.	股骨的
predominantly	[prɪˈdɒmɪnəntlɪ]	adv.	主要地,显著地
gamma	[ˈgæmə]	n.	γ(希腊字母)
vulnerability	[ˌvʌlnərəˈbɪlətɪ]	n.	易损性,弱点
aluminum	[əˈljuːmɪnəm]	n.	铝
element	[ˈelɪmənt]	n.	元素
hazard	[ˈhæzəd]	n.	危害
prosthesis	[prɒsˈθiːsɪs]	n.	假体(复数为 prostheses)
microstructure	[ˈmaɪkrəʊˌstrʌktʃə]	n.	微观结构,显微结构
electrochemical	[ɪˌlektrəʊˈkemɪkəl]	adj.	电化学的
synthetic polymer			合成高分子聚合物
naturally derived material			天然衍生材料
intervertebral disc			椎间盘
degenerative change			退行性病变
mechanical reliability			机械可靠性
biological stability			生物稳定性

tensile strength	抗拉强度
yield strength	屈服强度
elastic modulus	弹性模量
fatigue resistance	抗疲劳强度
resistance to implant wear	抗磨损
surface finish	表面光洁度
load-bearing	承重
stainless steel	不锈钢
iron-carbon alloy	铁碳合金
faced-centered cubic structure	面心立方结构
Young's modulus	杨氏模量
artificial joint	人工关节

Key sentences

(1) The definition of biomaterials itself is any material intended to perform the function of an organ or a tissue partly or completely in living organisms.

生物材料本身的定义是指在机体内部分或完全的执行器官或组织功能的材料。

(2) Biocompatibility, appropriate design, manufacturability, mechanical reliability, biological stabilities, special properties such as tensile strength, yield strength, elastic modulus, corrosion and fatigue resistance, surface finish, creep, hardness, resistance to implant wear, aseptic loosening, and corrosion of implants are all contributing to characteristic of implants.

生物相容性、合理的设计、可加工性、机械可靠性、生物稳定性,其他性能(如抗拉强度、屈服强度、弹性模量、耐腐蚀性、抗疲劳性、表面形貌、蠕变、硬度、耐磨性、无菌松动性和腐蚀性)都是植入材料的特性。

(3) The metallic implants that are widely used in orthopaedic surgery are: low carbon grade austenitic stainless steels, titanium(Ti) and its alloys, and cobalt-chromium(Co-Cr) alloys.

广泛用去骨外科的金属植入物主要有:低碳级的奥氏体不锈钢,钛及其合金,钴铬合金。

(4) Based on phase composition, these steels are classified into three categories: austenitic, martensitic or ferritic stainless steels.

根据金相组成,这些钢可分为三类:奥氏体、马氏体或铁素体不锈钢。

(5) However, the corrosion products of Co-Cr-Mo, especially the possible release of Cr ions is potentially more toxic than that of stainless steel 316L.

然而,钴-铬-钼的腐蚀产物,特别是其可能释放的铬离子比316L不锈钢具有更大的潜在毒性。

Learning test

Choose the following words below to fill in the blanks, change the form if necessary.

microstructure　　resist　　reliability　　electrochemical　　hazard resect　　　　　element　　　reconstruction　　　hardness　　　　defective

(1) Hip joint _____ and replacement is a very mature medical technology.

(2) A skilled doctor can _____ the tumor without even needing to remove the pancreas or any other organ.

(3) The product quality and _____ vary for different models.

(4) Relaxation is one of the treatments for _____ vision.

(5) The _____ of diamonds is highest in the earth.

(6) Some substances _____ corrosion by air or water.

(7) Hydrogen is a one-valence _____ .

(8) Smoking is a serious health _____ .

(9) _____ and composition of the material can be analyzed with an electronic microscope.

(10) _____ batteries work by converting stored chemical energy into electricity with high efficiency.

Translation

1. Translate the following words into Chinese.

(1) biomaterial　　　　　　　(2) organ

(3) fracture　　　　　　　　(4) corrosion

(5) element　　　　　　　　(6) orthopaedic

(7) titanium　　　　　　　　(8) ligament

(9) hardness　　　　　　　　(10) hazard

2. Translate the following phrases into Chinese.

(1) tensile strength　　　　　(2) elastic modulus

(3) stainless steel　　　　　　(4) intervertebral disc

(5) artificial joint　　　　　　(6) biological stability

Practice

1. Translate the following sentences into Chinese.

(1) Metallic biomaterials are one of the most commonly used biomaterial groups along with ceramics, synthetic polymers, and naturally derived materials.

(2) The main biomaterials used in orthopaedic surgery are divided into two groups: metals and non-metals.

(3) Metals have their main applications in load-bearing systems such as hip and knee prostheses and for the fixation of internal and external bone fractures.

(4) Alloying elements were added and contributed to the ability of stainless steel to resist corrosion.

(5) In term of corrosion resistance, Co-Cr alloys are considerably superior to stainless steels as it demonstrates better performance in chloride-rich environment.

2. Translate the following sentences into English.

(1) 人体承重部位的修复主要采用生物医用金属材料。

(2) 金属板被用于骨折的固定。

(3) 生物医用金属材料的毒性主要来自金属表面的离子溶出。

(4) 镁合金作为可降解医用材料被誉为第三代生物医用材料。

(5) 经过多年的临床应用，医用金属材料仍然存在许多问题。

Culture salon

Bioresorbable metal

After decades of developing strategies to minimize the corrosion of metallic biomaterials, there is now an increasing interest to use corrodible metals in a number of medical device applications. Bioresorbable (also called biodegradable or bioabsorbable) metals are metals or their alloys that degrade safely within the body. The primary metals in this category are magnesium-based and iron-based alloys, although recently zinc has also been investigated. Currently, the primary uses of bioresorbable metals are as stents for blood vessels (for example bioresorbable stents) and other internal ducts.

Chapter 2-Section 7-Text 3-看译文

Chapter 2-Section 7-习题

Chapter 3

Medical devices regulations

Learning objective

In this chapter, global regulations on medical devices, and its harmonization are introduced. Read the following passage, you should meet the following requirements:

(1) Grasp basic concepts, key points, involved professional terms and phrases.

(2) Acquire relevant regulations, regulatory agencies, and harmonization organizations on medical devices.

(3) Understand how to get a medical device product approved before enter the market in China

Section 1   Global medical device regulatory harmonization

Chapter 3-
Section 1-PPT

Text 1   Global regulations of medical devices

Lead-in questions:

(1) What's the purpose of regulations of medical devices?

(2) How many major phases in the lifespan of a medical device?

(3) Who are subject to the regulations of medical devices?

(4) How many methods of governmental regulation of medical devices?

(5) Who's the agency for regulating medical devices in China?

Chapter 3-
Section 1-
Text 1-听课文

Regulation is primarily concerned with enabling patient access to high quality, safe and effective medical devices, and restricting access to those products that are unsafe or have limited clinical use. When appropriately implemented, regulation ensures public health benefit and the safety of patients, health care workers and the community.

The objectives of regulatory controls are to safeguard the health and safety of patients and users by ensuring that manufacturers of medical devices follow specified procedures during the design, manufacture, and marketing of the medical device. Globally, medical devices are classified into three or four categories by level of risk, and the regulatory controls are proportional to the level of risk associated with the medical device accordingly.

To regulate the medical device market, there is a need to classify medical devices based on the potential risk to patients and users. The risk level of a medical device is attributed by a series of factors, including: the duration of the device's contact with the body; the degree of the invasiveness; whether the device is implantable or active; and whether or not the device contains a substance, which in its own right is considered to be a medicinal substance and has action ancillary to that of the device.

The major seven phases in the life span of a medical device include conception and development; manufacture; packaging and labeling; advertising; sale; use; and disposal. Persons who directly manage the different phases of medical devices are manufacturers; vendors including importers, distributors, retailers and manufacturers who sell medical equipment; and users who are usually patients or professionals in a health care facility.

Governmental regulations of medical devices usually include product control, vendor establishment control, post-market surveillance, and quality system requirements.

Generally, there are national healthcare policies and/or laws and regulations set in place to regulate medical devices in each country, and harmonized standards also could be used as references. Authorities acknowledge product clearance for the market in various ways. In Canada, a device license is awarded by the Therapeutic Products Directorate(TPD). In the European Union, after receiving the EC certificate from a notified body, the manufacturer places the CE mark on or with the device. In the United States, the manufacturer of the device receives a marketing clearance(510K) or an approval letter(PMA) from the FDA.

In China, National Medical products Administration(NMPA) is the Chinese agency for regulating drugs and medical devices. The predecessor to the NMPA was founded in 1998 to initially oversee drugs and medical devices. When it was given jurisdiction over food in 2003, it was renamed the State Food and Drug Administration(SFDA) and reported to the State Council. In 2008, the regulatory body was put under the supervision of the Ministry of Health(MoH). Formerly known as SFDA, the CFDA was restructured in March 2013 and elevated to a ministerial-level agency. In 2018, it was renamed the National Medical products Administration(NMPA). The NMPA is now part of the State Council of the People's Republic of China, the country's highest regulatory body that oversees the introduction of food, health products and cosmetics in Mainland China. Its responsibilities include drafting laws and regulations for food safety, drugs, medical devices and cosmetics as establishing medical device standards and classification systems. Medical devices are classified into Class Ⅰ, Ⅱ, and Ⅲ. Regulatory control increases from Class Ⅰ to Class Ⅲ. The device classification regulation defines the regulatory requirements for a general device type.

In the United States, FDA's Center for Devices and Radiological Health(CDRH) is responsible for regulating firms who manufacture, repackage, relabel, and/or import medical devices sold in the United States. In addition, CDRH regulates radiation-emitting electronic products(medical and non-medical) such as lasers, X-ray systems, ultrasound equipment, microwave ovens and color televisions. Medical devices are classified into Class Ⅰ, Ⅱ, and Ⅲ. Regulatory control increases from Class Ⅰ to Class Ⅲ. The device classification regulation defines the regulatory requirements for a general device type.

Most Class Ⅲ and new devices that are not substantially equivalent to a legally marketed product that does not require a Pre-Market Approval application, require clearance through the PMA processes. Most class Ⅱ and some class Ⅰ devices require pre-market entry notification(termed 510k), an information package for the FDA, which is subject to less stringent review than the PMA process. The 510k submission must demonstrate how the proposed medical device is substantially equivalent to a medical device that is already on the US market. Most class Ⅰ and some class Ⅱ(low-risk) devices are exempt from 510k submission before sale, but are still subject to general control requirements.

In the European system, manufacturers of devices of classes Ⅱ and Ⅲ, as well as devices of class Ⅰ with either measuring function or sterility requirements, must submit to the regulator: ①a Declaration of Conformity to the appropriate EC Directives; ②details of the conformity assessment procedure followed. In addition, for higher risk class devices that require design examination or type examination, the corresponding EC-Certificates issued by a notified body must also be submitted to the competent authority. Other medical devices of class Ⅰ are exempt from pre-market submissions, although they must follow the essential principles of safety and performance in their design, construction and labelling requirements.

In Canada, devices of classes Ⅲ and Ⅳ are subject to in-depth regulatory scrutiny, while class Ⅱ devices require only the manufacturer's declaration of device safety and effectiveness before sale. Class Ⅰ devices are exempted from pre-market submission, but they must still satisfy the safety, effectiveness and labelling requirements.

Words and phrases

regulation	[ˌregjʊˈleɪʃn]	n.	法规,规章
concern	[kənˈsɜːn]	vt.	涉及,关系到
quality	[ˈkwɒlɪtɪ]	n.	质量
effective	[ɪˈfektɪv]	adj.	有效的,生效的
restrict	[rɪˈstrɪkt]	vt.	限制,限定
implement	[ˈɪmplɪmənt]	vt.	贯彻,执行
community	[kəˈmjuːnɪtɪ]	n.	社会,公众
objective	[əbˈdʒektɪv]	n.	目标,目的
classify	[ˈklæsɪfaɪ]	vt.	分类,分等
category	[ˈkætɪgərɪ]	n.	种类,类别
risk	[rɪsk]	n.	风险
potential	[pəˈtenʃl]	adj.	潜在的,有可能的
attribute	[əˈtrɪbjuːt]	vt.	归因于…
duration	[djʊˈreɪʃn]	n.	期间,持续时间
invasiveness	[ɪnˈveɪsɪvnəs]	n.	有创,有创性

Chapter 3-
Section 1-
Text 1-听词汇

implantable	[ɪmˈplæntəbəl]	adj.	可植入的,可移植的
phase	[feɪz]	n.	阶段,时期
conception	[kənˈsepʃn]	n.	概念,构想,设想
label	[ˈleɪbl]	vt. &n.	贴标签于,标签
vendor	[ˈvendə]	n.	供应商
distributor	[dɪˈstrɪbjʊtə]	n.	批发商
retailer	[ˈriːteɪlə(r)]	n.	零售商
facility	[fəˈsɪlɪtɪ]	n.	机构
surveillance	[səˈveɪləns]	n.	监督,监管
reference	[ˈrefərəns]	n.	参考,参考资料
authority	[ɔːˈθɒrɪtɪ]	n.	官方,当局,机构
clearance	[ˈklɪərəns]	n.	准入,放行,批准
therapeutic	[ˌθerəˈpjuːtɪk]	adj.	治疗(学)的,疗法的
directorate	[dɪˈrektərət]	n.	委员会,理事会
certificate	[səˈtɪfɪkət]	n.	证书,执照
administration	[ədmɪnɪˈstreɪʃn]	n.	管理,管理局
predecessor	[ˈpriːdɪsesə]	n.	前身
agency	[ˈeɪdʒənsi]	n.	机构
jurisdiction	[ˌdʒʊərɪsˈdɪkʃn]	n.	管辖权,管辖范围,权限
supervision	[ˌsjuːpəˈvɪʒn]	n.	管理,监督
radiation	[ˌreɪdɪˈeɪʃn]	n.	辐射,放射
emit	[ɪˈmɪt]	vt.	发出,放射
electronic	[ɪlekˈtrɒnɪk]	adj.	电子的
microwave	[ˈmaɪkrəweɪv]	n.	微波
substantially	[səbˈstænʃəlɪ]	adv.	本质上,实质上
stringent	[ˈstrɪndʒənt]	adj.	严格的
review	[rɪˈvjuː]	n.	审查
submission	[səbˈmɪʃn]	n.	提交,呈递
demonstrate	[ˈdemənstreɪt]	vt.	证明,说明,演示,论证
exempt	[ɪgˈzempt]	adj.	被免除的,被豁免的
measure	[ˈmeʒə(r)]	vt.	测量
function	[ˈfʌŋkʃn]	n.	功能
sterility	[stəˈrɪlɪtɪ]	n.	灭菌
declaration	[dekləˈreɪʃn]	n.	声明,宣告
conformity	[kənˈfɔːmɪtɪ]	n.	符合

performance	[pəˈfɔːməns]	n.	性能
scrutiny	[ˈskruːtɪnɪ]	n.	审查
post-market			上市后
proportional to			与……成正比
pre-market			上市前
Therapeutic Products Directorate			治疗产品委员会
notified body			公告机构
marketing clearance			上市放行(证,书)
approval letter			批准书
radiation-emitting			辐射发射
equivalent to			相当于,等同于
exempt from			免除……,免于……
subject to			受……管制

Key sentences

(1) Regulation is primarily concerned with enabling patient access to high quality, safe and effective medical devices, and restricting access to those products that are unsafe or have limited clinical use.

监管主要关注的是使患者能够获得高质量、安全有效的医疗器械,并限制那些不安全或临床使用有限的产品的使用。

(2) To regulate the medical device market, there is a need to classify medical devices based on the potential risk to patients and users.

为规范医疗器械市场,有必要根据医疗器械对患者和使用者的潜在风险,对其进行分类。

(3) Governmental regulations of medical devices usually include product control, vendor establishment control, post-market surveillance, and quality system requirements.

医疗器械行政法规通常包括产品管制、供应商建立控制、上市后监督和质量体系要求。

(4) InChina, the National Medical products Administration(NMPA) is the Chinese agency for regulating drugs and medical devices.

在中国,国家药品监督管理局(NMPA)是药品和医疗器械监管的机构。

(5) In the United States, FDA's Center for Devices and Radiological Health(CDRH) is responsible for regulating firms who manufacture, repackage, relabel, and/or import medical devices sold in the United States.

在美国,FDA 的器械和放射健康中心(CDRH)负责监管制造,重新包装,重新贴标签,和/或进口在美国销售的医疗器械的公司。

Learning test

Choose the following words below to fill in the blanks, change the form if necessary.

regulation   function   requirement   performance   objective

submission   certificate   electronic   category   risk

(1) _____ computers are now in common use all over the world.

(2) The _____ of the human brain is well known.

(3) There's no _____ to control the working hours of employees.

(4) Patience is a _____ in teaching.

(5) Beingoverfat constitutes a health _____ .

(6) Are you satisfied with the _____ of your new car?

(7) Her _____ was to play golf and win.

(8) The application _____ system is quite easy to use.

(9) He was afforded a _____ upon completion of his course of study.

(10) Each medical device is assigned a _____ based on the risk.

Translation

1. Translate the following words into Chinese.

(1) quality   (2) regulation

(3) function   (4) label

(5) distributor   (6) classify

(7) clearance   (8) supervision

(9) measure   (10) radiation

2. Translate the following phrases into Chinese.

(1) post-market   (2) subject to

(3) pre-market   (4) equivalent to

(5) subject to   (6) radiation-emitting

Practice

1. Translate the following sentences into Chinese.

(1) When appropriately implemented, regulation ensures public health benefit and the safety of patients, health care workers and the community.

(2) Globally, medical devices are classified into three or four categories by level of risk, and the regulatory controls are proportional to the level of risk associated with the medical device accordingly.

(3) Its responsibilities include drafting laws and regulations for food safety, drugs, medical devices and cosmetics as establishing medical device standards and classification systems.

(4) The device classification regulation defines the regulatory requirements for a general device type.

(5) Class I devices are exempted from pre-market submission, but they must still satisfy the safety, ef-

fectiveness and labelling requirements.

2. Translate the following sentences into English.

(1) 在中国,医疗器械根据其风险级别分为三类。

(2) 医疗器械在其整个生命周期内都要受监管。

(3) 植入性医疗器械风险高,属于三类产品。

(4) 微波对人体是安全的。

(5) 医院更愿意购买安全、有效但是昂贵的外国产品。

Culture salon

Effectiveness/performance of medical devices

Every device has a designed purpose. A device is clinically effective when it produces the effect intended by the manufacturer relative to the medical condition. For example, if a device is intended for pain relief, one expects the device to actually relieve pain and would also expect the manufacturer to possess objective, scientific evidence, such as clinical test results, that the device does in fact relieve pain.

Clinical effectiveness is a good indicator of device performance. Performance, however, may include technical functions in addition to clinical effectiveness. For example, an alarm feature may not directly contribute to clinical effectiveness but would serve other useful purposes. Furthermore, it is easier to measure objectively and quantify performance than clinical effectiveness.

Chapter 3-Section 1-Text 1-看译文

Text 2 Global harmonization of medical devices

Lead-in questions:

(1) Can a medical device that has been approved by USFDA enter the China market?

(2) What is medical device harmonization?

(3) What's the purpose of global harmonization of medical devices?

(4) What's the purpose of IMDRF?

(5) Who is the official observer of IMDRF?

The medical device industry is a heterogeneous, innovative, and dynamic sector. The global market for medical devices is huge, and it will continue showing a significant growth in the future. There are differences between the regulatory systems and required documents for registration in different countries. There are around 60~65 countries which have implemented regulation for medical devices or will soon implement the regulations.

The manufacturers must clarify their target markets and comply with the regulations ac-

Chapter 3-Section 1-Text 2-听课文

cordingly. For example, a medical device that has been approved by the USFDA may not be able to enter the market in China without the CFDA's approval, even though it has undertaken the most stringent procedures in the world mandated by the USFDA. There is a need for global harmonization of medical devices, and the reasons include:

1) To minimize regulatory barriers;

2) To facilitate trade between different countries; and

3) To reduce the cost of implementing regulations for governments and industry.

But what is medical device harmonization? According to the World Health Organization (WHO), medical device harmonization is a process to encourage convergence in regulatory practices related to ensuring the safety, effectiveness/performance, and quality of medical devices, promoting technological innovation, and facilitating international trade. The European Union is a good example of the harmonization of medical devices, from which the advantages and benefits can be sensed—it was estimated that the European GDP had increased up to 1.5% between 1987 and 1993 due to the promoted completion of a single set of Europe requirements and regulations. Nowadays, different kinds of organizations are contributing to the global or regional harmonization of medical devices, and they will be introduced in this chapter.

1. The International Medical Device Regulators Forum (IMDRF)

IMDRF conceived in February 2011 as a forum to discuss future directions in medical device regulatory harmonization.

It is a voluntary group of medical device regulators from around the world, who have come together to build on the strong foundational work of the Global Harmonization Task Force (GHTF) on Medical Devices. The purpose of the IMDRF is to accelerate international medical device regulatory harmonization and convergence.

IMDRF was born in October 2011 in Ottawa. The current IMDRF regulatory members are: Australia, Brazil, Canada, China, Europe, Japan, Russia, and the U.S. The World Health Organization (WHO) is an official observer. The Asian Harmonization Working Party (AHWP) and APEC's Life Sciences Innovation Forum's regulatory harmonization steering committee are both affiliate organizations with IMDRF.

The IMDRF management committee comprises regulatory authority representatives from the following jurisdictions:

- Australia, Therapeutic Goods Administration
- Brazil, National Health Surveillance Agency (ANVISA)
- Canada, Health Canada
- China, China Food and Drug Administration
- European Union, European Commission Directorate-General for Internal Market, Industry, Entrepreneurship and SMEs
- Japan, Pharmaceuticals and Medical Devices Agency and theMinistry of Health, Labour and Welfare

- Russia, Ministry of Health of Russian Federation
- Singapore, Health Sciences Authority
- United States of America, US Food and Drug Administration

2. Asian Harmonization Working Party(AHWP)

AHWP is established as a non-profit organization. Its goals are to study and recommend ways to harmonize medical device regulations in the Asian and other regions and to work in coordination with the GHTF, APEC and other related international organizations aiming at establishing harmonized requirements, procedures and standards. The Working Party is a group of experts from the medical device regulatory authorities and the medical device industry. Membership is open to those representatives from the Asian and other regions that support the above stated goals.

The AHWP contains 24 member economies such as China, Cambodia, Indonesia and Laos.

3. Pan African Harmonization Working Party(PAHWP)

PAHWP is a voluntary body that aims to improve access to safe and affordable medical devices and diagnostics in Africa though harmonized regulation. Our first priority is in-vitro diagnostic devices. A new generation of diagnostic tests is being developed for use at the point of care that could save lives and stop the spread of infectious diseases. It is important that patients in Africa have access to these tests without delay.

PAHWP aims to study and recommend ways to ensure tests are safe and effective while minimizing costs and delays, allowing faster access to cheaper products. We shall undertake pilot projects in Africa using new point of care tests for CD4, viral load and early infant diagnosis of HIV as examples

4. Pan American Network for Drug Regulatory Harmonization(PANDRH)

PANDRH is an initiative of the national regulatory authorities within the region, and PAHO that supports the processes of pharmaceutical regulatory harmonization in the Americas, within the framework of national and sub-regional health policies and recognizing preexisting asymmetries.

The components of PANDRH are: ThePan American conference on Drug Regulatory Harmonization (PACDRH), the Steering Committee(SC), the Technical Working Groups(TWG) in the areas considered as priority by the conference and the secretariat.

Words and phrases

industry	[ˈɪndəstri]	n.	行业,产业
heterogeneous	[ˌhetərəˈdʒiːniəs]	adj.	多样化的,混杂的
innovative	[ˈɪnəveɪtɪv]	adj.	创新的,革新的
dynamic	[daɪˈnæmɪk]	adj.	充满活力的,动态的
significant	[sɪɡˈnɪfɪkənt]	adj.	显著的,可观的,重大的
registration	[ˌredʒɪˈstreɪʃn]	n.	注册
clarify	[ˈklærəfaɪ]	vt.	阐明,澄清
comply	[kəmˈplaɪ]	vi.	遵守

英文	音标	词性	中文
mandate	[ˈmændeɪt]	vi.	强制执行
harmonization	[ˌhɑːmənaɪˈzeɪʃn]	n.	协调
minimize	[ˈmɪnɪmaɪz]	vt.	最小化,使减到最少
barrier	[ˈbæriə(r)]	n.	壁垒,障碍
facilitate	[fəˈsɪlɪteɪt]	vt.	促进,帮助
trade	[treɪd]	n.	贸易
reduce	[rɪˈdjuːs]	vt.	降低,减少
cost	[kɒst]	n.	成本,代价,价格
convergence	[kənˈvɜːdʒəns]	n.	融合
promote	[prəˈməʊt]	vt.	提升,促进
technological	[ˌteknəˈlɒdʒɪkl]	adj.	技术的
benefit	[ˈbenɪfɪt]	n.	好处,利益
sense	[sens]	n.	领会到,体会到
estimate	[ˈestɪmət]	vt.	估计,估算
increase	[ɪnˈkriːs]	vi.	增长,提高
introduce	[ˌɪntrəˈdjuːs]	vt.	介绍
regulator	[ˈregjuleɪtə(r)]	n.	监管者
forum	[ˈfɔːrəm]	n.	论坛
conceive	[kənˈsiːv]	vi.	构想,设想
purpose	[ˈpɜːpəs]	n.	目的
accelerate	[əkˈseləreɪt]	vt.	加快,加速
current	[ˈkʌrənt]	adj.	目前的,现在的
official	[əˈfɪʃl]	adj.	官方的
steer	[ˈstɪə]	vi.	督导,指导
committee	[kəˈmɪti]	n.	委员会
affiliate	[əˈfɪlieɪt]	n.	分支机构,附属(机构)
comprise	[kəmˈpraɪz]	vt.	由……组成,由……构成
representative	[ˌreprɪˈzentətɪv]	n.	代表
coordination	[kəʊˌɔːdɪˈneɪʃn]	n.	协调,协作
standard	[ˈstændəd]	n.	标准
expert	[ˈekspɜːt]	n.	专家
diagnostic	[ˌdaɪəgˈnɒstɪk]	adj.	诊断的,诊断
infectious	[ɪnˈfekʃəs]	adj.	有传染性的,传染的
disease	[dɪˈziːz]	n.	疾病
infant	[ˈɪnfənt]	n.	婴儿

viral	[ˈvaɪrəl]	adj.	病毒的
pharmaceutical	[ˌfɑːməˈsuːtɪkl]	adj.	药品的
framework	[ˈfreɪmwɜːk]	n.	框架
asymmetry	[ˌeɪˈsɪmətrɪ]	n.	不对称,不对等
due to			由于,因为
contribute to			为…做出贡献,促成
life science			生命科学
non-profit			非营利
in-vitro			体外
point of care			床边,床旁
have access to			使用,接近,可以利用
preexisting			先前存在的
in coordination with			配合
The International Medical Device Regulators Forum (IMDRF)			国际医疗器械监管者论坛
Asian Harmonization Working Party (AHWP)			亚洲医疗器械法规协调组织
Pan African Harmonization Working Party (PAHWP)			泛非洲医疗器械法规协调组织
Pan American Network for Drug Regulatory Harmonization (PANDRH)			泛美洲药品注册协调组织

Key sentences

(1) The global market for medical devices is huge, and it will continue showing a significant growth in the future.

全球医疗器械市场庞大,未来将继续呈现显著增长。

(2) The manufacturers must clarify their target markets and comply with the regulations accordingly.

制造商必须明确其目标市场,并遵守相应法规。

(3) According to the World Health Organization (WHO), medical device harmonization is a process to encourage convergence in regulatory practices related to ensuring the safety, effectiveness/performance, and quality of medical devices, promoting technological innovation, and facilitating international trade.

根据世界卫生组织 WHO 的解释,医疗器械协调是指在确保医疗器械安全性、有效性/性能和质量的监管实践中,促进其融合的过程,以推动技术创新,促进国际贸易。

(4) It is a voluntary group of medical device regulators from around the world, who have come together to build on the strong foundational work of the Global Harmonization Task Force (GHTF) on Medical Devices.

它(IMDRF)是一个由来自全球的医疗器械监管者构建的志愿组织,他们已经共同建立了全球医疗器械协调工作组 GHTF 的坚实基础。

(5) PAHWP is a voluntary body that aims to improve access to safe and affordable medical devices and diagnostics in Africa though harmonized regulation.

PAHWP 是一个志愿机构,旨在通过法规协调提高非洲获得安全和负担得起的医疗器械和诊断。

Learning test

Choose the following words below to fill in the blanks, change the form if necessary.

industry   registration   organization   disease   standard
representative   increase   purpose   cost   trade

(1) The harmonization _____ now has around 18 members from all over the world.

(2) The electronic _____ is inappropriate to the region's present and future needs.

(3) Oversea _____ is the lifeblood of our country.

(4) Sales _____ is a stressful job.

(5) The _____ to develop a new product is too high for a small company.

(6) Medical devices are not allowed to enter the market before _____ .

(7) Their _____ is to build a fair society and a strong economy.

(8) The population in China continues to _____ .

(9) Sports reduce the risks of heart _____ .

(10) There will be new ISO _____ for this product soon.

Translation

1. Translate the following words into Chinese.

(1) reduce   (2) registration

(3) forum   (4) industry

(5) standard   (6) expert

(7) disease   (8) estimate

(9) trade   (10) cost

2. Translate the following phrases into Chinese.

(1) life science   (2) in-vitro

(3) due to   (4) point of care

(5) in coordination with (6) non-profit

Practice

1. Translate the following sentences into Chinese.

(1) There are differences between the regulatory systems and required documents for registration in different countries.

(2) A medical device that has been approved by the US FDA may not be able to enter the market in China without the NMPA's approval, even though it has undertaken the most stringent procedures in the world mandated by the US FDA.

(3) The European Union is a good example of the harmonization of medical devices, from which the ad-

vantages and benefits can be sensed—it was estimated that the European GDP had increased up to 1.5% between 1987 and 1993 due to the promoted completion of a single set of Europe requirements and regulations.

(4) The purpose of the IMDRF is to accelerate international medical device regulatory harmonization and convergence.

(5) PAHWP aims to study and recommend ways to ensure tests are safe and effective while minimizing costs and delays, allowing faster access to cheaper products.

2. Translate the following sentences into English.

(1)全球法规协调将有效降低产品价格。

(2)这个监管者论坛每年开一次。

(3)产品上市前注册是强制实施的。

(4)中国积极参与全球法规协调工作。

(5)这个组织有20个成员。

Culture salon

The role of medical device users

The user should make sure that he/she has qualifications and training in the proper use of the device, and is familiar with the indications, contra-indications and operating procedures recommended by the manufacturer. It is crucial that experience gained with medical devices should be shared with other users, the vendor and manufacturer to prevent future problems. This can be done by reporting any incidents to a coordinating centre from which warnings can be issued.

When using medical devices, users should always bear in mind that the safety and health of the patients are in their hands. The user has the responsibility to employ the medical device only for the intended indications(or to assure that any non-indicated use of the medical device does not compromise the safety of the patient and other users). The user also has the responsibility to ensure proper maintenance of medical devices during active use and safe disposal of obsolete medical devices.

Chapter 3-Section 1-Text 2-看译文

Chapter 3-Section 1-习题

Section 2　US FDA regulations

Chapter 3-Section 2-PPT

Lead-in questions:

(1) How many steps for marketing a medical device in US?

(2) How to find the regulation number of your medical device?

(3) How to choose correct premarketing submission?

(4) Which types of medical devices are exempt from premarket notification 510(k)?

(5) Which types of medical devices are subject to premarket approval(PMA)?

Text 1　How to market your medical devices in U.S.?

Chapter 3-Section 2-Text 1-听课文

In U.S., Medical devices are classified into Class Ⅰ, Ⅱ, and Ⅲ. The class to which your device is assigned determines the type of premarketing submission/application required for FDA clearance to market.

If your device is classified as Class Ⅰ or Ⅱ, and if it is not exempt, a premarket notification(PMN) 510(k) will be required for marketing. All devices classified as exempt are subject to the limitations on exemptions. Limitations of device exemptions are covered under 21 CFR xxx.9, where xxx refers to Parts 862~892.

For Class Ⅲ devices, a premarket approval(PMA) application will be required unless your device is a preamendments device(on the market prior to the passage of the medical device amendments in 1976, or substantially equivalent to such a device) and PMAs have not been called for. In that case, a 510(k) will be the route to market.

The 510(k), otherwise known as PMN, is unique to the FDA and its clearance permits manufacturers to market their device on the basis that the device to be marketed is as safe and effective as another previously legally marketed device. PMA is a more stringentpremarket submission process, which requires clinical evidence to prove the safety and effectiveness of the device to be marketed. Generally, there are five steps for marketing a medical device as described below.

1. Classify your device(Class Ⅰ, Ⅱ, Ⅲ)

Classification of the device depends on the risk level of the medical devices and its intended use. To find the classification of your device, as well as whether any exemptions may exist, you need to find the regulation number that is the classification regulation for your device. There are two methods for accomplishing this: go directly to the FDA classification database and search for a part of the device name, or, if you know the device panel(medical specialty) to which your device belongs, go directly to the listing for that panel and identify your device and the corresponding regulation.

The Food and Drug Administration(FDA) has established classifications for approximately 1,700 different generic types of devices and grouped them into 16 medical specialties referred to as panels. Each classification panel in the CFR begins with a list of devices classified in that panel. Each classified device has a 7-digit number associated with it, e. g. , 21 CFR 880. 2920-clinical mercury thermometer. Once you find your device in the panel's beginning list, go to the section indicated: in this example, 21 CFR 880. 2920. It describes the device and says it is Class Ⅱ.

2. Choose correctpremarket submission(exempted, 510(k) , PMA)

Choosing correctpremarket submission may refer to either 510(k) or PMA, whichever the medical device classification specifies in the CFR. Class Ⅰand Ⅱ medical devices are not as crucial to human life sustaining as Class Ⅲ medical devices and hence, only require "clearance" for 510(k) before marketing or an exemption if stated in the CFR. For the latter, it requires more stringent controls and PMA "approval" before marketing.

(1) Premarket Notification 510(k) -21 CFR Part 807 Subpart E

If your device requires the submission of apremarket notification 510(k) , you cannot commercially distribute the device until you receive a letter of substantial equivalence from FDA authorizing you to do so. A 510(k) must demonstrate that the device is substantially equivalent to one legally in commercial distribution in the United States: ① before May 28, 1976; or ② to a device that has been determined by FDA to be substantially equivalent.

Most Class Ⅰ devices and some Class Ⅱ devices are exempt from the Premarket Notification 510(k) submission. A list of exempt devices can be found on FDA website.

If you plan to send a 510(k) application to FDA for a Class Ⅰ or Class Ⅱ device, you may find 510(k) review by an accredited person beneficial. FDA accredited 12 organizations to conduct a primary review of 670 types of devices. By law, FDA must issue a final determination within 30 days after receiving a recommendation from an accredited person. Please note that 510(k) review by an accredited person is exempt from any FDA fee; however, the third party may charge a fee for its review.

(2) Premarket Approval(PMA) -21 CFR Part 814

Product requiringPMAs are Class Ⅲ high-risk devices that pose a significant risk of illness or injury, or devices found not substantially equivalent to Class Ⅰ and Ⅱ through the 510(k) process. The PMA process is more involved and includes the submission of clinical data to support claims made for the device.

3. Prepare documents for submission to appropriate authority

Preparation of a dossier for FDA submission can be a tedious process. However, with the understanding that all the information FDA requires is to ultimately establish that the device is safe and effective for use, the preparation of the dossier then becomes a logical process.

4. Sendpremarket submission for review

After preparation of all relevant documents, submitter must first submit user fees before submitting any application to the FDA. If application is received without payment(either before or during submission), application is deemed incomplete and will not be reviewed or processed. Besideshardcopies, submissions must also include an electronic copy. After submission, the FDA will conduct an administrative filing review to check if the submission is sufficient to be reviewed. During review, the FDA staff will communicate through an Interactive Review to increase the efficiency of the review process.

5. Complete establishment registration and device listing

The FDA requires all establishments/facilities and its manufactured medical devices to be registered and listed annually online at the device registration and listing module(DRLM) through the FDA Unified Registration and listing system(FURLS) Web site. Before registration can be done, establishments are required to pay the user fee at the "device facility user fee" website, between 1st October and 31st December each year. After confirmation of payment, a PIN and payment confirmation number(PCN) will be obtained for registration and listing in DRLM. Any changes in establishment/facility information and details may be updated in DRLM. The annual registration and listing will have to be done even if there are no changes. This is to make sure that establishment and device information are reviewed and kept up to date yearly. If a device is pending notification clearance orpremarket approval, registration of establishment and listing will have to be done after the device has been cleared or approved.

Words and phrases

Chapter 3-
Section 2-
Text 1-听词汇

market	[ˈmɑːkɪt]	vt.&n.	使上市,市场
classify	[ˈklæsɪfaɪ]	vt.	分类
assign	[əˈsaɪn]	vt.	分配,指定
determine	[dɪˈtɜːmɪn]	vt.	决定,确定
submission	[səbˈmɪʃn]	n.	递交,提交
application	[ˌæplɪˈkeɪʃn]	n.	申请
clearance	[ˈklɪərəns]	n.	准入,放行
notification	[ˌnəʊtɪfɪˈkeɪʃn]	n.	公告,通知
exempt	[ɪgˈzempt]	adj.	被免除的,被豁免的
limitation	[ˌlɪmɪˈteɪʃn]	n.	限制
approval	[əˈpruːvl]	n.	批准
route	[ruːt]	n.	途径
stringent	[ˈstrɪndʒənt]	adj.	严格的
process	[ˈprəʊses]	n.	过程
clinical	[ˈklɪnɪkl]	adj.	临床的
evidence	[ˈevɪdəns]	n.	证据
effectiveness	[ɪˌfekˈtɪvnɪs]	n.	有效性
method	[ˈmeθəd]	n.	方法

accomplish	[əˈkʌmplɪʃ]	vt.	完成,达到
database	[ˈdeɪtəbeɪs]	n.	数据库
panel	[ˈpænl]	n.	组
specialty	[ˈspeʃəlti]	n.	专业
identify	[aɪˈdentɪfaɪ]	vt.	确定,识别
approximately	[əˈprɒksɪmətli]	adv.	大约
mercury	[ˈmɜːkjəri]	n.	水银
thermometer	[θəˈmɒmɪtə(r)]	n.	温度计
indicate	[ˈɪndɪkeɪt]	vt.	标示,指示
specify	[ˈspesɪfaɪ]	vt.	指定,规定
crucial	[ˈkruːʃl]	adj.	决定性的,关键性的
sustain	[səˈsteɪn]	vt.	维持
state	[steɪt]	vt.	规定,陈述
subpart	[sʌbˈpɑːt]	n.	子部分
receive	[rɪˈsiːv]	vt.	收到
authorize	[ˈɔːθəraɪz]	vt.	批准,授权
substantially	[səbˈstænʃəli]	adv.	实质上
demonstrate	[ˈdemənstreɪt]	vt.	论证,证明
accredit	[əˈkredɪt]	vt.	委托,授权
conduct	[kənˈdʌkt]	vt.	实施,执行
primary	[ˈpraɪməri]	adj.	初步的
review	[rɪˈvjuː]	n.	审查
issue	[ˈɪʃuː]	vt.	发布
claim	[kleɪm]	n.	主张
document	[ˈdɒkjumənt]	n.	文件
dossier	[ˈdɒsieɪ]	n.	档案,资料
information	[ˌɪnfəˈmeɪʃn]	n.	信息
deem	[diːm]	vt.	认为,视为
hardcopy	[ˈhɑːdkɒpɪ]	n.	硬拷贝,纸质版
staff	[stɑːf]	n.	职员
interactive	[ˌɪntərˈæktɪv]	adj.	互动的
registration	[ˌredʒɪˈstreɪʃn]	n.	注册
confirmation	[ˌkɒnfəˈmeɪʃn]	n.	确认
subject to			受……管制
refer to			指的是

premarket notification(PMN)	上市前通告
premarket approval(PMA)	上市前批准
depend on	取决于
search for	搜索
associated with	与……相关
equivalent to	等同于……
electronic copy	电子版
thirty party	第三方

Key sentences

(1) The class to which your device is assigned determines the type of premarketing submission/application required for FDA clearance to market.

器械被指定的类别决定了FDA批准上市所需的上市前递交/申请的类型。

(2) Classification of the device depends on the risk level of the medical devices and its intended use.

设备的分类取决于医疗器械的风险级别及其预期用途。

(3) Choosing correctpremarket submission may refer to either 510(k) or PMA, whichever the medical device classification specifies in the CFR.

选择正确的上市前递交方式指的是510(k)或者PMA,无论哪种都在CFR(联邦规章典籍)中的医疗器械分类中有指定。

(4) If your device requires the submission of a Premarket Notification 510(k), you cannot commercially distribute the device until you receive a letter of substantial equivalence from FDA authorizing you to do so.

如果你的器械需要递交上市前通告510(k),你只有在收到FDA批准上市的一封实质等同函件后,才能上市销售该器械。

(5) Product requiringPMAs are Class III high-risk devices that pose a significant risk of illness or injury, or devices found not substantially equivalent to Class I and II through the 510(k) process.

要求PMA(上市前批准)的产品是Ⅲ类高危器械,是具有显著的疾病或损伤风险,或者通过510(k)过程发现不实质等同于Ⅰ类和Ⅱ类的器械。

Learning test

Choose the following words below to fill in the blanks, change the form if necessary.

review	clinical	market	hardcopy	information
document	authorize	specialty	receive	staff

(1) The _____ structure of vocational education is closely correlated with structural changes in industry.

(2) The annual _____ is still under review for later publishing.

(3) The company employs 20 development _____.

(4) The two big companies hold 50% of the _____.

(5) They will _____ their awards at a ceremony.

(6) Please contact me for any other _____ .

(7) I will _____ the transaction after I receive confirmation from my client.

(8) The foreign ministers of the two countries signed the _____ today.

(9) They have been treated with this drug in _____ trials.

(10) Please print the file for me, I need a _____ .

Translation

1. Translate the following words into Chinese.

(1) submission　　　　　(2) market

(3) review　　　　　　(4) hardcopy

(5) document　　　　　(6) database

(7) notification　　　　　(8) authorize

(9) clearance　　　　　(10) dossier

2. Translate the following phrases into Chinese.

(1) refer to　　　　　　(2) premarket notification

(3) premarket approval　　(4) depend on

(5) electronic copy　　　(6) thirty party

Practice

1. Translate the following sentences into Chinese.

(1) If your device is classified as Class I or II, and if it is not exempt, a Premarket Notification (PMN) 510(k) will be required for marketing.

(2) PMA is a more stringent premarket submission process, which requires clinical evidence to prove the safety and effectiveness of the device to be marketed.

(3) To find the classification of your device, as well as whether any exemptions may exist, you need to find the regulation number that is the classification regulation for your device.

(4) However, with the understanding that all the information FDA requires is to ultimately establish that the device is safe and effective for use, the preparation of the dossier then becomes a logical process.

(5) After preparation of all relevant documents, submitter must first submit user fees before submitting any application to the FDA.

2. Translate the following sentences into English.

(1) 任何一种医疗器械想要进入美国市场，必须先明确产品的管理类别和管理要求。

(2) Ⅲ类产品实施的是上市前许可，此类产品约占全部医疗器械的8%。

(3) 对Ⅰ类产品，企业向FDA递交资料后，FDA只进行公告，并无相关证件发给企业。

(4) 对Ⅱ类和Ⅲ类器械，企业须递交PMN或PMA，FDA在公告的同时，会给企业以正式的市场准入函件。

(5)FDA 是根据专家委员会的建议来最终决定医疗器械产品的类别的。

> **Culture salon**
>
> Medical device regulations of FDA Modernization Act
>
> The basic framework governing the regulation of medical devices is established in the Medical Device Amendments to the Federal Food, Drug, and Cosmetic(FFD&C) Act. The Medical Device Amendments were enacted on May 28, 1976. The FFD&C Act was again amended with respect to the regulation of medical devices by the Safe Medical Devices Act of 1990 and the Medical Device Amendments of 1992. New provisions governing the export of FDA regulated products, including medical devices, were established in the FDA Export Reform and Enhancement Act of 1996. The FFD&C Act was amended by the Food and Drug Administration Modernization Act of 1997(the Modernization Act). Signed into law on November 21, 1997, the Modernization Act contained provisions related to all products under FDA's jurisdiction.

Chapter 3-Section 2-Text 1-看译文

Text 2　FDA：Premarket notification (PMN) 510 (k)

Lead-in questions:

(1) Who must submit a FDA 510(k) premarket notification?

(2) How to classify a medical device into one of the three regulatory classes?

(3) When should device manufacturers notify FDA according to PMN?

(4) How many types of 510(k)?

(5) Is there any template of contents and documents for a 510(k) submission?

Are you ready to sell your medical devices in the United States? Who must submit a FDA 510(k) premarket notification? In general, manufacturers wishing to introduce Class Ⅱ medical devices(and a small number of Class Ⅰ and Ⅲ devices) or in-vitro diagnostic products(IVDs) to the US market must submit a 510(k) to the FDA. A 510(k) premarket notification is also required for manufacturers changing the intended use of their medical device, or changing the technology of a cleared device in such a way that it may significantly affect the device's safety or effectiveness.

Chapter 3-Section 2-Text 2-听课文

Federal law[Federal Food, Drug, and Cosmetic(FFD&C) Act, section 513], established the risk-based device classification system for medical devices. Each device is assigned to one of three regulatory classes: Class Ⅰ, Class Ⅱ or Class Ⅲ, based on the level of control necessary to provide reasonable assurance of its safety and effectiveness.

Section 510(k) of the FFD&C Act requires device manufacturers who must register, to notify FDA of their intent to market a medical device at least 90 days in advance. This is known as premarket notification-also called PMN or 510(k). This allows FDA to determine whether the device is equivalent to a device already placed into one of the three classification categories. Thus, "new" devices(not in commercial distribution prior to May 28, 1976) that have not been classified can be properly identified.

Specifically, medical device manufacturers are required to submit apremarket notification if they intend to introduce a device into commercial distribution for the first time or reintroduce a device that will be significantly changed or modified to the extent that its safety or effectiveness could be affected. Such change or modification could relate to the design, material, chemical composition, energy source, manufacturing process, or intended use.

The FDA developed a guidance document in March 1998 called, "The new 510(k) paradigm-alternate approaches to demonstrating substantial equivalence inpremarket notifications" in order to make the PMN review process more efficient and provides two other alternatives to the original Traditional 510(k) submission: special and abbreviated 510(k). "The new 510(k) paradigm" guidance also helps in the decision making process for choosing the right 510(k) submission method for a desired purpose by providing questions to direct submitters to the right method.

1. Traditional 510(k)

It is the original complete submission as provided in the 21 CFR 807 and may be used for any type of 510(k) submission.

2. Special 510(k)

Special 510(k) submission is used for device modification of a company's own medical device that has previously been cleared under the 510(k) process. It is also used when the device design complies with the QSR design controls(21 CFR 820.30) and allows manufacturers to declare conformance to design controls without providing the data.

3. Abbreviated 510(k)

The abbreviated 510(k) submission method may be used when a guidance document exists, a special control has been established, or when the FDA has recognized a relevant consensus standard.

Premarket notification, or 510(k), is submitted to the FDA to prove that the device to be marketed is substantially equivalent (SE) to a previously legally marketed device, also known as a predicate device. According to the FDA, a device is SE if, compared to a predicate, it:

- has the same intended use; and
- has the same technological characteristics;

or

- has the same intended use; and
- has different technological characteristics and does not raise new questions of safety and effectiveness; and

• the information submitted to FDA demonstrates that the device is at least as safe and effective as the legally marketed device.

Submitters will have to compare their device with similar predicate devices and support their claims, however, they need not be identical. The elements including intended use(anatomical site, human factors, sterility, electrical/chemical/ mechanical/thermo/radiation safety, etc.), indications for use(location of usage, materials, biocompatibility, etc.), and target population (performance, effectiveness, manufacturing process, etc.) should be taken into consideration for substantial equivalence. "Intended Use" refers to the functional capability of the device. It also refers to the objective intent of the person legally responsible for the labeling of the device. "Indications" are the disease(s) or condition(s) the device will diagnose, treat, prevent, cure, or mitigate and a description of the target population.

The submitter may not market a device until a letter from the FDA declaring that the device is SE is received. The device may be marketed immediately after the 510(k) clearance is obtained. After the 510(k) clearance, the facility in which the medical device is manufactured may be inspected at anytime. If the FDA determines that a device is not SE, the applicant may choose toresubmit another 510(k) with new data, request a Class I or II designation through the de novo process, file a reclassification petition, or submit a PMA.

There is no fixed template of contents and documents for a 510(k) submission. The relevant submission requirements could be found in the 21 CFR 807 Subpart E. Each 510(k) submission must be accompanied with a MDUFA cover sheet, which can be found at the FDA user fee website that will only be available after a disclaimer and login. All outstanding establishment registration fees have to be paid in order to proceed, otherwise, the FDA will not accept the submission. There would be a series of questions and options to answer before a draft of the MDUFA cover sheet would be generated.

During submission of the MDUFA cover sheet and payment of fees to the FDA, a payment identification number(PIN) would be obtained and that would be used to track the payment. If a 510(k) application is reviewed by an accredited third party, this may be omitted because the third party may charge review fees. Accredited persons, also known as accredited third parties, are people that are authorized by the FDA to conduct the primary review of 510(k) for eligible devices. This accredited persons program was created by the FDA Modernization Act of 1997(FDAMA). For more information, please refer to the FDA guideline "Implementation of Third Party Programs under the FDA Modernization Act of 1997; Final Guidance for Staff, Industry and Third Parties".

Words and phrases

technology	[tek'nɒlədʒi]	n.	技术
affect	[ə'fekt]	vt.	影响
drug	[drʌg]	n.	药品
cosmetic	[kɒz'metɪk]	n.	化妆品

Chapter 3- Section 2- Text 2-听词汇

英文	音标	词性	中文
act	[ækt]	n.	法案
regulatory	[ˈregjələtəri]	adj.	监管的
class	[klɑːs]	n.	类别,种类
assurance	[əˈʃʊərəns]	n.	保证
intent	[ɪnˈtent]	n.	意图,目的
category	[ˈkætɪɡ(ə)rɪ]	n.	种类,类别
determine	[dɪˈtɜːmɪn]	vt.	决定,确定
reintroduce	[ˌriːɪntrəˈdjuːs]	vt.	再次引入,重新引入
significantly	[sɪɡˈnɪfɪkənt]	adv.	显著地,重大地
modify	[ˈmɒdɪfaɪ]	vt.	决定,确定
extent	[ɪkˈstent]	n.	改变
design	[dɪˈzaɪn]	n.	设计
material	[məˈtɪəriəl]	n.	材料,原料
chemical	[ˈkemɪkl]	adj.	化学的
composition	[ˌkɒmpəˈzɪʃn]	n.	成分
energy	[ˈenədʒi]	n.	能量
source	[sɔːs]	n.	来源
guidance	[ˈɡaɪdns]	n.	指导
paradigm	[ˈpærədaɪm]	n.	模式
alternate	[ɔːlˈtɜːnət]	adj.	代替的
approach	[əˈprəʊtʃ]	n.	方法
original	[əˈrɪdʒənl]	adj.	最初的,原来的
abbreviate	[əˈbriːvieɪt]	vt.	缩简,简略
complete	[kəmˈpliːt]	adj.	完整的
comply	[kəmˈplaɪ]	vi.	符合
declare	[dɪˈkleə(r)]	vt.	声明,声称
conformance	[kənˈfɔːməns]	n.	一致性,符合性
consensus	[kənˈsensəs]	n.	共识,公认
element	[ˈelɪmənt]	n.	要素
anatomical	[ˌænəˈtɒmɪkl]	adj.	结构(上)的,解剖的
factor	[ˈfæktə(r)]	n.	参数
electrical	[ɪˈlektrɪkl]	adj.	电气的
chemical	[ˈkemɪkl]	adj.	化学的
mechanical	[məˈkænɪkl]	adj.	机械的
thermo	[θɜːməʊ]	adj.	热的

radiation	[ˌreɪdɪˈeɪʃ(ə)n]	n.	辐射
indication	[ˌɪndɪˈkeɪʃn]	n.	适应证
biocompatibility	[biːəʊkəmpætəˈbɪlɪtɪ]	n.	生物相容性
diagnose	[ˈdaɪəɡnəʊz]	vt.	诊断
treat	[triːt]	vt.	治疗
prevent	[prɪˈvent]	vt.	预防
cure	[kjʊə(r)]	vt.	治愈
mitigate	[ˈmɪtɪɡeɪt]	vt.	减轻
inspect	[ɪnˈspekt]	vt.	检查
petition	[pəˈtɪʃn]	n.	请求
designation	[ˌdezɪɡˈneɪʃn]	n.	指定
template	[ˈempleɪt]	n.	模板
available	[əˈveɪləbl]	adj.	可获得的,能找到的
disclaimer	[dɪsˈkleɪmə(r)]	n.	免责声明
outstanding	[aʊtˈstændɪŋ]	adj.	未完成的
draft	[drɑːft]	n.	草稿
generate	[ˈdʒenəreɪt]	vt.	生成
track	[træk]	vt.	跟踪,追踪
omit	[əˈmɪt]	vt.	忽略
eligible	[ˈelɪdʒəbl]	adj.	合适的

premarket notification	上市前公告
Federal Food, Drug, and Cosmetic (FFD&C) Act	联邦食品药品和化妆品法案
risk-based	基于风险的
notify... of...	通知……(某事)
in advance	提前
prior to	在……之前
chemical composition	化学成分
energy source	能源
special 510(k)	特殊 510(k)
abbreviated 510(k)	简略 510(k)
comply with	符合
substantially equivalent	实质等同,实质等效
take into consideration	考虑到
de novo process	重新归类程序
accompanied with	伴随着

Key sentences

(1) In general, manufacturers wishing to introduce Class II medical devices (and a small number of Class I and III devices) or IVDs to the US market must submit a 510(k) to the FDA.

一般来说,希望将Ⅱ类医疗器械(以及少量Ⅰ类和Ⅲ类器械)或体外诊断试剂引入美国市场的制造商必须向FDA提交510(k).

(2) Section 510(k) of the FFD&C Act requires device manufacturers who must register, to notify FDA of their intent to market a medical device at least 90 days in advance.

FFD&C法案的510(k)部分要求要注册的器械制造商必须至少提前90天通告FDA其将医疗器械上市的意愿。

(3) Special 510(k) submission is used for device modification of a company's own medical device that has previously been cleared under the 510(k) process.

特殊510(k)提交用于公司自己的曾经通过510(k)程序批准上市的器械修改。

(4) The abbreviated 510(k) submission method may be used when a guidance document exists, a special control has been established, or when the FDA has recognized a relevant consensus standard.

如果存在指南文件,特殊控制已建立,或FDA已经认可相关的公认标准,那么可以采用简略510(k)提交方式。

(5) There is no fixed template of contents and documents for a 510(k) submission. The relevant submission requirements could be found in the 21 CFR 807 Subpart E.

没有可供510(k)提交的内容和文件的固定范本。有关的提交要求可在21 CFR 807的子部分E中找到。

Learning test

Choose the following words below to fill in the blanks, change the form if necessary.

drug   draft   design   chemical   factor
diagnose   affect   modify   source   inspect

(1) High blood pressure is a major risk _____ in many diseases.

(2) Physical therapy is an important adjunct to _____ treatment.

(3) You never allow personal problems to _____ your performance.

(4) I faxed the final _____ of this article to him.

(5) She came to London in 1960 to study fashion _____ .

(6) Plants are the ultimate _____ of all food stuff.

(7) We are exposed to an overwhelming number of _____ contaminants every day in our air, water and food.

(8) ECG is the most commonly used clinical detection to _____ heart disease.

(9) The user can then edit and _____ these settings of the machine.

(10) They have the right to _____ the workshop of the company.

Translation

1. Translate the following words into Chinese.

(1) act (2) drug

(3) intent (4) approach

(5) diagnose (6) treat

(7) inspect (8) conformance

(9) biocompatibility (10) guidance

2. Translate the following phrases into Chinese.

(1) in advance (2) prior to

(3) chemical composition (4) energy source

(5) substantially equivalent (6) take into consideration

Practice

1. Translate the following sentences into Chinese.

(1) A 510(k) premarket notification is also required for manufacturers changing the intended use of their medical device, or changing the technology of a cleared device in such a way that it may significantly affect the device's safety or effectiveness.

(2) Specifically, medical device manufacturers are required to submit a premarket notification if they intend to introduce a device into commercial distribution for the first time or reintroduce a device that will be significantly changed or modified to the extent that its safety or effectiveness could be affected.

(3) Premarket notification, or 510(k), is submitted to the FDA to prove that the device to be marketed is substantially equivalent(SE) to a previously legally marketed device, also known as a predicate device.

(4) The submitter may not market a device until a letter from the FDA declaring that the device is SE is received.

(5) After the 510(k) clearance, the facility in which the medical device is manufactured may be inspected at anytime.

2. Translate the following sentences into English.

(1)器械初次投放美国市场前需要通过上市前公告或者上市前批准。

(2)除了传统的510(k)提交方式,还有特殊510(k)和简略510(k)两种。

(3)如果你的器械材料和化学成分已经改变,则应该重新提交上市前公告。

(4)产品注册是否收费呢?

(5)注册文件的准备是一件相当繁琐的工作。

Culture salon

Premarket approval(PMA)

Premarket approval(PMA) is the FDA process of regulatory review to evaluate the safety and effectiveness of Class Ⅲ medical devices. The regulation governing premarket approval is located in Title 21 Code of Federal Regulations(CFR) Part 814. FDA regulations provide 180 days to review the PMA and make a determination. In reality, the review time is normally longer. Before approving or denying a PMA, the appropriate FDA advisory committee may review the PMA at a public meeting and provide FDA with the committee's recommendation on whether FDA should approve the submission. After FDA notifies the applicant that the PMA has been approved or denied, a notice is published on the Internet: ① announcing the data on which the decision is based; ② providing interested persons an opportunity to petition FDA within 30 days for reconsideration of the decision.

Chapter 3-Section 2-Text 2-看译文

Chapter 3-Section 2-习题

Section 3　NMPA regulations

Chapter 3-Section 3-PPT

Text 1　Regulations on supervisory management of medical devices

Lead-in questions:

(1) What's the purpose to formulate the Regulations?

(2) Who should abide by these Regulations?

(3) Who is responsible for the supervisory management of medical devices all over the country?

(4) How to categorize medical devices?

(5) What's the management system for marketing medical devices?

Chapter Ⅰ　General rules

Article 1　These regulations are formulated in order to protect the safety and effectiveness of medical devices and safeguard the physical health and life safety of the public.

Article 2　Any unit or individual engaged in the research, manufacture, operation, use, supervisory management of medical devices within the territory of the People's

Chapter 3-Section 3-Text 1-听课文

Republic of China should abide by these regulations.

Article 3 The State Food and Drug Administration should be responsible for the supervisory management of medical devices all over the country. Relevant departments under the State Council should be responsible for the work relating to the supervisory management of medical devices within their respective scopes of responsibility.

The food and drug supervision departments of various local people's governments above the county level should be responsible to supervise and manage the medical devices within their respective administrative regions. Relevant departments of various local people's governments above the county level should be responsible for the work relating to supervision and management of medical devices within their respective scopes of responsibility.

The Food and Drug Administration Department of the State Council should cooperate with relevant departments of the State Council to carry out and implement the planning for and policies on the medical devices industry.

Article 4 The State should carry out classified management for medical devices according to its potential risk.

The medical devices with lower risks are categorized as Class I medical devices; their safety and effectiveness can be guaranteed by implementing general management.

The medical devices with moderate risks are categorized as Class II medical devices; their safety and effectiveness can be guaranteed by implementing stricter management.

The medical devices with higher risks are categorized as Class III medical devices; their safety and effectiveness can be guaranteed by taking special measures and by stricter management.

When evaluating the risks of any medical devices, their intended use, structural feature, use method and other factors must be taken into account.

The Food and Drug Administration Department of the State Council should be responsible for formulating classification rules and classified catalogue for medical devices, and should, depending on the production, operation and application of medical devices, analyze and evaluate their risk changes in a timely manner and then adjusts the classified catalogue. When formulating and adjusting the classified catalogue, the opinions and suggestions from manufacturers, users and various medical devices associations should be taken into account sufficiently, while referring to international medical device classification practice. The classification catalogue of medical devices should be published to the society.

Article 5 The research and manufacture of medical devices should follow the principles of safety, effectiveness and cost conservation. The State encourages various manufacturers to research innovative medical devices and take full advantage of the role of market mechanism to promote popularization and application of new technologies in medical devices so as to boost the development of the medical devices industry.

Article 6 All medical devices and products should comply with the mandatory standards of the state for medical devices. If there is no mandatory national standards, the compulsory standards of medical devices industry should be executed.

The catalogue of single-use medical devices should be formulated, adjusted and publicized by the food & drug administration department of the State council together with the health and family planning departments of the State Council. Any medical devices whose safety and effectiveness can be guaranteed in case of reuse should not be listed in the catalogue of single-use medical devices. Any medical devices whose safety and effectiveness can be guaranteed in case of reuse after having their designs, production processes, disinfection & sterilization technologies improved should be deleted from the catalogue of single-use medical devices.

Article 7 Various industrial associations in the field of the medical device industry should strengthen self-discipline, promote the construction of a credit system, urge enterprises to carry out various production and operation activities in compliance with laws and guide enterprises to be honest and trustworthy.

Chapter II   Registration and Notification of Medical Devices

Article 8 Class I medical devices are for notification-based management and Class II and Class III are for registration-based management.

Article 9 When Class I medical devices are submitted for notification and Class II and Class III medical devices are submitted for applying for registration, the following information should be submitted:

(I) Analysis on risks of the product;

(II) Technical requirements for the product;

(III) Test report of the product;

(IV) Clinical evaluation data;

(V) The Instructions for Use and the sample of product labels for the product;

(VI) Quality Assurance System documents related to product research & development and manufacture;

(VII) Other documents necessary to prove the safety and effectiveness of the medical device.

The registration applicant and the person to submit the medical device for notification should be responsible for the authenticity of all the documents submitted.

Article 10 For Class I medical devices, the person to submit the product for notification should submit all the required documents to the food & drug administration department of the local municipal-level people's government. The test report therein may be an internal test report of the applicant; the clinical evaluation data should not include the clinical trial report and may be the document that uses the data obtained from references or during clinical application of similar products to prove their safety and effectiveness.

An overseas manufacturing enterprise that exports Class I medical devices to China may have their documents required for notification and the official marketability certificates issued by state/local authority submitted by their representative office or legal representative in China to the Food and Drug Administration Department of the State Council.

If any items stated in any document submitted for notification change, the applicant should submit it to the original competent department for notification.

Words and phrases

Chapter 3-Section 3-Text 1-听词汇

regulation	[ˌregjuˈleɪʃn]	n.	条例,规章
supervisory	[ˌsjuːpəˈvaɪzəri]	adj.	监督的
decree	[dɪˈkriː]	n.	法令
state	[steɪt]	n.	国家
formulate	[ˈfɔːmjuleɪt]	vt.	制定
public	[ˈpʌblɪk]	n.	公众,民众
operation	[ˌɒpəˈreɪʃn]	n.	经营
territory	[ˈterətri]	n.	领土,领地
abide	[əˈbaɪd]	vi.	遵守
respective	[rɪˈspektɪv]	adj.	各自的,分别的
scope	[skəʊp]	n.	范围
local	[ˈləʊkl]	adj.	地方的,本地的
county	[ˈkaʊnti]	n.	县
administrative	[ədˈmɪnɪstrətɪv]	adj.	管理的,行政的
region	[ˈriːdʒən]	n.	区域
cooperate	[kəʊˈɒpəreɪt]	vi.	配合,合作
implement	[ˈɪmplɪment]	vt.	实施,执行
potential	[pəˈtenʃl]	adj.	潜在的
moderate	[ˈmɒdərət]	adj.	中等的,适度的
measures	[ˈmeʒəz]	n.	措施
catalogue	[ˈkætəlɒg]	n.	目录
production	[prəˈdʌkʃn]	n.	生产,制作
analyze	[ˈænəlaɪz]	vt.	分析
timely	[ˈtaɪmli]	adj.	及时的,适时的
manner	[ˈmænə(r)]	n.	方式,方法
adjust	[əˈdʒʌst]	vt.	调整
association	[əˌsəʊʃiˈeɪʃn]	n.	协会
principle	[ˈprɪnsəpl]	n.	原则
conservation	[ˌkɒnsəˈveɪʃn]	n.	节约
mechanism	[ˈmekənɪzəm]	n.	机制
popularization	[ˌpɒpjələraɪˈzeɪʃn]	n.	推广
boost	[buːst]	vt.	促进,推动
mandatory	[ˈmændətəri]	adj.	强制性的

compulsory	[kəmˈpʌlsəri]	*adj.*	强制性的
delete	[dɪˈliːt]	*vt. & vi.*	删除
disinfection	[ˌdɪsɪnˈfekʃən]	*n.*	消毒
sterilization	[ˌsterəlaɪˈzeɪʃn]	*n.*	灭菌
self-discipline	[selfˈdɪsəplɪn]	*n.*	自律
promote	[prəˈməʊt]	*vt.*	促进,推进
credit	[ˈkredɪt]	*n.*	信誉,信用
system	[ˈsɪstəm]	*n.*	体系,系统
trustworthy	[ˈtrʌstwɜːði]	*adj.*	值得信赖的,可靠的
notification	[ˌnəʊtɪfɪˈkeɪʃn]	*n.*	备案
registration	[ˌredʒɪˈstreɪʃn]	*n.*	注册
authenticity	[ˌɔːθenˈtɪsəti]	*n.*	真实性,可靠性
proof	[pruːf]	*n.*	证明
State Council			国务院
general rule			通则,总则
physical health			体格健康
engaged in			从事
abide by			遵守
cooperate with			配合……
carry out			进行,执行
take into account			考虑在内
refer to			参考
mandatory standard			强制性标准
in the field of			在……领域
in compliance with			遵守……
apply for			申请……
post-market			上市后

Key sentences

(1) These regulations are formulated in order to protect the safety and effectiveness of medical devices and safeguard the physical health and life safety of the public.

为了保证医疗器械的安全、有效,保障人体健康和生命安全而制定本条例。

(2) The National Medical products Administration should be responsible for the supervisory management of medical devices all over the country.

国家药品监督管理局负责全国医疗器械监督管理工作。

(3) The State should carry out classified management for medical devices according to its potential risk.

国家对医疗器械按照风险程度实行分类管理。

(4) The research and manufacture of medical devices should follow the principles of safety, effectiveness and cost conservation.

医疗器械的研制应当遵循安全、有效和节约的原则。

(5) Class Ⅰ medical devices are for notification-based management and Class Ⅱ and Class Ⅲ are for registration-based management.

第一类医疗器械实行产品备案管理,第二类、第三类医疗器械实行产品注册管理。

Learning test

Choose the following words below to fill in the blanks, change the form if necessary.

catalogue   association   manner   local   operation
promote   production   credit   delete   scope

(1) _____ is everything to a trader.

(2) The lawyer speaks in a professional _____ .

(3) We visited the _____ markets and bought wonderful fruit and vegetables.

(4) China will further expand the _____ of its opened-up areas.

(5) The company starts to make profits under his _____ .

(6) _____ and invalidation date should be indicated in product instructions.

(7) I _____ the files from the computer system.

(8) The meeting discussed how to _____ cooperation between the two countries.

(9) China _____ for medical devices industry is a non-profit social organization.

(10) The sales give me an electronic product _____ for reference.

Translation

1. Translate the following words into Chinese.

(1) registration   (2) notification

(3) decree   (4) operation

(5) analyze   (6) measures

(7) popularization   (8) catalogue

(9) state   (10) production

2. Translate the following phrases into Chinese.

(1) physical health   (2) cooperate with

(3) abide by   (4) engaged in

(5) carry out   (6) in the field of

Practice

1. Translate the following sentences into Chinese.

(1) Relevant departments under the State Council should be responsible for the work relating to the su-

pervisory management of medical devices within their respective scopes of responsibility.

(2) The medical devices with moderate risks are categorized as Class Ⅱ medical devices.

(3) All medical devices and products should comply with the mandatory standards of the state for medical devices.

(4) For Class Ⅰ medical devices, the person to submit the product for notification should submit all the required documents to the food & drug administration department of the local municipal-level people's government.

(5) If any items stated in any document submitted for notification change, the applicant should submit it to the original competent department for notification.

2. Translate the following sentences into English.

(1)申请第三类医疗器械产品注册,注册申请人应当向国务院药品监督管理部门提交注册申请资料。

(2)备案资料载明的事项发生变化的,应当向原备案部门变更备案。

(3)医疗器械注册证有效期为5年。

(4)第一类医疗器械产品备案,不需要进行临床试验。

(5)免于进行临床试验的医疗器械目录由国务院药品监督管理部门制定、调整并公布。

> **Culture salon**
>
> Medical device user
>
> Medical device user refers to the unit or individual that uses medical device to provide such technical services as medical treatment to others, including medical institutions that have obtained the Medical Institution Practicing License, family planning related technical service institutions that have obtained the Family Planning Related Technical Service Institution Practicing License, and blood station, single plasma station, rehabilitation supporting device fitting institution, etc. that do not need to obtain Medical Institution Practicing License.

Text 2 Provisions for medical device registration

Lead-in questions:

(1) What's the purpose to formulate the provisions?

(2) Which kind of medical devices should apply for product registration?

(3) What's the process of medical device filing?

(4) Who should take legal responsibilities for the marked product?

(5) What's the requirement for the persons undertaking registration or filing for medical devices?

Chapter I General provisions

Article 1 The Provisions is formulated in accordance with the Regulations on Supervision and Administration of Medical Devices, with a view to standardizing the administration of medical device registration and filing and guaranteeing the safety and effectiveness of medical devices.

Article 2 All medical devices sold and used within the territory of the People's Republic of China shall apply for registration or filing according to the Provisions.

Article 3 Medical device registration refers to the prescribed procedures conducted by the food and drug regulatory department upon an application submitted by the registration applicant to decide whether the medical device to be marketed can be sold based on a comprehensive evaluation of the research and results of its safety and effectiveness.

Medical device filing is a process that the filing entity submits the filing documents to the food and drug regulatory department and the food and drug regulatory department files the filing documents submitted by the filing entity.

Article 4 The medical device registration and filing shall be conducted under the principles of publicity, equity and justice.

Article 5 Class I medical device is subject to filing administration and class II and class III medical devices are subject to registration administration.

To apply for filing of class I domestic medical devices, the filing entity shall submit the filing documents to the food and drug regulatory department of the city consisting of districts.

Class II domestic medical devices shall be reviewed by the food and drug regulatory department of the provinces, autonomous regions and municipalities directly under the central government, and the medical device registration certificate shall be issued after approval.

Class III domestic medical devices shall be reviewed by China Food and Drug Administration, and the medical device registration certificate shall be issued after approval.

To apply for filing of import class I medical devices, the filing entity shall submit the filing documents to China Food and Drug Administration.

Import class II and class III medical devices shall be reviewed by China Food and Drug Administration, and the medical device registration certificate shall be issued after approval.

The medical devices from Hong Kong, Macao and Taiwan shall be registered and filed by reference to the import medical devices.

Article 6 Where a registration applicant or filing entity of a medical device brings the products to the market in his own name, he shall take responsibilities for the product.

Article 7 The food and drug regulatory department shall publicize relevant information on medical device registration and filing promptly in accordance with laws. The applicant can look up the approval progress and relevant results as well as the public can look up the result of approval.

Article 8 The State encourages the research and innovation of medical devices, implements special review for innovative medical devices, accelerates the popularization and application for new technologies of medical devices and promotes the development of medical device industry.

Chapter II   Essential Requirements

Article 9 The registration applicant and filling entity of medical devices shall establish a quality management system related to product research, development and manufacture and shall keep its effective operation.

Where applying for registration of a domestic medical device subject to the special review procedures for innovative medical devices and its sample production is entrusted to another enterprise, the applicant shall entrust a medical device manufacturer having corresponding production range. Where the domestic medical device applied for registration is not subject to the special review procedures for innovative medical devices, the sample production shall not be entrusted to another enterprise.

Article 10 The persons undertaking registration or filing for medical devices shall have corresponding specialized knowledge and be familiar with the laws, rules, regulations and technical requirements of the administration on registration or filing of medical devices.

Article 11 When applying for registration or filing, the applicant or filing entity shall complete the research and development of medical devices in accordance with the essential requirements for safety and effectiveness of medical devices and make sure the research and development process is true and standardized and all the data is authentic, complete and traceable.

Article 12 The documents for registration application or filing shall be in Chinese. Where they are translated from another language, the original documents shall also be provided at the same time. When referring to unpublished literature, the applicant shall provide documents proving the owner's permission to use the information.

The applicant and filing entity shall take full responsibility for the authenticity of the documents submitted.

Article 13 An import medical device being applied for registration shall be one which has already got permission for distribution in the country(region) where the applicant or filing entity is registered or the manufacture is carried out.

If the product is not managed as medical device in the country(region) where the applicant or filing entity is registered or the manufacture is carried out, the applicant or filing entity shall provide relevant proof documents, including the permission for distribution of sold product in the country(region) where the applicant or filing entity is registered or the manufacture is carried out.

Article 14 An overseas applicant or filing entity shall conduct related work with the support of its representative office established in China or an enterprise legal person in China designated by it as its agent.

The agent shall undertake the following liabilities in addition to conducting medical device registration or filing:

(Ⅰ) Contacting with corresponding food and drug regulatory department and overseas applicant or filing entity.

(Ⅱ) Delivering related laws and regulations and technical requirements to the applicant of filing entity in a truthful and faithful manner.

(Ⅲ) Collecting post-market adverse event information of medical devices and sending feedback to overseas registrants or filing entity, meanwhile, reporting to relevant food and drug regulatory department.

(Ⅳ) Coordinating the medical device post-market product recall and reporting to relevant food and drug regulatory department.

(Ⅴ) Other joint liabilities related to product quality and after-sales service.

Words and phrases

provision	[prəˈvɪʒn]	n.	办法,规定,条款
registration	[ˌredʒɪˈstreɪʃn]	n.	注册
supervision	[ˌsjuːpəˈvɪʒn]	n.	监督
administration	[ədˌmɪnɪˈstreɪʃn]	n.	管理
standardize	[ˈstændədaɪz]	vt.	规范,使标准化
filing	[ˈfaɪlɪŋ]	n.	备案
conduct	[kənˈdʌkt]	vt.	进行,实施,办理
comprehensive	[ˌkɒmrɪˈhensɪv]	adj.	综合的,系统的,全面的
procedure	[prəˈsiːdʒə(r)]	n.	程序,过程,步骤
department	[dɪˈpɑːtmənt]	n.	部门
application	[ˌæplɪˈkeɪʃn]	n.	申请
applicant	[ˈæplɪkənt]	n.	申请人
evaluation	[ɪˌvæljʊˈeɪʃn]	n.	评价,评估
process	[ˈprəʊses]	n.	程序,过程
entity	[ˈentəti]	n.	(法人)实体
file	[faɪl]	vt.	把……归档,把……存档
principle	[ˈprɪnsəpl]	n.	原则
publicity	[pʌbˈlɪsəti]	n.	公开
equity	[ˈekwəti]	n.	公平
justice	[ˈdʒʌstɪs]	n.	公正
district	[ˈdɪstrɪkt]	n.	行政区
domestic	[dəˈmestɪk]	adj.	国内的,境内的
province	[ˈprɒvɪns]	n.	省份
autonomous	[ɔːˈtɒnəməs]	adj.	自治的

municipality	[mjuːˌnɪsɪˈpæləti]	n.	市
certificate	[səˈtɪfɪkət]	n.	证书
issue	[ˈɪʃuː]	vt.	颁发,发布
approval	[əˈpruːvl]	n.	批准
import	[ˈɪmpɔːt]	n.	进口
progress	[ˈprəʊgres]	n.	进度,进展
accelerate	[əkˈseləreɪt]	vt.	促进,加快
essential	[ɪˈsenʃl]	adj.	基本的,必要的
operation	[ˌɒpəˈreɪʃn]	n.	实施,运作
entrust	[ɪnˈtrʌst]	vt.	委托
enterprise	[ˈentəpraɪz]	n.	企业
range	[reɪndʒ]	n.	范围
authentic	[ɔːˈθentɪk]	adj.	真实的,可信的,可靠的
traceable	[ˈtreɪsəbl]	adj.	可追踪的,可追溯
permission	[pəˈmɪʃn]	n.	允许,批准
distribution	[ˌdɪstrɪˈbjuːʃn]	n.	分销,上市销售
proof	[pruːf]	n.	证明
designate	[ˈdezɪgneɪt]	vt.	指定,指派
undertake	[ˌʌndəˈteɪk]	vt.	承担
liability	[ˌlaɪəˈbɪləti]	n.	责任
deliver	[dɪˈlɪvə(r)]	vt.	传达、递送
faithful	[ˈfeɪθfl]	adj.	诚实的
coordinate	[kəʊˈɔːdɪneɪt]	vt.	协调
recall	[rɪˈkɔːl]	n.	召回
in accordance with			根据
with a view to			为了……
consist of			包括,由……组成
autonomous region			自治区
municipality directly under the central government			直辖市
registration certificate			注册证
by reference to			参考……,参照……
look up			查询
quality management system			质量管理体系
post-market			上市后
after-sales			售后

Key sentences

(1) The Provisions is formulated in accordance with the Regulations on Supervision and Administration of Medical Devices, with a view to standardizing the administration of medical device registration and filing and guaranteeing the safety and effectiveness of medical devices.

为规范医疗器械的注册与备案管理,保证医疗器械的安全、有效,根据《医疗器械监督管理条例》,制定本办法。

(2) The medical device registration and filing shall be conducted under the principles of publicity, equity and justice.

医疗器械注册与备案应当遵循公开、公平、公正的原则。

(3) Where a registration applicant or filing entity of a medical device brings the products to the market in his own name, he shall take responsibilities for the product.

医疗器械注册人、备案人以自己名义把产品推向市场,对产品负法律责任。

(4) The registration applicant and filling entity of medical devices shall establish a quality management system related to product research, development and manufacture and shall keep its effective operation.

医疗器械注册申请人和备案人应当建立与产品研制、生产有关的质量管理体系,并保持有效运行。

(5) The documents for registration application or filing shall be in Chinese.

申请注册或者办理备案的资料应当使用中文。

Learning test

Choose the following words below to fill in the blanks, change the form if necessary.

applicant   procedure   department   principle   issue
entrust   coordinate   permission   progress   designate

(1) You must _____ what you say with what you do.

(2) The school has received various grants from the education _____ .

(3) The job _____ applied for the sales position.

(4) The company did not follow the correct _____ in applying for a registration certificate.

(5) Bill never did anything that went against his _____ .

(6) We must _____ a company to conduct safety inspections regularly.

(7) Each employee will be _____ with a written employment statement before appointment.

(8) The road construction work is in _____ .

(9) We are grateful to you for _____ to copy this article.

(10) Smoking is allowed in _____ areas.

Translation

1. Translate the following words into Chinese.

(1) recall   (2) filing

(3) provision　　　　　　　　(4) liability

(5) traceable　　　　　　　　(6) provision

(7) import　　　　　　　　　(8) proof

(9) deliver　　　　　　　　　(10) evaluation

2. Translate the following phrases into Chinese.

(1) after-sales　　　　　　　　(2) by reference to

(3) registration certificate　　　(4) quality management system

(5) consist of　　　　　　　　(6) with a view to

Practice

1. Translate the following sentences into Chinese.

(1) Medical device registration refers to the prescribed procedures conducted by the food and drug regulatory department upon an application submitted by the registration applicant to decide whether the medical device to be marketed can be sold based on a comprehensive evaluation of the research and results of its safety and effectiveness.

(2) To apply for filing of class Ⅰ domestic medical devices, the filing entity shall submit the filing documents to the food and drug regulatory department of the city consisting of districts.

(3) The State encourages the research and innovation of medical devices, implements special review for innovative medical devices, accelerates the popularization and application for new technologies of medical devices and promotes the development of medical device industry.

(4) Where applying for registration of a domestic medical device subject to the special review procedures for innovative medical devices and its sample production is entrusted to another enterprise, the applicant shall entrust a medical device manufacturer having corresponding production range.

(5) An overseas applicant or filing entity shall conduct related work with the support of its representative office established in China or an enterprise legal person in China designated by it as its agent.

2. Translate the following sentences into English.

(1)申请第二类、第三类医疗器械注册,应当进行注册检验。

(2)临床评价资料是指申请人或者备案人进行临床评价所形成的文件。

(3)办理第一类医疗器械备案,不需进行临床试验。

(4)第一类医疗器械生产前,应当办理产品备案。

(5)国家药品监督管理局负责全国医疗器械注册与备案的监督管理工作。

Culture salon

The forms of operation of medical devices

Medical devices are designated into different forms of operation in accordance with their intended purposes.

(1) Passive devices in terms of their form of operation can be classified as device used for transportation and storage of pharmaceutical liquid, device for alteration of blood, body fluids, medical dressing, surgical instruments; reusable surgical instruments, disposable aseptic device, implantable device, device for contraception and birth control, device for sterilization and cleaning, patient care device, in vitro diagnostic reagent, as well as other passive contacting device or passive supplementary device.

(2) Active devices in terms of their form of operation can be classified as device for treatment through energy, diagnostic monitoring, body fluids transportation and ionized radiation, laboratory instruments and medical sterilizer; as well as other active contacting device or active supplementary device.

Chapter 3-
Section 3-
Text 2-看译文

Chapter 3-Section 3-习题

Section 4 Quality management system (QMS) for medical device

Text 1 Quality management system (QMS) for medical devices

Chapter 3-
Section 4-PPT

Lead-in questions:

(1) What is the title of ISO 13485?

(2) What is the purpose of QMS regulation?

(3) What are the two major quality management system regulations?

(4) Is ISO 13485 certification acceptable in US?

(5) What are the essential elements in the QMS regulation?

The quality of medical devices has a direct impact on their effectiveness as well as the safety of patients and users. Quality management system requirements for medical devices have been internationally recognized as a way to assure product safety.

To prevent low-quality medical devices from entering the market, an effective quality

Chapter 3-
Section 4-
Text 1-听课文

management system is mandatory for medical device manufacturers before their products can enter the target market. The requirements vary in different countries according to their different regulations on medical devices. The two major regulations/standards for medical device manufacture are the ISO 13485 "Medical devices—Quality management systems: requirements for regulatory purposes" and the US FDA 21 CFR 820 regulations. The two have a great deal of similarity, and most countries have adopted modified versions of the ISO 13485 and/or the USFDA regulations.

In Europe, the requirements of ISO 13485 have been harmonized with the essential requirements of the EU's Medical Device Directive(93/42/EEC), the Directive for In-Vitro Diagnostic Medical Devices(98/79/EC) and the Directive for Active Implantable Medical Devices(90/385/EEC). Certification to ISO 13485 by an accredited certification body provides a presumption of conformity with the essential requirements of these important directives. In the U.S., the Food and Drug Administration(FDA) permits device manufacturers to submit ISO 13485 audit reports as a substitute of the evidence of compliance with its quality systems regulations(QSR). Canada also requires medical device manufacturers marketing their products to have their QMS certified to ISO 13485.

Manufacturers must establish and follow quality systems to help ensure that their products consistently meet applicable requirements and specifications. The quality systems for FDA-regulated products (food, drugs, biologics, and devices) are known as current good manufacturing practices(CGMP).

The FDA has identified in the QS regulation the essential elements that a quality system shall embody design, production and distribution, without prescribing specific ways to establish these elements. These elements include:

- training and qualification of personnel;
- product design controls;
- documentation controls;
- purchasing controls;
- identification and traceability throughout production;
- production and process;
- definitions and controls for inspection, measurement and testing equipment;
- process validation;
- product acceptance;
- nonconforming product controls;
- corrective and preventive actions and controls;
- labeling and packaging controls;
- handling, storage, distribution, and installation;
- records;
- servicing;
- statistical techniques

1. Flexibility of the QS regulation

The QS regulation embraces the same "umbrella" approach to the CGMP regulation that was the underpinning of the original CGMP regulation. Because the regulation must apply to so many different types of devices, the regulation does not prescribe in detail how a manufacturer must produce a specific device. Rather, the regulation provides the framework that all manufacturers must follow by requiring that manufacturers develop and follow procedures and fill in the details that are appropriate to a given device according to the current state-of-the-art manufacturing for that specific device.

Manufacturers should use good judgment when developing their quality system and apply those sections of the QS regulation that are applicable to their specific products and operations, 21 CFR 820. 5 of the QS regulation. Operating within this flexibility, it is the responsibility of each manufacturer to establish requirements for each type or family of devices that will result in devices that are safe and effective, and to establish methods and procedures to design, produce, distribute, etc. Devices that meet the quality system requirements. The responsibility for meeting these requirements and for having objective evidence of meeting these requirements may not be delegated even though the actual work may be delegated.

Because the QS regulation covers a broad spectrum of devices, production processes, etc. , it allows some leeway in the details of quality system elements. It is left to manufacturers to determine the necessity for, or extent of, some quality elements and to develop and implement specific procedures tailored to their particular processes and devices.

2. Applicability of the QS regulation

The QS regulation applies to finished device manufacturers who intend to commercially distribute medical devices. A finished device is defined in 21 CFR 820. 3(l) as any device or accessory that is suitable for use or capable of functioning, whether or not it is packaged, labeled, or sterilized.

Certain components such as blood tubing and diagnosticX-ray components are considered by FDA to be finished devices because they are accessories to finished devices. A manufacturer of accessories is subject to the QS regulation.

3. GMP exemptions

FDA has determined that certain types ofmedical devices are exempt from GMP requirements. These devices are exempted by FDA classification regulations published in the Federal Register and codified in 21 CFR 862 to 892. Exemption from the GMP requirements does not exempt manufacturers of finished devices from keeping complaint files(21 CFR 820. 198) or from general requirements concerning records(21 CFR 820. 180).

Medical devices manufactured under an investigational device exemption(IDE) are not exempt from design control requirements under 21 CFR 820. 30 of the QS regulation.

Words and phrases

impact	['ɪmpækt]	n.	影响
recognize	['rekəgnaɪz]	vt.	认可,承认
assure	[ə'ʃʊə(r)]	vt.	保证,确保

vary	['veəri]	vi.	变化
major	['meɪdʒə(r)]	adj.	主要的
similarity	[ˌsɪmə'lærəti]	n.	相似性
adopt	[ə'dɒpt]	vt.	采用
version	['vɜːʃn]	n.	版本
directive	[dɪ'rektɪv]	n.	指令
certification	[ˌsɜːtɪfɪ'keɪʃn]	n.	认证
accredit	[ə'kredɪt]	vt.	认可
presumption	[prɪ'zʌmpʃn]	n.	认定,推定
conformity	[kən'fɔːməti]	n.	符合,一致
permit	[pə'mɪt]	vt.	允许
audit	['ɔːdɪt]	n.	审核,审查
substitute	['sʌbstɪtjuːt]	n.	替代
evidence	['evɪdəns]	n.	证据
compliance	[kəm'plaɪəns]	n.	符合,合规
establish	[ɪ'stæblɪʃ]	vt.	建立
follow	['fɒləʊ]	vt.	遵从,遵照
consistently	[kən'sɪstəntlɪ]	adv.	一贯地,持续地
meet	[miːt]	vt.	满足,符合
specification	[ˌspesɪfɪ'keɪʃn]	n.	规格,规范
biologics	[baɪə'lɒdʒɪkz]	n.	生物制剂
element	['elɪmənt]	n.	要素
embody	[ɪm'bɒdi]	vt.	包含
prescribe	[prɪ'skraɪb]	vt.	指定,规定
training	['treɪnɪŋ]	n.	培训
qualification	[ˌkwɒlɪfɪ'keɪʃn]	n.	资格鉴定
personnel	[ˌpɜːsə'nel]	n.	人员,员工
purchase	['pɜːtʃəs]	vt.	购买,采购
traceability	[ˌtreɪsə'bɪlətɪ]	n.	可追溯性
validation	[ˌvælɪ'deɪʃn]	n.	验证,确认
embrace	[ɪm'breɪs]	vt.	包括,包含
underpinning	['ʌndəˌpɪnɪŋ]	n.	基础
framework	['freɪmwɜːk]	n.	框架
leeway	['liːweɪ]	n.	回旋余地
determine	[dɪ'tɜːmɪn]	vt.	决定,确定

tailored	['teɪləd]	adj.	定制的
applicability	[ˌæplɪk'bɪlətɪ]	n.	适用性
suitable	['suːtəbl]	adj.	合适的,适当的
codify	['kəʊdɪfaɪ]	vt.	编列,编纂
concerning	[kən'sɜːnɪŋ]	prep.	关于
quality management system			质量管理体系
a great deal of			大量的
in-vitro			体外的
low-quality			劣质的
certification body			认证机构
audit report			审核报告,审查报告
compliance with			符合,遵守
current good manufacturing practice (CGMP)			现行生产质量管理规范
state-of-the-art			最新的,最先进的
result in			引起,招致
suitable for			适合……的
such as			例如

Key sentences

(1) To prevent low-quality medical devices from entering the market, an effective quality management system is mandatory for medical device manufacturers before their products can enter the target market.

为防止低质量的医疗器械进入市场,有效的质量管理体系是医疗器械制造商在产品进入目标市场必须执行的。

(2) Manufacturers must establish and follow quality systems to help ensure that their products consistently meet applicable requirements and specifications.

制造商必须建立和遵循质量体系,以确保其产品始终符合适用的要求和规定。

(3) The FDA has identified in the QS regulation the essential elements that a quality system shall embody design, production and distribution, without prescribing specific ways to establish these elements.

FDA在质量体系法规中确定了基本要素,质量体系应包括设计,生产和销售,但并没有规定建立这些要素的具体方法。

(4) Because the QS regulation covers a broad spectrum of devices, production processes, etc., it allows some leeway in the details of quality system elements.

由于质量体系法规涵盖广泛的器械,生产流程等,因此允许在质量体系要素的细节上有一些余地。

(5) The QS regulation applies to finished device manufacturers who intend to commercially distribute medical devices.

质量体系法规适用于打算销售医疗器械的成品器械制造商。

Learning test

Choose the following words below to fill in the blanks, change the form if necessary.

determine   vary   assure   permit   meet
training   purchase   framework   adopt   follow

(1) I _____ an air ticket to Spain.

(2) Prices _____ according to the quantity ordered.

(3) These camera factors _____ the depth of focus.

(4) My boss _____ me to send him the business report next month.

(5) The salary only _____ my basic living needs, and is not acceptable.

(6) I need to _____ that my pet dog is well cared for.

(7) The company requires all the employees to pass an exam of safety _____ .

(8) Please note to read and _____ the instructions carefully.

(9) Bill has drafted a rough _____ for their story.

(10) I finally _____ his suggestions to accept the offer.

Translation

1. Translate the following words into Chinese.

(1) directive   (2) compliance

(3) element   (4) validation

(5) applicability   (6) personnel

(7) element   (8) qualification

(9) purchase   (10) establish

2. Translate the following phrases into Chinese.

(1) quality management system   (2) compliance with

(3) in-vitro   (4) current good manufacturing practice

(5) suitable for   (6) such as

Practice

1. Translate the following sentences into Chinese.

(1) The requirements vary in different countries according to their different regulations on medical devices.

(2) In the U. S., the Food and Drug Administration(FDA) permits device manufacturers to submit ISO 13485 audit reports as a substitute of the evidence of compliance with its quality systems regulations(QSR).

(3) Operating within this flexibility, it is the responsibility of each manufacturer to establish requirements for each type or family of devices that will result in devices that are safe and effective, and to establish methods and procedures to design, produce, distribute, etc. devices that meet the quality system requirements.

(4) It is left to manufacturers to determine the necessity for, or extent of, some quality elements and to

develop and implement specific procedures tailored to their particular processes and devices.

(5) Certain components such as blood tubing and diagnostic X-ray components are considered by FDA to be finished devices because they are accessories to finished devices.

2. Translate the following sentences into English.

(1) ISO 13485 为医疗器械的质量达到安全有效起到了很好的促进作用。

(2) ISO 13485 认证有利于增强产品的竞争力，提高产品的市场占有率。

(3) 医疗器械生产企业的员工应定期进行培训，提高职业技能。

(4) 在提出认证申请前的一年内，申请组织的产品无重大顾客投诉及质量事故。

(5) ISO 13485 最新版已于 2016 年发布。

> ### Culture salon
>
> #### What is a quality management system?
>
> A quality management system, often called a QMS, is a set of internal rules that are used to define a collection of policies, processes, documented procedures and records. This system defines how a company will achieve the manufacture and delivery of the product or service they provide to their customers. When implemented in your company, the QMS needs to be specific to the product or service you provide, so it is important to tailor it to your needs. However, in order to help ensure that you do not miss elements of a good system, some general guidelines exist in the form of ISO 13485, which is intended to help standardize how a QMS is designed.

Chapter 3-Section 4-Text 1-看译文

Text 2   ISO 13485：2003 Medical devices —quality management systems — requirements for regulatory purposes

Lead-in questions：

(1) What's the scope of this standard?

(2) What are the possible factors to influence the design and implementation of an organization's quality management system?

(3) What's the relationship of ISO 13485 with ISO 9001?

(4) What are the general requirements of quality management system?

(5) What are the documentation requirements for the content?

1. General

This international standard specifies requirements for a quality management system that can be used by an organization for the design and development, production, installation and servicing of medical devices, and the design, development, and provision of related services.

It can also be used by internal and external parties, including certification bodies, to assess the organization's ability to meet customer and regulatory requirements.

It is emphasized that the quality management system requirements specified in this international standard are complementary to technical requirements for products.

Chapter 3-
Section 4-
Text 2-听课文

The adoption of a quality management system should be a strategic decision of an organization. The design and implementation of an organization's quality management system is influenced by varying needs, particular objectives, the products provided, the processes employed and the size and structure of the organization. It is not the intent of this international standard to imply uniformity in the structure of quality management systems or uniformity of documentation.

There is a wide variety of medical devices and some of the particular requirements of this international standard only apply to named groups of medical devices. These groups are defined in Clause 3.

2. Process approach

This international standard is based on a process approach to quality management.

Any activity that receives inputs and converts them to outputs can be considered as a process.

For an organization to function effectively, it has to identify and manage numerous linked processes.

Often the output from one process directly forms the input to the next. The application of a system of processes within an organization, together with the identification and interactions of these processes, and their management, can be referred to as the "process approach".

3. Relationship with other standards

(1) **Relationship with ISO 9001**

While this is a stand-alone standard, it is based on ISO 9001. Those clauses or subclauses that are quoted directly and unchanged from ISO 9001 are in normal font. The fact that these subclauses are presented unchanged is noted in Annex B.

Where the text of this international standard is not identical to the text of ISO 9001, the sentence or indent containing that text as a whole is shown in italics (in blue italics for electronic versions). The nature and reasons for the text changes are noted in Annex B.

(2) **Relationship with ISO/TR 14969**

ISO/TR 14969 is a technical report intended to provide guidance for the application of ISO 13485.

(3) **Compatibility with other management systems**

This international standard follows the format of ISO 9001 for the convenience of users in the medical device community.

This international standard does not include requirements specific to other management systems, such as those particular to environmental management, occupational health and safety management, or financial management.

However, this International Standard enables an organization to align or integrate its own quality management system with related management system requirements. It is possible for an organization to adapt its existing management system(s) in order to establish a quality management system that complies with the requirements of this International Standard.

4. Quality management system

(1) General requirements

The organization shall establish, document, implement and maintain a quality management system and maintain its effectiveness in accordance with the requirements of this international standard.

The organization shall

1) identify the processes needed for the quality management system and their application throughout the organization;

2) determine the sequence and interaction of these processes;

3) determine criteria and methods needed to ensure that both the operation and control of these processes are effective;

4) ensure the availability of resources and information necessary to support the operation and monitoring of these processes;

5) monitor, measure and analyze these processes;

6) implement actions necessary to achieve planned results and maintain the effectiveness of these processes.

These processes shall be managed by the organization in accordance with the requirements of this International Standard.

Where an organization chooses to outsource any process that affects product conformity with requirements, the organization shall ensure control over such processes. Control of suchoutsourced processes shall be identified within the quality management system.

NOTE Processes needed for the quality management system referred to above should include processes for management activities, provision of resources, product realization and measurement.

(2) Documentation requirements

The quality management system documentation shall include:

1) documented statements of a quality policy and quality objectives;

2) a quality manual;

3) documented procedures required by this International Standard;

4) documents needed by the organization to ensure the effective planning, operation and control of its processes;

5) records required by this International Standard;

6) any other documentation specified by national or regional regulations.

Where this international standard specifies that a requirement, procedure, activity or special arrangement be "documented", it shall, in addition, be implemented and maintained.

For each type or model of medical device, the organization shall establish and maintain a file either containing or identifying documents defining product specifications and quality management system requirements. These documents shall define the complete manufacturing process and, if applicable, installation and servicing.

NOTE 1 The extent of the quality management system documentation can differ from one organization to another due to

1) the size of the organization and type of activities;

2) the complexity of processes and their interactions;

3) the competence of personnel.

NOTE 2 The documentation can be in any form or type of medium.

Words and phrases

Chapter 3-Section 4-Text 2-听词汇

specify	[ˈspesɪfaɪ]	vt.	规定,详述
installation	[ˌɪnstəˈleɪʃn]	n.	安装
service	[ˈsɜːvɪs]	vt.	向……提供服务
provision	[prəˈvɪʒn]	n.	供应,提供
internal	[ɪnˈtɜːnl]	adj.	内部的
external	[ɪkˈstɜːnl]	adj.	外部的
emphasize	[ˈemfəsaɪz]	vt.	强调
complementary	[ˌkɒmplɪˈmentri]	adj.	互补的,补充的
influence	[ˈɪnfluəns]	vt.	影响
objective	[əbˈdʒektɪv]	n.	目标
size	[saɪz]	n.	规模
uniformity	[ˌjuːnɪˈfɔːməti]	n.	统一,一致性
named	[neɪmd]	adj.	指定的
define	[dɪˈfaɪn]	vt.	定义,规定
clause	[klɔːz]	n.	条款
input	[ˈɪnpʊt]	n.	输入
convert	[kənˈvɜːt]	vt.	转变,转化
output	[ˈaʊtpʊt]	n.	输出
function	[ˈfʌŋkʃn]	vi.	运作,运转
identify	[aɪˈdentɪfaɪ]	vt.	识别
numerous	[ˈnjuːmərəs]	adj.	很多的,众多的
interaction	[ˌɪntərˈækʃn]	n.	相互作用
relationship	[rɪˈleɪʃnʃɪp]	n.	关系
quote	[kwəʊt]	vt.	引用
identical	[aɪˈdentɪkl]	adj.	相同的
guidance	[ˈɡaɪdns]	n.	指南

compatibility	[kəmˌpætəˈbɪləti]	n.	兼容性,相容性
convenience	[kənˈviːniəns]	n.	方便
community	[kəˈmjuːnəti]	n.	团体
occupational	[ˌɒkjuˈpeɪʃənl]	adj.	职业的
align	[əˈlaɪn]	vt.	对齐
integrate	[ˈɪntɪɡreɪt]	vt.	整合,结合
existing	[ɪɡˈzɪstɪŋ]	adj.	现存的,目前的
sequence	[ˈsiːkwəns]	n.	顺序
criteria	[kraɪˈtɪəriə]	n.	标准,准则
availability	[əˌveɪləˈbɪləti]	n.	可获得性,可利用性
resource	[rɪˈsɔːs]	n.	资源
monitor	[ˈmɒnɪtə(r)]	n.	监控,监视
achieve	[əˈtʃiːv]	vt.	实现,取得
outsource	[ˈaʊtsɔːs]	vt.	外包
documentation	[ˌdɒkjumenˈteɪʃn]	n.	文档,形成文件
maintain	[meɪnˈteɪn]	vt.	保持
certification body			认证机构
complementary to			与……互补
identical to			与……相同
for the convenience of			为了方便……
align with			与……结合
integrate with			与……整合
comply with			遵守
in accordance with			根据,按照
in addition			此外,另外

Key sentences

(1) This international standard specifies requirements for a quality management system that can be used by an organization for the design and development, production, installation and servicing of medical devices, and the design, development, and provision of related services.

本标准规定了质量管理体系要求,组织可依此要求进行医疗器械的设计和开发、生产、安装和服务以及相关服务的设计、开发和提供。

(2) This international standard is based on a process approach to quality management.

本国际标准以质量管理的过程方法为基础。

(3) While this is a stand-alone standard, it is based on ISO 9001.

本标准是一个以 ISO 9001 为基础的独立标准。

（4）The organization shall establish, document, implement and maintain a quality management system and maintain its effectiveness in accordance with the requirements of this international standard.

组织应按本标准的要求建立,形成文件,实施和保持质量管理体系,并保持其有效性。

（5）For each type or model of medical device, the organization shall establish and maintain a file either containing or identifying documents defining product specifications and quality management system requirements.

组织应对医疗器械的每一类型或型号建立和保持一套文件,包含或识别规定产品规格和质量体系要求的文件定义。

Learning test

Choose the following words below to fill in the blanks, change the form if necessary.

maintain   service   influence   guidance   resource
criteria   relationship   convenience   size   monitor

（1）The principles outlined above provide technical _____ for system development teams.

（2）Push yourself to make friends and to _____ the friendships.

（3）There are now 100 staff at headquarters to _____ our regional work.

（4）I wanted to build a close _____ with my team.

（5）Water is becoming an increasingly precious _____ .

（6）The sales' suggestions _____ my final decision.

（7）Most mainland medical device companies are small in _____ .

（8）You need to _____ your blood pressure at home regularly.

（9）The productperformance _____ should be high enough to guarantee their quality.

（10）Please reply at your earliest _____ .

Translation

1. Translate the following words into Chinese.

（1）specify   （2）uniformity

（3）compatibility   （4）outsource

（5）input   （6）identify

（7）documentation   （8）input

（9）define   （10）achieve

2. Translate the following phrases into Chinese.

（1）certification body   （2）complementary to

（3）identical to   （4）for the convenience of

（5）in addition   （6）comply with

Practice

1. Translate the following sentences into Chinese.

（1）It can also be used by internal and external parties, including certification bodies, to assess the organization's ability to meet customer and regulatory requirements.

(2) It is emphasized that the quality management system requirements specified in this International Standard are complementary to technical requirements for products.

(3) This international standard does not include requirements specific to other management systems, such as those particular to environmental management, occupational health and safety management, or financial management.

(4) Processes needed for the quality management system referred to above should include processes for management activities, provision of resources, product realization and measurement.

(5) Where this international standard specifies that a requirement, procedure, activity or special arrangement be "documented", it shall, in addition, be implemented and maintained.

2. Translate the following sentences into English.

(1)质量管理体系所要求的文件应予以控制。

(2)组织应确保文件的更改得到原审批部门的评审和批准。

(3)生产企业应确保顾客的要求得到满足。

(4)本标准的所有要求是针对提供医疗器械的组织,不论组织的类型或规模。

(5)所有文件应标注日期和版本。

Culture salon

About ISO standard

ISO (the International Organization for Standardization) is a worldwide federation of national standards bodies (ISO member bodies). The work of preparing International Standards is normally carried out through ISO technical committees. Each member body interested in a subject for which a technical committee has been established has the right to be represented on that committee. International organizations, governmental and non-governmental, in liaison with ISO, also take part in the work. ISO collaborates closely with the International Electrotechnical Commission (IEC) on all matters of electrotechnical standardization.

International Standards are drafted in accordance with the rules given in the ISO/IEC Directives, Part 2. The main task of technical committees is to prepare International Standards. Draft International Standards adopted by the technical committees are circulated to the member bodies for voting. Publication as an International Standard requires approval by at least 75% of the member bodies casting a vote.

Chapter 3-Section 4-Text 2-看译文

Chapter 3-Section 4-习题

Chapter 4

Medical device standards and practices

Learning objective

In this chapter, the use of medical device standards and their applications in practices are introduced. Read the following passage, you should meet the following requirements:

(1) Grasp basic concepts, key points, involved professional terms and phrases.

(2) Acquire representative biological, chemical and electrical test methods involved in relevant standards.

(3) Understand how to use relevant standards in variant application cases.

Section 1 Medical device standards

Chapter 4-Section 1-PPT

Text 1 ISO 10993-5: Test for in vitro cytotoxicity

Lead-in questions:

(1) What is the definition of negative control?

(2) What are the extraction vehicles for mammalian cell assays?

(3) What are the extraction conditions used?

(4) What are the two means to determine cytotoxic effects?

(5) How to determine cytotoxicity by qualitative evaluation?

Chapter 4-Section 1-Text 1-听课文

This test allows both qualitative and quantitative assessment of cytotoxicity. A minimum of three replicates shall be used for test samples and controls. Neutral red uptake (NRU) cytotoxicity test, colony formation cytotoxicity test, MTT cytotoxicity test and XTT cytotoxicity test may be used for the quantitative determination of cytotoxicity of extracts. The numerous methods used and endpoints measured in cytotoxicity determination can be grouped into the following categories of evaluation: assessments of cell damage by morphological means, measurements of cell damage, measurements of cell growth, measurements of specific aspects of cellular metabolism. Generally the test shall be performed on an extract of the test sample and/or the test sample itself.

1. Terms and definitions

(1) culture vessels:

Vessels appropriate for cell culture including glass petri dishes, plastic culture flasks or plastic multi-wells and microtiter plates.

(2) positive control material

Material which, when tested in accordance with this part of ISO 10993, provides a reproducible cytotoxic response.

(3) blank

Extraction vehicle not containing the test sample, retained in a vessel identical to that which holds the test sample and subjected to conditions identical to those to which the test sample is subjected during its extraction.

(4) negative control

Material which, when tested in accordance with this part of ISO 10993, does not produce a cytotoxic response.

(5) test sample

Material, device, device portion, component, extract or portion thereof that is subjected to biological or chemical testing or evaluation.

(6) subconfluency

Approximately 80% confluency, i. e. the end of the logarithmic phase of growth.

2. Preparation of liquid extracts of material

(1) Principles of extraction

Extracting conditions should attempt to simulate or exaggerate the clinical use conditions so as to determine the potential toxicological hazard without causing significant changes in the test sample, such as fusion, melting or any alteration of the chemical structure, unless this is expected during clinical application. Due to the nature of certain materials(e. g. biodegradable materials), alteration of the chemical structure can occur during the extraction procedure.

For devices that involve mixing two or more components in the patient to arrive at the final device(for example bone cement), the final device should not be washed prior to extraction. Washing the test sample can reduce or remove residuals present on the device. If the test sample is to be used in a sterile environment, a sterilized test sample should be used to extract chemical constituents.

(2) Extraction vehicle

The choice of the extraction vehicle(s) taking into account the chemical characteristics of the test sample shall be justified and documented. For mammalian cell assays one or more of the following vehicles shall be used:

1) culture medium with serum;

2) physiological saline solution;

3) other suitable vehicle.

The choice of vehicle should reflect the aim of the extraction. Consideration shall be given to the use of both a polar and a non-polar vehicle. Culture medium with serum is the preferred extraction vehicle. The use of culture medium with serum is preferred for extraction because of its ability to support cellular growth as well as extract both polar and non-polar substances. In addition to culture medium with serum, use of medium without serum should be considered in order to specifically extract polar substances(e. g. ionic compounds). Other suitable vehicles include purified water and dimethyl sulfoxide(DMSO). DMSO is cytotoxic in selected assay systems at greater than 0. 5%(volume fraction). The cellular exposure concentration of extractables in DMSO will be lower due to the greater dilution as compared to extraction in culture medium with serum.

NOTE 1 Different types of serum(e. g. foetal, bovine/calf serum, newborn calf serum) might be used and the choice of the serum is dependent on the cell type.

NOTE 2 It is important to recognize that serum/proteins are known to bind, to some extent, extractables.

(3) Extraction conditions

With the exception of circumstances given below, the extraction shall be conducted under one of the following conditions and shall be applied according to the device characteristics and specific conditions for use:

1) (24 ± 2) h at (37 ± 1) ℃;

2) (72 ± 2) h at (50 ± 2) ℃;

3) (24 ± 2) h at (70 ± 2) ℃;

4) (1 ± 0.2) h at (121 ± 2) ℃.

Cell culture medium with serum should only be used in accordance with 1) because extraction temperatures greater than (37 ± 1) ℃ can adversely impact chemistry and/or stability of the serum and other constituents in the culture medium. For polymeric test samples, the extraction temperature should not exceed the glass transition temperature as the higher temperature can change the extractant composition.

3. Determination of cytotoxicity

Determine cytotoxic effects by either qualitative or quantitative means. Quantitative evaluation of cytotoxicity is preferable. Qualitative means are appropriate for screening purposes.

Qualitative evaluation: Examine the cells microscopically using cytochemical staining if desired. Assess changes in, for example, general morphology, vacuolization, detachment, cell lysis and membrane integrity.

Quantitative evaluation: Measure cell death, inhibition of cell growth, cell proliferation or colony formation. The number of cells, amount of protein, release of enzymes, release of vital dye, reduction of vital dye or any other measurable parameter may be quantified by objective means. Reduction of cell viability by more than 30% is considered a cytotoxic effect. Other criteria, including different cut-off points or an acceptable ratio of test-to-control result shall be justified for alternate cell lines or multi-layered tissue constructs. The criteria shall be justified and documented.

Words and phrases

standard	[ˈstændəd]	n.	标准
test	[test]	n.	试验,测试
cytotoxicity	[sɪtəʊtɒkˈsɪsɪtɪ]	n.	细胞毒性
qualitative	[ˈkwɒlɪtətɪv]	adj.	定性的
quantitative	[ˈkwɒntɪtətɪv]	adj.	定量的
assessment	[əˈsesmənt]	n.	评定
replicate	[ˈreplɪkeɪt]	n.	重复
numerous	[ˈnjuːmərəs]	adj.	大量的,许多的
endpoint	[ˈendpɔɪt]	n.	终点
evaluation	[ɪˌvæljuˈeɪʃn]	n.	评价
morphological	[ˌmɔːfəˈlɒdʒɪkl]	adj.	形态学的,形态的
means	[miːnz]	n.	方法
damage	[ˈdæmɪdʒ]	n.	损伤
aspect	[ˈæspekt]	n.	方面,方向
cellular	[ˈseljʊlə]	adj.	细胞的
metabolism	[məˈtæbəlɪzəm]	n.	代谢
extract	[ˈekstrækt]	n.	提取物
culture	[ˈkʌltʃə(r)]	n.	培养
vessel	[ˈvesl]	n.	器皿
appropriate	[əˈprəʊprɪət]	adj.	适当的
reproducible	[ˌriːprəˈdjuːsəbl]	adj.	可重现的
subconfluency	[sʌbˈkɒnfluənsɪ]	n.	近汇合
vehicle	[ˈviːəkl]	n.	介质
preparation	[ˌprepəˈreɪʃn]	n.	制备
principle	[ˈprɪnsəpl]	n.	原则
simulate	[ˈsɪmjuleɪt]	vt.	模拟
exaggerate	[ɪɡˈzædʒəreɪt]	vt.	使扩大,使增大
fusion	[ˈfjuːʒn]	n.	熔化
sterile	[ˈsteraɪl]	adj.	无菌的
mammalian	[mæˈmeɪlɪən]	adj.	哺乳动物的

Chapter 4-
Section 1-
Text 1-听词汇

dilution	[daɪˈluːʃn]	n.	稀释,冲淡
foetal	[ˈfiːtl]	adj.	胎儿的
bovine	[ˈbəʊvaɪn]	adj.	牛的
cytochemical	[saɪtəˈkemɪkəl]	adj.	细胞化学的
morphology	[mɔːˈfɒlədʒɪ]	n.	形态
vacuolization	[vækjuəʊlaɪˈzeɪʃən]	n.	空泡形成
detachment	[dɪˈtætʃmnt]	n.	脱落
integrity	[ɪnˈtegrəti]	n.	完整性
inhibition	[ˌɪnhɪˈbɪʃn]	n.	抑制
proliferation	[prəˌlɪfəˈreɪʃn]	n.	繁殖
document	[ˈdɒkjumənt]	vt.	记录

in vitro	体外
culture vessel	培养器皿
petri dish	培养皿
plastic culture flask	塑料培养瓶
multiwall plate	多孔培养板
microtiter plate	微量滴定板
positive control material	阳性对照材料
negative control	阴性对照
reagent control	试剂对照
test sample	试样
logarithmic phase of growth	对数生长期
liquid extract	浸提液
toxicological harzard	毒理学危害
biodegradable materials	生物降解材料
culture medium	培养基
physiological saline solution	生理盐水溶液
dimethyl sulfoxide(DMSO)	二甲基亚砜
volume fraction	体积分数
assay system	测试系统
Newborn calf serum	新生牛血清
qualitative evaluation	定性评价
Quantitative evaluation	定量评价
cell lysis	细胞溶解
vital dye	活体染料

Key sentences

(1) This test allows both qualitative and quantitative assessment of cytotoxicity. A minimum of three replicates shall be used for test samples and controls.

本测试方法可进行细胞毒性的定性和定量评价。测试样品和对照物应至少重复三次。

(2) Extracting conditions should attempt to simulate or exaggerate the clinical use conditions so as to determine the potential toxicological hazard without causing significant changes in the test sample, such as fusion, melting or any alteration of the chemical structure, unless this is expected during clinical application.

为了测定潜在的毒理学危害,浸提条件应模拟或严于临床使用条件,但不应导致试验材料发生诸如熔化、溶解或化学结构改变等明显变化,除非这是临床应用中所希望的。

(3) The use of culture medium with serum is preferred for extraction because of its ability to support cellular growth as well as extract both polar and non-polar substances.

应优先选择含血清培养基来浸提,因为它具有支持细胞生长、同时提取极性和非极性物质的能力。

(4) For polymeric test samples, the extraction temperature should not exceed the glass transition temperature as the higher temperature can change the extractant composition.

对高分子试验样品,浸提温度不应超过其玻璃化转变温度,因为高温会改变提取的成分。

(5) Qualitative evaluation: Examine the cells microscopically using cytochemical staining if desired. Assess changes in, for example, general morphology, vacuolization, detachment, cell lysis and membrane integrity.

定性评价:用显微镜检查细胞(如果需要,使用细胞化学染色),评价诸如一般形态、空泡形成、脱落、细胞溶解和膜完整性等方面的变化。

Learning test

Choose the following words below to fill in the blanks, change the form if necessary.

qualitative   test   proliferation   cytotoxicity   means
numerous   simulate   document   principle   dilution

(1) Analytical chemistry encompasses _____ analysis and quantitative analysis.

(2) The standard GB/T 16886.11 describes systemic toxicity _____ methods.

(3) Many genes take part in the regulation of cell _____ and differentiation.

(4) They also carried out _____ tests using human cells.

(5) All possible _____ have been tried to solve the problems.

(6) _____ marriages now end in divorce.

(7) Games may _____ reality to attract players.

(8) He wrote a book _____ his working experiences in IBM.

(9) We have to defend the _____ of freedom and equality.

(10) Serial _____ is performed to reduce the sample concentration.

Translation

1. Translate the following words into Chinese.

(1) culture   (2) morphology

(3) numerous   (4) quantitative

(5) cytotoxicity   (6) inhibition

(7) dilution   (8) proliferation

(9) evaluation   (10) vehicle

2. Translate the following phrases into Chinese.

(1) cell lysis   (2) negative control

(3) culture vessel   (4) liquid extract

(5) volume fraction   (6) physiological saline solution

Practice

1. Translate the following sentences into Chinese.

(1) Determine cytotoxic effects by either qualitative or quantitative means.

(2) The choice of vehicle should reflect the aim of the extraction.

(3) Washing the test sample can reduce or remove residuals present on the device.

(4) Generally thetest shall be performed on an extract of the test sample and/or the test sample itself.

(5) The criteria shall be justified and documented.

2. Translate the following sentences into English.

(1)无菌医疗器械的细胞毒性测试应符合要求。

(2)实验条件和结果应该详细记录。

(3)一般来说定量测试比定性测试复杂。

(4)浸提介质的选择会直接影响实验的结果。

(5)浸提液不能与被测样品发生化学反应。

Culture salon

Sample required for direct-contact tests

Once the direct-contact tests have been chose, sterility of the test sample shall be taken into account. The preferred test sample of a solid material should have at least one flat surface. If not, adjustments shall be made to achieve flat surfaces. Liquid test samples shall be tested by either direct deposition or deposition on a biologically inert absorbent matrix. If appropriate, test samples that are absorbent shall be soaked with culture medium prior to testing to prevent adsorption of the culture medium in the testing vessel.

No matter which kind of tests were chose, controls should be selected so that they can be prepared by the same procedure as the test sample.

Chapter 4-
Section 1-
Text 1-看译文

Text 2 Heavy metals impurities test methods in United States pharmacopeia

Lead-in questions:

(1) What is the purpose of heavy metals impurities test?

(2) How many test methods to determine the amount of heavy metals?

(3) When can Method I be used to determine the amount of heavy metals?

(4) How to prepare Lead nitrate stock solution?

(5) How to prepare pH 3.5 Acetate buffer?

This test is provided to demonstrate that the content of metallic impurities that are colored by sulfide ion, under the specified test conditions, does not exceed the *Heavy metals* limit specified in the individual monograph in percentage (by weight) of lead in the test substance, as determined by concomitant visual comparison (see *Visual Comparison* in the section *Procedure* under *Spectrophotometry and Light-Scattering* <851>) with a control prepared from a *Standard Lead Solution*. (NOTE—Substances that typically will respond to this test are lead, mercury, bismuth, arsenic, antimony, tin, cadmium, silver, copper, and molybdenum.)

Chapter 4-
Section 1-
Text 2-听课文

Determine the amount of heavy metals by *Method I*, unless otherwise specified in the individual monograph. *Method I* is used for substances that yield clear, colorless preparations under the specified test conditions. *Method II* is used for substances that do not yield clear, colorless preparations under the test conditions specified for *Method I*, or for substances that, by virtue of their complex nature, interfere with the precipitation of metals by sulfide ion, or for fixed and volatile oils. *Method III*, a wet-digestion method, is used only in those cases where neither *Method I* nor *Method II* can be used.

Special Reagents

Lead Nitrate Stock Solution—Dissolve 159.8 mg of lead nitrate in 100 ml of water to which has been added 1 ml of nitric acid, then dilute with water to 1000 ml. Prepare and store this solution in glass containers free from soluble lead salts.

Standard Lead Solution—On the day of use, dilute 10.0 ml of *Lead Nitrate Stock Solution* with water to 100.0 ml. Each mL of *Standard Lead Solution* contains the equivalent of 10 μg of lead. A comparison solution prepared on the basis of 100 μL of *Standard Lead Solution* per g of substance being tested contains the equivalent of 1 part of lead per million parts of substance being tested.

Method I

pH 3.5 Acetate Buffer — Dissolve 25.0 g of ammonium acetate in 25 ml of water, and add 38.0 ml of 6 N hydrochloric acid. Adjust, if necessary, with 6 N ammonium hydroxide or 6 N hydrochloric acid to a pH of 3.5, water to 100 ml, and mix.

Standard Preparation — Into a 50-ml color-comparison tube pipet 2 ml of *Standard Lead Solution* (20 μg of Pb), and dilute with water to 25 ml. Using a pH meter or short-range pH indicator paper as external indicator, adjust with 1 N acetic acid or 6 N ammonium hydroxide to a pH between 3.0 and 4.0, dilute with water to 40 ml, and mix.

Test Preparation — Into a 50-ml color-comparison tube place 25 ml of the solution prepared for the test as directed in the individual monograph; or, using the designated volume of acid where specified in the individual monograph, dissolve in and dilute with water to 25 ml the quantity, in g, of the substance to be tested, as calculated by the formula:

$$2.0/(1000L)$$

in which L is the *Heavy metals* limit, as a percentage. Using a pH meter or short-range pH indicator paper as external indicator, adjust with 1 N acetic acid or 6 N ammonium hydroxide to a pH between 3.0 and 4.0, dilute with water to 40 ml, and mix.

Monitor Preparation — Into a third 50-ml color-comparison tube place 25 ml of a solution prepared as directed for *Test Preparation*, and add 2.0 ml of *Standard Lead Solution*. Using a pH meter or short-range pH indicator paper as external indicator, adjust with 1 N acetic acid or 6 N ammonium hydroxide to a pH between 3.0 and 4.0, dilute with water to 40 ml, and mix.

Procedure — To each of the three tubes containing the *Standard Preparation*, the *Test Preparation*, and the *Monitor Preparation*, add 2 ml of pH 3.5 *Acetate Buffer*, then add 1.2 ml of thioacetamide-glycerin base TS, dilute with water to 50 ml, mix, allow to stand for 2 minutes, and view downward over a white surface: the color of the solution from the *Test Preparation* is not darker than that of the solution from the *Standard Preparation*, and the color of the solution from the *Monitor Preparation* is equal to or darker than that of the solution from the *Standard Preparation*. (NOTE—If the color of the *Monitor Preparation* is lighter than that of the *Standard Preparation*, use *Method* II instead of *Method* I for the substance being tested.)

Words and phrases

pharmacopedia	[fɑ:məkəʊˈpedɪə]	n.	药典
provide	[prəˈvaɪd]	vt.	提供
demonstrate	[ˈdemənstreɪt]	vt.	确证,证明,证实
content	[ˈkɒntent]	n.	含量
metallic	[mɪˈtælɪk]	adj.	金属的
impurity	[ɪmˈpjʊərɪtɪ]	n.	杂质
color	[ˈkʌlə(r)]	vt.	显色
sulfide	[ˈsʌlfaɪd]	n.	硫化物
specify	[ˈspesɪfaɪ]	vt.	规定,指定
exceed	[ɪkˈsi:d]	vt.	超过

Chapter 4-
Section 1-
Text 2-听词汇

limit	[ˈlɪmɪt]	n.	限度,限量
individual	[ˌɪndɪˈvɪdʒuəl]	adj.	单个的,独个的
monograph	[ˈmɒnəgrɑːf]	n.	专论,专题
percentage	[pəˈsentɪdʒ]	n.	百分比,百分率
lead	[liːd]	n.	铅
concomitant	[kənˈkɒmɪt(ə)nt]	adj.	相伴的,共存的
visual	[ˈvɪʒjuəl]	adj.	视觉的,视力的
procedure	[prəˈsiːdʒə]	n.	步骤
spectrophotometry	[spektrəʊfəʊˈtɒmɪtrɪ]	n.	分光光度法
prepare	[prɪˈpeə(r)]	vt.	制备
bismuth	[ˈbɪzməθ]	n.	铋
arsenic	[ˈɑːsnɪk]	n.	砷
antimony	[ˈæntɪmənɪ]	n.	锑
tin	[tɪn]	n.	锡
cadmium	[ˈkædmɪəm]	n.	镉
molybdenum	[məˈlɪbdənəm]	n.	钼
yield	[jiːld]	vt.	产生
precipitation	[prɪˌsɪpɪˈteɪʃn]	n.	沉淀
volatile	[ˈvɒlətaɪl]	adj.	挥发性的,易挥发的
digestion	[daɪˈdʒestʃən]	n.	消化
nitrate	[ˈnaɪtreɪt]	n.	硝酸盐
dissolve	[dɪˈzɒlv]	vt.	溶解
dilute	[daɪˈluːt]	v.	稀释
soluble	[ˈsɒljəbl]	adj.	可溶的
equivalent	[ɪˈkwɪvələnt]	n.	当量
buffer	[ˈbʌfə(r)]	n.	缓冲液
ammonium	[əˈməʊnɪəm]	n.	铵,氨盐基
acetate	[ˈæsɪteɪt]	n.	醋酸盐
pipet	[pɪˈpet]	vt.	用移液管移,精密量取
indicator	[ˈɪndɪkeɪtə]	n.	指示剂
designate	[ˈdezɪgneɪt]	vt.	规定,指定
thioacetamide	[θaɪəʊæsɪˈtæmɪd]	n.	硫代乙酰胺
downward	[ˈdaʊnwəd]	adv.	向下地
sulfide ion			硫离子
test condition			试验条件

visual comparison	目视比较
light-scattering	光散射
by virtue of	由于,凭借
wet-digestion method	湿消化法
stock solution	储备溶液,储溶液
free from	无……,不含
ammonium acetate	醋酸铵
hydrochloric acid	盐酸
standard lead solution	铅标准溶液
acetate buffer	醋酸盐缓冲液
color-comparison tube	比色管
pH meter	pH 计
acetic acid	乙酸,醋酸
ammonium hydroxide	氢氧化铵

Key sentences

(1) Substances that typically will respond to this test are lead, mercury, bismuth, arsenic, antimony, tin, cadmium, silver, copper, and molybdenum.

对本试验有相应的典型物质有:铅、汞、铋、砷、锑、锡、镉、银、铜和钼等。

(2) Dissolve 159.8mg of lead nitrate in 100ml of water to which has been added 1ml of nitric acid, then dilute with water to 1000ml.

将 1ml 硝酸加入 100ml 水中,然后取 159.8mg 的硝酸铅溶解于水中,用水稀释至 1000ml。

(3) Using a pH meter or short-range pH indicator paper as external indicator, adjust with 1 N acetic acid or 6 N ammonium hydroxide to a pH between 3.0 and 4.0, dilute with water to 40ml, and mix.

采用 pH 计或短程 pH 指示剂作为外指示剂,用 1 N 的醋酸或 6 N 的氢氧化铵调节 pH 到 3.0 至 4.0 之间,加水稀释至 40ml,混匀。

(4) Into a third 50-ml color-comparison tube place 25ml of a solution prepared as directed for *Test Preparation*, and add 2.0ml of *Standard Lead Solution*.

取 25ml 按规定制备的供试液加入到第 3 支 50ml 比色管中,再加入 2.0ml 铅标准溶液。

(5) If the color of the *Monitor Preparation* is lighter than that of the *Standard Preparation*, use *Method II* instead of *Method I* for the substance being tested.

如果供试品溶液的颜色比铅标准溶液的颜色浅,用方法二代替方法一进行测定供试品。

Learning test

Choose the following words below to fill in the blanks, change the form if necessary.

prepare	specify	downward	pipet	dilute
content	comparison	dissolve	percentage	exceed

(1) Could you please _____ the hydrochloric acid solution for me?

(2) The _____ of heavy metal impurities in the liquid extract can be tested first.

(3) Do you know the _____ test temperature for the reaction?

(4) To prepare the physiological saline solution, _____ 0.9g of NaCl in 100 ml of water.

(5) _____ the stock solution with water to reduce its concentration before use.

(6) Look _____, you will view a wonderful scenery.

(7) Visual _____ is not accurate, it's just a semi-quantitative method.

(8) Can you teach me how to _____ 2 ml of water to the tube?

(9) Exports yearly _____ imports.

(10) The content of NaCl in physiological saline solution is 0.9% in _____ by weight.

Translation

1. Translate the following words into Chinese.

(1) impurity   (2) prepare

(3) lead   (4) soluble

(5) buffer   (6) indicator

(7) designate   (8) dissolve

(9) acetate   (10) limit

2. Translate the following phrases into Chinese.

(1) test condition   (2) acetate buffer

(3) standard lead solution (4) visual comparison

(5) nitric acid   (6) stock solution

Practice

1. Translate the following sentences into Chinese.

(1) Method Ⅰ is used for substances that yield clear, colorless preparations under the specified test conditions.

(2) Prepare and store this solution in glass containers free from soluble lead salts.

(3) On the day of use, dilute 10.0ml of *Lead Nitrate Stock Solution* with water to 100.0ml.

(4) Each mL of *Standard Lead Solution* contains the equivalent of $10\mu g$ of lead.

(5) Dissolve 25.0g of ammonium acetate in 25ml of water, and add 38.0ml of 6 N hydrochloric acid.

2. Translate the following sentences into English.

(1) 称取2g氯化钠,加入到100ml水中溶解。

(2) 5分钟后,在白色背景下将样品溶液与标准溶液进行对比。

(3) 用精密pH计测试溶液的pH.

(4) 铅标准溶液比色法是一种半定量的重金属含量测试方法。

(5) 分光光谱法可以定量测试溶液中重金属离子的浓度。

Culture salon

A brief introduction to USP

The United States Pharmacopeia (USP) is a pharmacopeia (compendium of drug information) for the United States published annually by the United States Pharmacopeial Convention (usually also called the USP), a nonprofit organization that owns the trade mark and copyright. The USP is published in a combined volume with the National Formulary (a formulary) as the USP-NF. If a drug ingredient or drug product has an applicable USP quality standard (in the form of a USP-NF monograph), it must conform in order to use the designation "USP" or "NF." Drugs subject to USP standards include both human drugs (prescription, over-the-counter, or otherwise), as well as animal drugs. USP-NF standards also have a role in U. S. federal law; a drug or drug ingredient with a name recognized in USP-NF is deemed adulterated if it does not satisfy compendial standards for strength, quality or purity. USP also sets standards for dietary supplements, and food ingredients (as part of the Food Chemicals Codex). USP has no role in enforcing its standards; enforcement is the responsibility of FDA and other government authorities in the U. S. and elsewhere.

Chapter 4-
Section 1-
Text 2-看译文

Text 3   IEC 60601-1-2-2014: Medical electrical equipment-General requirements for basic safety and essential performance

Lead-in questions: √

(1) What is the scope of this standard?

(2) Are the requirements and tests applicable to what kind of medical devices?

(3) What are the purposes of the electromagnetic emission standards?

(4) Are all electromagnetic disturbance phenomena covered by this standard?

(5) When should the manufacturers address the electromagnetic disturbance?

1. General guidance and rationale— safety and performance

The scope of this collateral standard includes safety (basic safety and essential performance) with regard to electromagnetic disturbances, which is also called EMC for safety.

The words "Electromagnetic compatibility" have been deleted from the title of this collateral standard based on the following text from IEC/TS 61000-1-2: 2001:

Chapter 4-
Section 1-
Text 3-听课文

Whether a test on the influence of an electromagnetic phenomenon on the behaviour of an equipment should be included in an EMC standard (or clause) or in a safety standard (or clause) is dependent on the approval criterion:

1) If it is required that during or after the test the equipment continue to operate as intended, the test should be included in an EMC immunity standard (or clause) of a product (product family).

2) If it is required that during or after the test no unsafe situation occurs(performance maybe degraded incidentally or permanently, but not resulting in an unsafe situation), the test should be included in a safety standard(or clause). It is obvious that for products with safety functions the immunity levels may be chosen to be higher than in the generic standards for that environment.

Because this collateral standard is a safety standard, it is clear that the term "EMC" should not be used without qualification to refer to the requirements.

2. Introduction

The need for establishing specific standards for basic safety and essential performance with regard to electromagnetic disturbances for medical electrical equipment and medical electrical systems is well recognized.

The requirements and tests specified by this collateral standard are generally applicable to medical electrical equipment and medical electrical systems as defined in 3. 63 and 3. 64 in the general standard. For certain types of medical electrical equipment and medical electrical systems, these requirements might need to be modified by the special requirements of a particular standard. Writers of particular standards are encouraged to refer to Annex D for guidance in the application of this collateral standard.

Medical electrical equipment and medical electrical systems are expected to provide their basic safety and essential performance without interfering with other equipment and systems in the electromagnetic environments in which they are intended by their manufacturer to be used. The application of electromagnetic emission standards is essential for the protection of:

1) safety services;

2) other medical electrical equipment and medical electrical systems;

3) non-ME equipment(e. g. computers);

4) telecommunications(e. g. radio/TV, telephone, radio-navigation).

Of even more importance, the application of electromagnetic immunity standards is essential to ensure safety of medical electrical equipment and medical electrical systems. To ensure safety, medical electrical equipment and medical electrical systems are expected to provide their basic safety and essential performance in the electromagnetic environments of intended use throughout their expected service life.

This collateral standard specifies immunity test levels for safety for ME equipment and ME systems intended by their manufacturer for use in the professional healthcare facility environment or the home healthcare environment. It recognizes that RF wireless communications equipment can no longer be prohibited from most patient environments because in many cases it has become essential to the efficient provision of healthcare. This collateral standard also recognizes that, for certain special environments, higher or lower immunity test levels than those specified for the professional healthcare facility environment and the home healthcare environment might be appropriate. This collateral standard provides guidance in determining appropriate immunity test levels for special environments.

The immunity test levels specified for basic safety and essential performance are based on the reasonably foreseeable maximum of the electromagnetic disturbance phenomena in the applicable environments of intended use.

Not all electromagnetic disturbance phenomena are covered by this collateral standard, as it is not practical to do so. Manufacturers of medical electrical equipment and medical electrical systems need to address this during their risk assessment and evaluate if other electromagnetic disturbance phenomena could make their product unsafe. This evaluation should be based on the environments of intended use and the reasonably foreseeable maximum levels of electromagnetic disturbances expected throughout the expected service life.

This collateral standard recognizes that the manufacturer has the responsibility to design and perform verification of medical electrical equipment and medical electrical systems to meet the requirements of this collateral standard and to disclose information to the responsible organization or operator so that the medical electrical equipment or medical electrical system will remain safe throughout its expected service life.

This collateral standard provides guidance in incorporating considerations regarding electromagnetic disturbances into the risk management process.

This collateral standard is based on existing IEC standards prepared by subcommittee 62A, technical committee 77(electromagnetic compatibility between electrical equipment including networks), ISO(International standards organization), and CISPR(International special committee on radio interference).

Words and phrases

essential	[ɪˈsenʃl]	adj.	基本的
rationale	[ˌræʃəˈnɑːl]	n.	基本原理,基础理论
guidance	[ˈgaɪdns]	n.	指导
collateral	[kəˈlætərəl]	adj.	并列的,并行的
scope	[skəʊp]	n.	范围
electromagnetic	[ɪˌlektrəʊmægˈnetɪk]	adj.	电磁的
disturbance	[dɪˈstɜːbəns]	n.	干扰
compatibility	[kəmˌpætəˈbɪləti]	n.	兼容性
phenomenon	[fɪˈnɒmɪnən]	n.	现象
clause	[klɔːz]	n.	条款
approval	[əˈpruːvəl]	n.	批准,认可
criterion	[kraɪˈtɪərɪən]	n.	标准,准则
immunity	[ɪˈmjuːnəti]	n.	抗干扰
generic	[dʒɪˈnerɪk]	adj.	一般的
incidentally	[ˌɪnsɪˈdentəlɪ]	adv.	偶然地
permanently	[ˈpɜːmənəntlɪ]	adv.	长期地

define	[dɪˈfaɪn]	vt.	规定
modify	[ˈmɒdɪfaɪ]	vt.	修改
emission	[ɪˈmɪʃn]	n.	发射
ensure	[ɪnˈʃʊə(r)]	vt.	确保,保证
throughout	[θruːˈaʊt]	prep.	贯穿,遍及
facility	[fəˈsɪləti]	n.	设备
address	[əˈdres]	vt.	解决,处理
assessment	[əˈsesmənt]	n.	评估,评定
evaluate	[ɪˈvæljueɪt]	vt.	评价
subcommittee	[ˈsʌbkəmɪti]	n.	下属委员会
interference	[ɪntəˈfɪərəns]	n.	干扰
network	[ˈnetwɜːk]	n.	网络
communication	[kəˌmjuːnɪˈkeɪʃn]	n.	通信
wireless	[ˈwaɪələs]	n.	无线的
prohibit	[prəˈhɪbɪt]	vt.	禁止
foreseeable	[fɔːˈsiːəbl]	adj.	可预见的
healthcare	[ˈhelθkeə]	n.	医疗保健
responsibility	[rɪˌspɒnsəˈbɪləti]	n.	责任
verification	[ˌverɪfɪˈkeɪʃn]	n.	验证,确认
meet	[miːt]	vt.	满足
disclose	[dɪsˈkləʊz]	vt.	公开
operator	[ˈɒpəreɪtə]	n.	经营者,操作员
incorporate	[ɪnˈkɔːpəreɪt]	vt.	包含
regarding	[rɪˈɡɑːdɪŋ]	prep.	关于
medical electrical equipment			医用电气设备
with regard to			关于
collateral standard			并列标准
electromagnetic disturbance			电磁干扰
electromagnetic compatibility(EMC)			电磁兼容性
international standards organization(ISO)			国际标准化组织
radio-navigation			无线电导航
wireless communication			无线通信
risk assessment			风险评估
risk management process			风险管理过程

Key sentences

(1) The scope of this collateral standard includes safety (basic safety and essential performance) with regard to electromagnetic disturbances, which is also called EMC for safety.

本并列标准的范围包括有关电磁干扰的安全性(基本安全和基本性能),也称为 EMC 安全性。

(2) The need for establishing specific standards for basic safety and essential performance with regard to electromagnetic disturbances for medical electrical equipment and medical electrical systems is well recognized.

对于医用电气设备和医用电气系统的电磁干扰,需要制定基本安全和基本性能的专门标准是公认的。

(3) Medical electrical equipment and medical electrical systems are expected to provide their basic safety and essential performance without interfering with other equipment and systems in the electromagnetic environments in which they are intended by their manufacturer to be used.

医用电气设备和医用电气系统应能够在其预期使用的电磁环境中提供基本的安全性和基本性能,而不干扰其他设备和系统。

(4) The immunity test levels specified for basic safety and essential performance are based on the reasonably foreseeable maximum of the electromagnetic disturbance phenomena in the applicable environments of intended use.

为保证其基本安全性和基本性能所需要的抗扰度试验水平是基于在预期使用环境中可预期的最大电磁干扰来规定的。

(5) This evaluation should be based on the environments of intended use and the reasonably foreseeable maximum levels of electromagnetic disturbances expected throughout the expected service life.

评估应基于预期使用环境和设备有效期内可预见的最大电磁干扰水平。

Learning test

Choose the following words below to fill in the blanks, change the form if necessary.

electromagnetic   foreseeable   phenomenon   interference   scope
address   meet   guidance   prohibit   wireless

(1) Visible light, X-rays, and microwaves are all natural _____ waves.

(2) This content is beyond the _____ of this article.

(3) The moon eclipse is an interesting _____ .

(4) This _____ mouse is very convenient to use.

(5) This problem should be _____ as soon as possible.

(6) In the _____ future, Alibaba will dominate China's electronic payment market.

(7) I couldn't hear even a word because of too much _____ .

(8) Please follow the _____ of the user manual and try to solve the problems.

(9) The job's salary is high enough to _____ your requirement.

(10) I'm sorry that smoking is _____ here.

Translation

1. Translate the following words into Chinese.

(1) electromagnetic (2) immunity

(3) rationale (4) disturbance

(5) facility (6) compatibility

(7) responsibility (8) verification

(9) disclose (10) network

2. Translate the following phrases into Chinese.

(1) collateral standard (2) electromagnetic disturbance

(3) electromagnetic compatibility (4) risk assessment

(5) medical electrical equipment (6) wireless communication

Practice

1. Translate the following sentences into Chinese.

(1) Of even more importance, the application of electromagnetic immunity standards is essential to ensure safety of medical electrical equipment and medical electrical systems.

(2) To ensure safety, medical electrical equipment and medical electrical systems are expected to provide their basic safety and essential performance in the electromagnetic environments of intended use throughout their expected service life.

(3) Manufacturers of medical electrical equipment and medical electrical systems need to address this during their risk assessment and evaluate if other electromagnetic disturbance phenomena could make their product unsafe.

(4) Not all electromagnetic disturbance phenomena are covered by this collateral standard, as it is not practical to do so.

(5) This collateral standard provides guidance in incorporating considerations regarding electromagnetic disturbances into the risk management process.

2. Translate the following sentences into English.

(1) 本标准不适用于植入式医用电气设备。

(2) 厂商在产品设计阶段就应考虑电磁兼容性要求。

(3) 我们生活的环境中充斥着大量不同频率的电磁波。

(4) 各种电子产品在使用的过程中存在相互干扰的现象。

(5) 电磁干扰是指任何能使设备或系统性能降级的电磁现象。

Culture salon

Introduction of IEC 60601

IEC 60601 list a series of technical standards for the safety and effectiveness of medical electrical equipment, published by the International Electrotechnical Commission. First published in 1977 and regularly updated and restructured, as of 2011 it consists of a general standard, about 10 collateral standards, and about 60 particular standards.

The general standard IEC 60601-1-Medical electrical equipment-Part 1: General requirements for basic safety and essential performance give general requirements of the series of standards. IEC 60601 is a widely accepted benchmark for medical electrical equipment and compliance with IEC 60601-1 has become a requirement for the commercialization of electrical medical equipment in many countries. Many companies view compliance with IEC 60601-1 as a requirement for most markets.

Chapter 4-Section 1-Text 3-看译文

Chapter 4-Section 1-习题

Section 2 Application cases

Chapter 4-Section 2-PPT

Text 1 Classes and types of medical electrical equipment

Lead-in questions:

(1) How to categorize the electrical equipment?

(2) What are the two levels of protection used for mains powered electrical equipment?

(3) What is the basic means of protection for Class I equipment?

(4) What is the method of protectionagainst for Class II equipment?

(5) What is the definition of safety extra low voltage(SELV)?

All electrical equipment is categorized into classes according to the method of protection against electric shock that is used. For mains powered electrical equipment there are usually two levels of protection used, called "basic" and "supplementary" protection. The supplementary protection is intended to come into play in the event of failure of the basic protection.

Chapter 4-Section 2-Text 2-听课文

1. Class I Equipment

Class I equipment has a protective earth. The basic means of protection is the insulation between live parts and exposed conductive parts such as the metal enclosure. In the event of a fault that would otherwise cause an exposed conductive part to become live, the supplementary protection(i. e. the protective earth) comes into effect. A large fault current flows from the mains part to earth via the protective earth conductor, which causes a protective device(usually a fuse) in the mains circuit to disconnect the equipment from the supply.

It is important to realize that not all equipment having an earth connection is necessarily class I. The earth conductor may be for functional purposes only such as screening. In this case the size of the conductor may not be large enough to safely carry a fault current that would flow in the event of a main short to earth for the length of time required for the fuse to disconnect the supply.

Class I medical electrical equipment should have fuses at the equipment end of the mains supply lead in both the live and neutral conductors, so that the supplementary protection is operative when the equipment is connected to an incorrectly wired socket outlet.

Further confusion can arise due to the use of plastic laminates for finishing equipment. A case that appears to be plastic does not necessarily indicate that the equipment is not Class I.

There is no agreed symbol in use to indicate that equipment is Class I and it is not mandatory to state on the equipment itself that it is Class I. Where any doubt exists, reference should be made to equipment manuals.

The symbols below may be seen on medical electrical equipment adjacent to terminals as shown in Figure 4-1.

Figure 4-1 Symbols seen on earthed equipment.

2. Class II Equipment

The method of protection against electric shock in the case of Class II equipment is either double insulation or reinforced insulation. In double insulated equipment the basic protection is afforded by the first layer of insulation. If the basic protection fails then supplementary protection is provided by a second layer of insulation preventing contact with live parts.

In practice, the basic insulation may be afforded by physical separation of live conductors from the equipment enclosure, so that the basic insulation material is air. The enclosure material then forms the supplementary insulation.

Reinforced insulation is defined in standards as being a single layer of insulation offering the same degree of protection against electric shock as double insulation.

Class II medical electrical equipment should be fused at the equipment end of the supply lead in either mains conductor or in both conductors if the equipment has a functional earth.

The symbol for Class II equipment is two concentric squares illustrating double insulation as shown below in Figure 4-2.

Figure 4-2 Symbol for class II equipment.

3. Class III Equipment

Class III equipment is defined in some equipment standards as that in which protection against electric shock relies on the fact that no voltages higher than safety extra low voltage(SELV) are present. SELV is defined in turn in the relevant standard as a voltage not exceeding 25V ac or 60V dc.

In practice such equipment is either battery operated or supplied by a SELV transformer.

If battery operated equipment is capable of being operated when connected to the mains(for example, for battery charging) then it must be safety tested as either Class I or Class II equipment. Similarly, equipment powered from a SELV transformer should be tested in conjunction with the transformer as Class I or Class II equipment as appropriate.

It is interesting to note that the current IEC standards relating to safety of medical electrical equipment do not recognize Class III equipment since limitation of voltage is not deemed sufficient to ensure safety of the patient. All medical electrical equipment that is capable of mains connection must be classified as Class I or Class II. Medical electrical equipment having no mains connection is simply referred to as "internally powered".

4. Equipment Types

As described above, the class of equipment defines the method of protection against electric shock. The degree of protection for medical electrical equipment is defined by the type designation. The reason for the existence of type designations is that different pieces of medical electrical equipment have different areas of application and therefore different electrical safety requirements. For example, it would not be necessary to make a particular piece medical electrical equipment safe enough for direct cardiac connection if there is no possibility of this situation arising.

Table4-1 and Figure 4-3 shows the symbols and definitions for each type classification of medical electrical equipment.

Table4-1 Medical electrical equipment types.

Type	Definition
B	Equipment providing a particular degree of protection against electric shock, particularly regarding allowable leakage currents and reliability of the protective earth connection(if present).
BF	As type B but with isolated or floating(F-type) applied part or parts.
CF	Equipment providing a higher degree of protection against electric shock than type BF, particularly with regard to allowable leakage currents, and having floating applied parts.

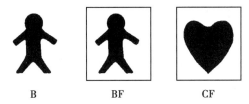

Figure 4-3　Symbols of B-type, BF-type and CF-type medical electrical equipment.

All medical electrical equipment should be marked by the manufacturer with one of the type symbols above.

Words and phrases

categorize	[ˈkætəɡəraɪz]	vt.	分类
protection	[prəˈtekʃn]	n.	保护,防护
shock	[ʃɒk]	n.	电击
mains	[meɪnz]	n.	网电源
supplementary	[ˌsʌplɪˈmentri]	adj.	补充的,辅助的
means	[miːnz]	n.	方法
insulation	[ˌɪnsjuˈleɪʃn]	n.	绝缘,隔离
live	[lɪv]	adj.	带电的
conductive	[kənˈdʌktɪv]	adj.	导电的
enclosure	[ɪnˈkləʊʒə]	n.	外壳
fault	[fɔːlt]	n.	故障
current	[ˈkʌrənt]	n.	电流
fuse	[fjuːz]	n. &vt.	保险丝,装保险丝
disconnect	[ˌdɪskəˈnekt]	vt.	断开,切断
short	[ʃɔːt]	n.	短路
neutral	[ˈnjuːtrəl]	adj.	中性的
operative	[ˈɒpərətɪv]	adj.	起作用的,工作的
confusion	[kənˈfjuːʒn]	n.	混淆,混乱
laminate	[ˈlæmɪneɪt]	n.	薄片
symbol	[ˈsɪmbl]	n.	符号
mandatory	[ˈmændətri]	adj.	强制的
manual	[ˈmænjʊl]	n.	手册
adjacent	[əˈdʒeɪsnt]	adj.	邻近的
terminal	[ˈtɜːmɪnl]	n.	电路端子
reinforced	[riːɪnˈfɔːsd]	adj.	加强的
illustrate	[ˈɪləstreɪt]	vt.	说明

battery	[ˈbætri]	n.	电池
voltage	[ˈvəʊltɪdʒ]	n.	电压
transformer	[trænsˈfɔːmə]	n.	变压器
deem	[diːm]	vt.	认为,视作
existence	[ɪɡˈzɪstns]	n.	存在
designation	[dezɪɡˈneɪʃn]	n.	指定,名称
electric shock			电击
medical electrical equipment			医用电气设备
come into play			起作用
in the event of			在……情况下
protective earth			保护接地
live part			带电体,带电部件
comes into effect			开始生效
fault current			故障电流
exposed conductive part			外露导电部分
metal enclosure			金属外壳
earth conductor			接地导体
double insulation			双重绝缘
reinforced insulation			加强绝缘
physical separation			物理隔离
functional earth			功能接地

Key sentences

(1) For mains powered electrical equipment there are usually two levels of protection used, called "basic" and "supplementary" protection. The supplementary protection is intended to come into play in the event of failure of the basic protection.

对于网电源供电的电气设备,通常有两个保护级别,称为"基本"和"辅助"保护。辅助保护旨在在基本保护失效的情况下发挥作用。

(2) Class Ⅰ equipment has a protective earth. The basic means of protection is the insulation between live parts and exposed conductive parts such as the metal enclosure.

Ⅰ类设备有保护接地。基本的保护方法是带电部件与暴露的导电部件之间的绝缘,例如金属外壳。

(3) The method of protection against electric shock in the case of Class Ⅱ equipment is either double insulation or reinforced insulation. In double insulated equipment the basic protection is afforded by the first layer of insulation.

Ⅱ类设备的防电击保护方法为双重绝缘或加强绝缘。双重绝缘设备的基本保护由第一层绝缘提供。

(4) Reinforced insulation is defined in standards as being a single layer of insulation offering the same

degree of protection against electric shock as double insulation.

加强绝缘在标准中定义为能够提供与双层绝缘相同程度的防电击保护的单层绝缘。

(5) Class Ⅲ equipment is defined in some equipment standards as that in which protection against electric shock relies on the fact that no voltages higher than safety extra low voltage(SELV) are present.

Ⅲ类设备在一些设备标准中定义为,防电击保护依赖于没有电压高于安全特低电压(SELV)这样一个情况。安全特低电压在相关标准中定义为不超过交流 25V 或直流 60V 的电压。

Learning test

Choose the following words below to fill in the blanks, change the form if necessary.

insulation   conductive   fault   short   disconnect

protection   symbol   mandatory   operative   deem

(1) Every citizen may claim the _____ of the law.

(2) Plastics are good _____ materials with very high resistance.

(3) The _____ current will flow to the earth through the protective earth resistor.

(4) What's the chemical _____ for mercury?

(5) I _____ it a great honor to receive your invitation.

(6) The fuse will _____ the circuit when the current is over 1A.

(7) The timer is _____ when the green light is on.

(8) The 10Ω resistor is burnt because of the power supply _____ to earth.

(9) Our sweating skin is electrically _____ and the resistance is about 1KΩ.

(10) This standard(GB 8368-2005) is a _____ standard in China.

Translation

1. Translate the following words into Chinese.

(1) protection   (2) fuse

(3) symbol   (4) neutral

(5) deem   (6) insulation

(7) fault   (8) battery

(9) voltage   (10) disconnect

2. Translate the following phrases into Chinese.

(1) come into play   (2) live part

(3) electric shock   (4) fault current

(5) protective earth   (6) double insulation

Practice

1. Translate the following sentences into Chinese.

(1) All electrical equipment is categorized into classes according to the method of protection against electric shock that is used.

(2) In the event of a fault that would otherwise cause an exposed conductive part to become live, the supplementary protection (i.e. the protective earth) comes into effect.

(3) If the basic protection fails then supplementary protection is provided by a second layer of insulation preventing contact with live parts.

(4) Class II medical electrical equipment should be fused at the equipment end of the supply lead in either mains conductor or in both conductors if the equipment has a functional earth.

(5) If battery operated equipment is capable of being operated when connected to the mains (for example, for battery charging) then it must be safety tested as either Class I or Class II equipment.

2. Translate the following sentences into English.

(1) 无论什么类型的医用电气设备，用电安全是最重要的。

(2) 通过人体的最低安全电流是 10 毫安。

(3) 变压器是利用电磁感应的原理来改变交流电压的装置。

(4) 医用电气设备应该有防电击保护措施，保证使用安全。

(5) 三类医用电气设备一般使用电池供电。

> **Culture salon**
>
> ### Electric shock
>
> Electric shock is the physiological reaction, sensation, or injury caused by electric current passing through the (human) body. It occurs upon contact of a (human) body part with any source of electricity that causes a sufficient current through the skin, muscles, or hair.
>
> Very small currents can be imperceptible. Stronger current passing through the body may make it impossible for a shock victim to let go of an energized object. Still larger currents can cause fibrillation of the heart and damage to tissues. Death caused by an electric shock is called electrocution. Wiring or other metal work at a hazardous voltage which can constitute a risk of electric shock is called "live," as in "live wire".

Text 2 The AK 96 dialysis machine

Lead-in questions:

(1) What are the main parts for AK 96 dialysis machine?

(2) What is the usage of the blood unit?

(3) What is the relevant standard used to design the machine?

(4) How to reduce the risk of blood contamination?

(5) What is the usage of the venous pressure measuring system?

1. Structure of AK 96

The AK 96 dialysis machine is designed to be used as a single patient machine to perform hemodialysis treatments upon prescription by a physician. Patient counseling and teaching of treatment techniques are directly under the supervision and discretion of the physician.

Chapter 4-
Section 2-
Text 2-听课文

The AK 96 dialysis machine can be divided into following parts:

- Fluid unit
- Blood unit
- Power supply
- Operator's panel

(1) Fluid unit

The fluid unit is used to produce the dialysis fluid (with correct temperature and correct composition) from reverse osmosis water and concentrate, and to transport the dialysis fluid through the dialyzer.

The fluid unit maintains the dialysis fluid flow through the dialyzer with controlled ultrafiltration. If a fault occurs, the fluid unit bypasses the dialyzer.

(2) Blood unit

The blood unit is designed to control and supervise the extracorporeal blood circuit. Single needle treatment can be performed with one pump (double clamp function). To prevent coagulation, anticoagulants may be administered by means of the syringe pump.

(3) Power supply

The main voltage is fed to an AC/DC converter which generates different DC supplies to the monitor.

(4) Operator's panel

Both the blood unit and the fluid unit are controlled and supervised from the operator's panel. The panel consists of a number of buttons, flow paths, a time display and an information display. The information in the display can be set to different languages.

2. Safety Philosophy

The AK 96 dialysis machine is designed according to the current standards for hemodialysis equipment, IEC 60601-2-16. This means that safety under so-called Single Fault Conditions is granted. In practice this means that controllable treatment parameters (i. e. conductivity, temperature and ultrafiltration) are controlled by one system, the control system, and monitored by another completely separate protective system, utilizing its own sensors, electrical circuits and microprocessors. The functionality of the protective system is checked by the AK 96 dialysis machine before each treatment. A fault detection during the pre-treatment tests will make it impossible to start the treatment.

In order to verify that the corresponding control and protective systems are operating with the correct input values, the user is instructed to compare the readings from these systems before connecting to the patient. Check that the calculated conductivity values (C/P) displayed in the conductivity menu is in agreement. If this comparison is not satisfactory, call an authorized technician.

The protective system will, when a parameter (measured by the protective system) is outside the alarm limits, put the AK 96 dialysis machine into a patient-safe condition. This means that the protective system can stop the blood pump, close the venous clamp, prevent the dialysis fluid from reaching the dialyzer and alert the operator with sound and light.

There is a risk that the blood of the patient may be contaminated with bacteria and endotoxins due to transport of undesired substances from the dialysis fluid compartment to the blood compartment of the dialyzer. This risk is reduced by using intact dialyzers, high inlet water quality, high quality of concentrates and by using the ultrafilter for dialysis fluid (option).

For the ultrafiltration control system, the transmembrane pressure (TMP) is used as the protective system. Alarm limits for TMP related to the dialyzer UF coefficient and the expected UF rate, are to be set around the actual TMP value when starting treatment. The TMP alarm limit correspond to a UF-deviation limit described by $TMP_{Alarm\ limit} \times UF_{coefficient}$. Example: If the alarm window is set to ±50 mmHg and the UF-coefficient is 10 ml/mmHg/h, the maximum weight deviations without any alarm is ±500 g/h. Default the alarm window is set to ±100 mmHg. It is essential to ensure that the alarm window is set as close as possible to the working TMP. As an additional precaution it is recommended that the blood pressure is checked regularly.

To protect the patient against a hazardous blood loss to the environment AK 96 dialysis machine incorporates a venous pressure monitoring system. This system will react on a change in the venous pressure, i. e. when the pressure falls below the low alarm limit. It must be observed that under certain pressure/flow conditions a blood loss to the environment may not be able to cause the venous pressure to fall below the low alarm limit. To avoid blood loss to the environment it is essential to ensure that all connections in the extracorporeal blood circuit are tight and secured, that the fistula needle is correctly positioned and secured and that the low alarm limit is set as close as possible to the working venous pressure.

The venous pressure measuring system is the protection against blood loss to the environment. This measuring system is automatically checked before each treatment. A failure will make it impossible to start the treatment.

The supervision of the stop time of the blood pump is the protection system against patient blood loss due to coagulation during treatment. The operator will be notified via an alarm that the blood pump stop time has been exceeded.

The blood leak detector system, which utilizes an optical sensor, is automatically tested before each treatment for being able to detect transparency (no blood) and non-transparency (blood) before each treatment. If the system cannot detect these states, it is impossible to start the treatment.

The air detector utilizes an ultrasonic sound sensing system in which the transmitter is handled by one microprocessor and the receiver is handled by both microprocessors in the protective system. The system is tested pre-treatment for parameter deviation in terms of sensitivity change.

Words and phrases

Chapter 4-
Section 2-
Text 2-听词汇

dialysis	[ˌdaɪˈæləsɪs]	n.	透析
hemodialysis	[ˌhiːmədaɪˈælɪsɪs]	n.	血液透析
prescription	[prɪˈskrɪpʃn]	n.	指示,处方
physician	[fɪˈzɪʃn]	n.	医师
counsel	[ˈkaʊnsəlɪŋ]	vt.	提供专业咨询
reverse	[rɪˈvɜːs]	adj.	反向
osmosis	[ɒzˈməʊsɪs]	n.	渗透
concentrate	[ˈkɒnsntreɪt]	vi. &n.	浓缩(物)
syringe	[sɪˈrɪndʒ]	n.	注射器
pump	[pʌmp]	n.	泵
ultrafiltration	[ˌʌltrəfɪlˈtreɪʃn]	n.	超滤
bypass	[ˈbaɪpɑːs]	vt.	绕开
supervise	[ˈsuːpəvaɪz]	vt.	监控,管理
extracorporeal	[ˌekstrəkɔːˈpɔːriəl]	adj.	体外的
circuit	[ˈsɜːkɪt]	n.	回路
pump	[pʌmp]	n.	泵
coagulation	[kəʊˌæɡjʊˈleɪʃən]	n.	凝固,凝结
anticoagulant	[ˌæntikəʊˈæɡjələnt]	n.	抗凝血剂
administer	[ədˈmɪnɪstə(r)]	vt.	给(药)
syringe	[sɪˈrɪndʒ]	n.	注射器,注射筒
converter	[kənˈvɜːtə(r)]	n.	转换器
monitor	[ˈmɒnɪtə]	n.	显示屏
panel	[ˈpænl]	n.	面板
transmembrane	[ˌtrænzˈmembreɪn]	adj.	跨膜
parameter	[pəˈræmɪtə(r)]	n.	参数
conductivity	[ˌkɒndʌkˈtɪvəti]	n.	电导率
sensor	[ˈsensə(r)]	n.	传感器
functionality	[ˌfʌŋkʃəˈnæləti]	n.	功能(性)
check	[tʃek]	vt.	检查
verify	[ˈverɪfaɪ]	vt.	验证,核实
input	[ˈɪnpʊt]	n.	输入
reading	[ˈriːdɪŋ]	n.	读数
menu	[ˈmenjuː]	n.	菜单

contaminate	[kənˈtæmɪneɪt]	vt.	污染
endotoxin	[ˌendəʊˈtɒksɪn]	n.	内毒素
coefficient	[ˌkəʊɪˈfɪʃnt]	adj.	系数
deviation	[ˌdiːvɪˈeɪʃn]	n.	偏差,误差
satisfactory	[ˌsætɪsˈfæktərɪ]	adj.	符合要求的
precaution	[prɪˈkɔːʃn]	n.	预警,预防措施
incorporate	[ɪnˈkɔːpəreɪt]	vt.	包含,整合
venous	[ˈviːnəs]	adj.	静脉的
leak	[liːk]	n.	泄漏
transparency	[trænsˈpærənsi]	n.	透明度
fistula	[ˈfɪstjʊlə]	n.	瘘管
sensitivity	[ˌsensəˈtɪvəti]	n.	灵敏度
reverse osmosis			反渗透,逆向渗透
syringe pump			注射泵
by means of			通过,用,借助
input value			输入值
in agreement			一致
blood pump			血液泵
dialysis machine			血液透析机
due to			由于
safety philosophy			安全理念
transmembrane pressure			跨膜压差
venous pressure			静脉压
fistula needle			瘘管缝合针
blood leak detector			血漏检测器
in terms of			在…方面

Key sentences

(1) The AK 96 dialysis machine is designed to be used as a single patient machine to perform hemodialysis treatments upon prescription by a physician.

AK 96 透析机是为一个患者而设计的机器,在医生的处方下进行血液透析治疗。

(2) To prevent coagulation, anticoagulants may be administered by means of the syringe pump.

为了防止凝血,可通过注射泵添加抗凝剂。

(3) The AK 96 dialysis machine is designed according to the current standards for hemodialysis equipment, IEC 60601-2-16.

AK 96 血液透析机是根据现行的血液透析设备的标准(IEC 60601-2-16)设计的。

(4) In order to verify that the corresponding control and protective systems are operating with the correct input values, the user is instructed to compare the readings from these systems before connecting to the patient.

为了验证相应的控制和保护系统是在正确的输入值下运行,用户在连接到患者之前应当比较这些系统的读数。

(5) There is a risk that the blood of the patient may be contaminated with bacteria and endotoxins due to transport of undesired substances from the dialysis fluid compartment to the blood compartment of the dialyzer.

由于不需要的物质从透析液室运送到透析器的血液室,存在患者血液可能被细菌和内毒素污染的风险。

Learning test

Choose the following words below to fill in the blanks, change the form if necessary.

input   venous   deviation   reverse   check
osmosis   coagulation   sensor   precaution   parameter

(1) The wrong attitude will have exactly the _____ effect.

(2) The culture _____ such as temperature and culture medium, will affect the cell growth.

(3) Please _____ the input parameters before you press the "enter" buttom.

(4) I post the _____ notice on the operation panel of the machine, so that everyone can read it before using it.

(5) This camera has an optical _____ in it.

(6) Semi-permeable membranes are often used for ultra filtration and reverse _____ process.

(7) Once you specify the _____ and output types, click the Next button.

(8) During operation the vital sign and blood _____ were monitored.

(9) The measurement _____ is only one millimetre.

(10) The _____ pressure is monitored during hemodialysis.

Translation

1. Translate the following words into Chinese.

(1) leak   (2) transparency
(3) sensitivity   (4) ultrafiltration
(5) panel   (6) conductivity
(7) functionality   (8) osmosis
(9) coagulation   (10) deviation

2. Translate the following phrases into Chinese.

(1) syringe pump   (2) input value
(3) dialysis machine   (4) venous pressure
(5) reverse osmosis   (6) due to

Practice

1. Translate the following sentences into Chinese.

(1) The fluid unit maintains the dialysis fluid flow through the dialyzer with controlled ultra filtration.

(2) The blood unit is designed to control and supervise the extracorporeal blood circuit.

(3) Both the blood unit and the fluid unit are controlled and supervised from the operator's panel.

(4) The venous pressure measuring system is the protection against blood loss to the environment.

(5) The system is tested pre-treatment for parameter deviation in terms of sensitivity change.

2. Translate the following sentences into English.

(1)血液透析机的设计要符合医用电气设备的相关标准。

(2)血泵部分往往具有转速检测功能。

(3)一般透析液温度控制在37℃左右,根据患者情况可适当调节。

(4)具有热消毒的机器,在进行热消毒时加热温度可达到100℃。

(5)抗凝剂的作用是防止血液凝固。

Culture salon

Hemoperfusion

Hemoperfusion is a medical procedure which is used to cleanse the blood of toxins. During this process the blood is passed through an adsorbent material which attracts toxic substances. The adsorbent material is usually charcoal or activated carbon fixed to a solid surface inside a column. During treatment, the patient's blood is passed through the column and toxins bind to the adsorbent material, allowing cleansed blood to flow out of the column. This process continues until as much toxic material as possible has been removed from the blood.

Chapter 4-Section 2-Text 2-看译文

Text 3 Maintenance manual of the AK 96 dialysis machine

Lead-in questions:

(1) How to disinfect the AK 96 dialysis machine?

(2) How to prevent cross-infection between patients?

(3) Can heat disinfection be replaced by chemical disinfection?

(4) How to maintain a machine if it is planned not to be in use for a period of time exceeding seven days?

(5) How many circulation phase cycles for a heat disinfection program?

In order to maintain a high microbiological quality of the dialysis fluid, it is important that the operator/user is attentive to the hygiene and maintenance of the machine. There are factors and procedures that affect the hygiene of the flow path, and consequently the quality of the prepared dialysis fluid. The following must be considered:

(1) The disinfection, decalcification and cleaning processes must be performed according to instructions in this chapter.

Chapter 4-
Section 2-
Text 3-听课文

(2) The frequency of disinfection, decalcification and cleaning of the machine should be carried out according to recommendations in this chapter.

(3) The quality of the inlet water entering the machine.

(4) The quality of the concentrates being used.

(5) The inlet water tube, which preferably should be disinfected by using integrated heat disinfection.

(6) The arrangement of the drain connection from the machine, i. e. there must be a sufficient air-gap between the drain tube of the machine and the drainage system to avoid back contamination from the drainage system.

The AK 96 dialysis machine may be disinfected using heat where the fluid flow path is heated up. The AK 96 dialysis machine may also be disinfected by filling the fluid path with chemical disinfectants. For information concerning description and handling instructions of the different disinfection programs, see further on in this chapter.

The test procedure by which the effectiveness of disinfection has been verified is available on request, contact your local Gambro representative.

To prevent any cross-infection between patients the following precautions must be considered:

(1) Wipe down the outside of the machine with proper disinfectant.

(2) To protect the pressure connections on the machine, use blood lines where hydrophobic filters are integrated. Especially note that if any of the hydrophobic filters have been filled with blood and show signs of damage, the machine components for pressure measurements must be replaced and cleaned by an authorized technician.

1. Schedule for hygiene and maintenance

The schedule below is recommended for use by the operator/user of the AK 96 dialysis machine. This recommendation is to keep a high level of performance of the machine. It is also a guideline for maintaining the hygiene of the fluid path and the exterior of the machine to ensure the safety of the patient.

Any heat disinfection program can be exchanged to a heat disinfection program in combination with CleanCart C cartridge or liquid citric acid. Heat disinfection can be replaced by chemical disinfection. But note that decalcification(CleanCart C cartridge or liquid citric acid) cannot be replaced by chemical disinfection.

An integrated heat disinfection program is recommended when possible since it also includes the inlet water tube. In order to enable the cleaning agents to remove fats, proteins etc. efficiently, cleaning(CleanCart A cartridge or sodium hypochlorite) should be carried out after decalcification(CleanCart C cartridge or liquid citric acid).

2. When the machine is planned not to be in use

If the machine is planned not to be in use for a period of time exceeding seven days, the microbiological standard of the flow path can be maintained with one of the following alternatives:

(1) Perform a heat disinfection program at least every seventh day and a heat disinfection program before treatment.

(2) Fill the machine with a proper chemical disinfectant in order to preserve it.

3. Heat disinfection-general

During the heat disinfection program the inlet water is heated up and flushed through the fluid monitor. The program begins with a rinse in order to eradicate possible residues from concentrates used in the preceding treatment. After this, the program consists of three consecutive phases; the fill up phase, the circulation phase and the drain phase. The actual disinfection takes place in the circulation phase. The circulation phase consists of a number of cycles, where each cycle is a sequence in which heated water is passed through all parts of the fluid path. The standard heat disinfection program consists of 12 such cycles.

The AK 96 dialysis machine fluid path is constructed in a way that prevents possible contaminants in the post-dialyzer fluid path from reaching the patient. The construction principle is called *single pass*, which means no recirculation in the fluid path, and a separation between the pre-dialyzer circuit and the post-dialyzer circuit. This is to ensure that no fluid from the downstream(post) dialyzer circuit comes into contact with the pre-dialyzer circuit. The construction principle is an effective obstacle to transmission of blood-borne viruses like Hepatitis B, from the fluid path to the patient, if a blood leak in the dialyzer occurs. In addition, the standard heat disinfection program has the capacity to make viruses inactive.

The combination of the standard heat disinfection program process(12 cycles) and the safe hygienic construction of the fluid path(single pass) described above provides high safety levels for the patient.

The standard heat disinfection program is validated by using microbiological challenge tests which gives a 5 log reduction of the following microorganisms: Pseudomonas aeruginosa(ATCC 9027), Staphylococcus aureus(ATCC 6538), Burkholderia cepacia(ATCC 17770) and Candida albicans(ATCC 10231).

The heat disinfection program can be extended for 22 cycles. The 22 cycles variant may be used in case of substantial contamination of the fluid path. This program provides different positions of the fluid path with at least 30% more heat. As a consequence, the duration of this heat disinfection program will be extended by approximately 10 minutes, compared to the standard program.

It is possible to run a heat disinfection program in combination with CleanCart cartridge:

(1) Decalcification program with CleanCart C cartridge

(2) Cleaning program with CleanCart A cartridge

If heat disinfection program with CleanCart cartridge is chosen, the contents of the CleanCart cartridge are dissolved and the solution is heated-up and flushed through the fluid monitor during the heat disinfection program as described above. This program is extended by a priming phase, a fill-up phase and a recirculation phase for the CleanCart cartridge, which leads to a longer time to complete the program.

The CleanCart C cartridge can be replaced by liquid citric acid in order to perform a decalcification program; heat disinfection program with liquid citric acid. The program is performed in a similar way as heat disinfection program with CleanCart cartridge.

Words and phrases

maintenance	[ˈmeɪntənəns]	n.	维护,保养
manual	[ˈmænjuəl]	n.	手册,指南
microbiological	[ˌmaɪkrəʊˌbaɪəˈlɒdʒɪkl]	adj.	微生物(学)的
hygiene	[ˈhaɪdʒiːn]	n.	卫生
procedure	[prəˈsiːdʒə]	n.	步骤
disinfection	[ˌdɪsɪnˈfekʃən]	n.	消毒
decalcification	[ˈdiːkælsɪfɪˈkeɪʃən]	n.	脱钙
recommendation	[ˌrekəmenˈdeɪʃn]	n.	推荐,建议
inlet	[ˈɪnlet]	n.	入口,进口
arrangement	[əˈreɪndʒmənt]	n.	布置
drain	[dreɪn]	n.	排水
drainage	[ˈdreɪnɪdʒ]	n.	排水
contamination	[kənˌtæmɪˈneɪʃən]	n.	污染
disinfectant	[ˌdɪsɪnˈfektənt]	n.	消毒剂
precaution	[prɪˈkɔːʃn]	n.	预警,预防措施
hydrophobic	[haɪdrəˈfəʊbɪk]	adj.	疏水的
integrate	[ˈɪntɪgreɪt]	vt.	集成
technician	[tekˈnɪʃn]	n.	技术员
schedule	[ˈʃedjuːl]	n.	时间表,计划表
hypochlorite	[ˌhaɪpəˈklɔːraɪt]	n.	次氯酸盐
alternative	[ɔːlˈtɜːnətɪv]	n.	替代,备选
rinse	[rɪns]	n.	冲洗,漂洗
eradicate	[ɪˈrædɪkeɪt]	vt.	去除,清除
residue	[ˈrezɪdjuː]	n.	残渣
consecutive	[kənˈsekjʊtɪv]	adj.	连贯的,连续的
cartridge	[ˈkɑːtrɪdʒ]	n.	匣,盒
circulation	[sɜːkjʊˈleɪʃn]	n.	循环
downstream	[ˈdaʊnˈstriːm]	adj.	下游
circuit	[ˈsɜːkɪt]	n.	回路

Chapter 4-
Section 2-
Text 3-听词汇

obstacle	[ˈɒbstəkl]	n.	障碍,干扰
virus	[ˈvaɪrəs]	n.	病毒
inactive	[ɪnˈæktɪv]	adj.	无活性的
hepatitis	[ˌhepəˈtaɪtɪs]	n.	肝炎
validate	[ˈvælɪdeɪt]	vt.	验证,证实
substantial	[səbˈstænʃl]	adj.	大量的,实质的
air-gap			空气隙
cross-infection			交叉感染
chemical disinfectant			化学消毒剂
sodium hypochlorite			次氯酸钠
citric acid			柠檬酸,枸橼酸
flow path			流路
dialysis fluid			透析液,透析液
blood-borne virus			血源性病毒
hepatitis B			乙型肝炎
Pseudomonas aeruginosa			铜绿假单胞菌
Staphylococcus aureus			金黄色葡萄球菌
Burkholderia cepacia			洋葱伯克霍尔德氏菌
Candida albicans			白色念珠菌

Key sentences

(1) There are factors and procedures that affect the hygiene of the flow path, and consequently the quality of the prepared dialysis fluid.

有很多因素和步骤会影响管路的卫生,并因此影响到制备的透析液的质量。

(2) The AK 96 dialysis machine may be disinfected using heat where the fluid flow path is heated up. The AK 96 dialysis machine may also be disinfected by filling the fluid path with chemical disinfectants.

AK 96 透析机通过流体流路加热时的热量进行消毒。AK 96 透析机也可以通过用化学消毒剂填充流体路径来消毒。

(3) Perform a heat disinfection program at least every seventh day and a heat disinfection program before treatment.

至少每七天对设备进行一次热消毒,而且治疗之前也要进行热消毒。

(4) During the heat disinfection program the inlet water is heated up and flushed through the fluid monitor. The program begins with a rinse in order to eradicate possible residues from concentrates used in the preceding treatment.

在热消毒期间,进入的水被加热,并通过流体监控器冲洗。该程序首先要冲洗,以清除之前治疗中使用的浓缩物可能存在的残留。

(5) The standard heat disinfection program is validated by using microbiological challenge tests which gives a 5 log reduction of the following microorganisms: Pseudomonas aeruginosa(ATCC 9027), Staphylococcus aureus(ATCC 6538), Burkholderia cepacia(ATCC 17770) and Candida albicans(ATCC 10231).

标准的热消毒程序用微生物挑战试验验证,可使下列微生物减少5个数量级:铜绿假单胞菌(ATCC 9027),金黄色葡萄球菌(ATCC 6538),洋葱伯克霍尔德氏菌(ATCC 17770)和白念珠菌(ATCC 10231).

Learning test

Choose the following words below to fill in the blanks, change the form if necessary.

technician   schedule   procedure   manual   validate
rinse   drainage   hygiene   inlet   inactive

(1) I'm a maintenance _____ in GE healthcare.

(2) Please _____ the disinfection effect of the dialysis machine before treatment.

(3) It's important that all possible viruses are _____ after disinfection.

(4) We successfully repaired the ECG monitor with the aid of the user _____ .

(5) I have prepared a time _____ for the meeting next week.

(6) Something gets into my eyes. I have to _____ it out immediately.

(7) The _____ system has collapsed because of too much rain.

(8) Close outlet valve and _____ valve of cooling water.

(9) There are so many operation _____ and I'm confused.

(10) Be extra careful about personal _____ .

Translation

1. Translate the following words into Chinese.

(1) virus   (2) disinfection

(3) recommendation   (4) manual

(5) hygiene   (6) disinfectant

(7) technician   (8) residue

(9) maintenance   (10) eradicate

2. Translate the following phrases into Chinese.

(1) cross-infection   (2) chemical disinfectant

(3) citric acid   (4) blood-borne virus

(5) Hepatitis B   (6) flow path

Practice

1. Translate the following sentences into Chinese.

(1) Heat disinfection can be replaced by chemical disinfection.

(2) An integrated heat disinfection program is recommended when possible since it also includes the in-

let water tube.

(3) Fill the machine with a proper chemical disinfectant in order to preserve it.

(4) In addition, the standard heat disinfection program has the capacity to make viruses inactive.

(5) As a consequence, the duration of this heat disinfection program will be extended by approximately 10 minutes, compared to the standard program.

2. Translate the following sentences into English.

(1) 热消毒分为干热消毒和湿热消毒。

(2) 大部分医疗器械出厂前应进行灭菌处理。

(3) 血液透析机不是一次性使用产品,所以它的灭菌处理非常重要。

(4) 血液透析是一种血液净化技术,可清除体内代谢产物和毒性物质。

(5) 血液透析机有血液监护警报系统和透析液供给系统两部分。

Culture salon

Heat disinfection

Heat is the most common physical agent used for the decontamination of pathogens. "Dry" heat, which is totally non-corrosive, is used to process many items of laboratory ware which can withstand temperatures of 160°C or higher for 2~4 hours. Burning or incineration (see below) is also a form of dry heat. "Moist" heat is most effective when used in the form of autoclaving. Boiling does not necessarily kill all microorganisms and/or pathogens, but it may be used as the minimum processing for disinfection where other methods (chemical disinfection or decontamination, autoclaving) are not applicable or available.

Chapter 4-Section 2-Text 3-看译文

Chapter 4-Section 2-习题

Chapter 5

Medical device marketing and management

Learning objective

In this chapter, we will learn how to do business in medical device market, which involves how to cooperate perfectly with customersto make them satisfied, how to sell medical device products successfully and how to negotiate with customers to gain a win-win outcome. Read the following passages, you should meet the following requirements:

(1) Grasp basic concepts of business etiquette, medical device sales jobs and business negotiations.

(2) Acquire key qualities and business strategies the medical device positions need.

(3) Understand the importance of business etiquette and negotiation skills in the medical device market.

Section 1 Business etiquette

Chapter 5-Section 1-PPT

Text 1 Business etiquette tips everyone should follow

Lead-in questions:

(1) What is business etiquette?

(2) Is business etiquette the same in every culture?

(3) What are the simple guidelines for one to start a new job in the business market?

(4) What should you do if you are going to be late for the meeting?

(5) What is the best choice if you are going to express your gratitude?

Chapter 5-Section 1-Text 1-听课文

Business etiquette is about building relationships with other people. Etiquette is not about rules & regulations, but is about providing basic social comfort and creating an environment where others feel comfortable and secure, this is possible through better communication.

Business etiquette differs from region to region and from country to country. This creates a complex situation for people as it is hard to balance the focus on both international business etiquette and other business activities at the same time. Therefore, a wise step is to focus on some key pillars of business etiquette. Whether you're just starting out in business or vying for a promotion, these simple guidelines will never fail you.

1. Turn off your gadgets in meetings. Turn our phones face down, close our laptops, and focus only on the meeting. It's difficult to make that happen, but when everyone is unplugged and focused, meetings are much more productive.

2. Arrive on time for meetings — face to face or virtual. If you're the meeting host, on time means at least five minutes early. If you're the guest, on time means on time. If you get held up and know you're going to be delayed, a quick email can keep the person on the other end from feeling stood up.

3. No nail trimming at work. This wouldn't be on the list if I hadn't witnessed it too many times to count. This also goes for toes. I've seen that, too, and hardly survived it. Just because most of us work in open work environments doesn't mean it's a green light to do things better done in the privacy of your own home/bathroom.

4. Express gratitude. Thank you notes go a long way. If possible, a handwritten note is best. If you prefer a paperless option, make sure your message is concise, thoughtful, and free of misspelled words. Fine stationery is an investment.

5. Whether it's a company-wide function or a team dinner, ask before inviting your significant other. If you bring your significant other and nobody else does, talking shop feels awkward to everyone(your guest included), and that's what business dinners are often about.

6. Don't say anything in email or instant messaging that you don't mind being broadcast to your entire organization. Watch what you're saying on instant messaging systems, too.

7. Don't dominate the "Questions" segment in meetings. How many questions should you ask? Frankly, I think one question per big meeting is usually enough; three is the maximum. For smaller meetings, just consider how much air time is available, and try not to dominate.

Following is an example for practice.

(Mark is the sales manager of ABC Company in Beijing. Jenny is his colleague. Mark talked about his uneasiness with her. He thought that Mr. Black, the vice-president, did not approve of him since Mr. Black kept spinning the pen and looked outside of the window constantly during his presentation in the meeting. Mark felt much more nervous after the sectary noted him that Mr. Black would have a talk with him later. That afternoon, Mark and Jenny met again.)

Jenny: How did things go with Mr. Black?

Mark: It is completely a false alarm. Mr. Black affirmed the achievement I made and praised me.

Jenny: Wow, that's great.

Mark: Were my worries uncalled-for? Think of his reactions during the meeting this morning.

Jenny: You were reading his body language this morning and he clearly wasn't sending a positive message.

Mark: Body language?

Jenny: Words aren't the only way we communicate. We send messages with our bodies.

Mark: But why Mr. Black's body language was quite different from his minds?

Jenny: Perhaps he was distracted or had something else on his mind. Evidently he was listening to you, but that is not the impression you got.

Mark: What should I do when I encounter this situation next time?

Jenny: Normally, you could have stopped at that point and asked for questions or opinions, specially addressing the person who didn't seem to be involved.

Mark: What? Stop and ask for questions? You know, Mr. Black is the big boss.

Jenny: In this case, you were in a tough spot since he is the vice-president.

Mark: Actually, I will pay much more attention to my body language after this matter.

Jenny: That's a good lesson for all of us. Too often we send nonverbal messages without being aware of how others interpret them.

Mark: Well, many mistakes would happen without attention.

Jenny: Body language can be used to impress people. It is directly related to good manners if you think about it.

Mark: How dose body language relate to a man's manners?

Jenny: Positive body language like giving someone your full attention and having a pleasant expression on your face are all parts of showing courtesy and respect.

Mark: Listening carefully itself shows respect for others.

Words and phrases

etiquette	[ˈetɪket]	n.	礼仪,礼节
tip	[tɪp]	n.	提示,技巧,窍门
relationship	[rɪˈleɪʃnʃɪp]	n.	关系,联系
region	[ˈriːdʒən]	n.	地区,区域
complex	[ˈkɒmpleks]	adj.	复杂的
balance	[ˈbæləns]	vt.	权衡,平衡
focus	[ˈfəʊkəs]	n.	焦点,中心
pillar	[ˈpɪlə(r)]	n.	核心
vying	[ˈvaɪɪŋ]	vi.	竞争(vie 的分词形式)
guideline	[ˈgaɪdlaɪn]	n	指南
gadget	[ˈgædʒɪt]	n.	小玩意,小器具
laptop	[ˈlæptɒp]	n.	笔记本电脑
unplugged	[ˌʌnˈplʌgd]	adj.	解脱的

productive	[prəˈdʌktɪv]	adj.	多产的
virtual	[ˈvɜːtʃuəl]	adj.	虚拟的
trim	[trɪm]	vt.	修剪
witness	[ˈwɪtnəs]	vt.	目击，证明
gratitude	[ˈɡrætɪtjuːd]	n.	谢意，感谢
handwritten	[ˌhændˈrɪtn]	adj.	手写的
paperless	[ˈpeɪpələs]	adj.	无纸的
concise	[kənˈsaɪs]	adj.	简明的，简洁的
thoughtful	[ˈθɔːtfl]	adj.	有思想的
misspell	[ˌmɪsˈspel]	vt.	拼错
stationery	[ˈsteɪʃənri]	n.	文具，信纸
dominate	[ˈdɒmɪneɪt]	vt.	控制，主导
segment	[ˈseɡmənt]	n.	环节，部分
maximum	[ˈmæksɪməm]	n.	最大限度，最大量
available	[əˈveɪləbl]	adj.	可得的，有空的
uneasiness	[ʌnˈiːzɪnəs]	n.	不安，担忧
affirm	[əˈfɜːm]	vt.	肯定
achievement	[əˈtʃiːvmənt]	n.	成就，成绩
distracted	[dɪˈstræktɪd]	adj.	思想不集中的，分心的
tough	[tʌf]	adj.	困难的，难办的
encounter	[ɪnˈkaʊntə(r)]	vt.	遭遇
nonverbal	[ˌnɒnˈvɜːbl]	adj.	非言语的
courtesy	[ˈkɜːtəsi]	n.	礼貌
business etiquette			商务礼仪
rules & regulations			规章制度
free of			无，没有……
instant messaging			即时通讯
vice-president			副总裁
stand up			失约，让人空等
go for			适用于
go a long way			对……大有帮助
talk shop			三句话不离本行，谈工作
air time			活动时间
approve of			满意，认同
uncalled-for			不必要的，多余的

tough spot	困境
body language	肢体语言

Key sentences

(1) Business etiquette is about building relationships with other people.

商务礼仪就是与他人建立好关系。

(2) Etiquette is not about rules & regulations, but is about providing basic social comfort and creating an environment where others feel comfortable and secure, this is possible through better communication.

商务礼仪不是规章制度,而是提供基本的社会舒适感,创造一个让他人感到舒适和安全的环境,这可以通过更好的交流来实现。

(3) Business etiquette differs from region to region and from country to country.

商务礼仪因地而异,各国不同。

(4) This creates a complex situation for people as it is hard to balance the focus on both international business etiquette and other business activities at the same time.

这给人们造成了复杂的局面,因为很难同时兼顾国际商务礼仪和其他商务活动。

(5) You were reading his body language this morning and he clearly wasn't sending a positive message.

今天上午你在读他的肢体语言,他显然没有传达积极的信息。

Learning test

Choose the following words below to fill in the blanks, change the form if necessary.

productive	distracted	complex	guideline	concise
focus	region	available	gratitude	Etiquette

(1) In theory, _____ is a way for everyone to express mutual respect for one another.

(2) It is hard to understate the consequences of Mr. Chen being _____ from his job.

(3) Guangdong is the most dynamic economic _____ in China.

(4) Negotiations become more _____ when each party acknowledges that the other may have legitimate concerns.

(5) I wish to express my _____ to Kathy Davis for her immense practical help.

(6) The UN's role in promoting peace is increasingly the _____ of international attention.

(7) Burton's text is _____ and informative.

(8) If you are _____ this afternoon, I want to talk to you.

(9) The only _____ it that the business had to sell a mobile-based service to help companies reach customers.

(10) The problem was more _____ than he supposed.

Translation

1. **Translate the following words into Chinese.**

   (1) tip   (2) gratitude

(3) witness   (4) concise
(5) courtesy   (6) etiquette
(7) virtual   (8) achievement
(9) handwritten   (10) encounter

2. Translate the following phrases into Chinese.

(1) talk shop   (2) uncalled-for
(3) go a long way   (4) tough spot
(5) business etiquette (6) body language

Practice

1. Translate the following sentences into Chinese.

(1) Turn our phones face down, close our laptops, and focus only on the meeting.

(2) If you're the meeting host, on time means at least five minutes early.

(3) If you prefer a paperless option, make sure your message is concise, thoughtful, and free of misspelled words.

(4) Don't say anything in email or instant messaging that you don't mind being broadcast to your entire organization.

(5) Don't dominate the "Questions" segment in meetings.

2. Translate the following sentences into English.

(1) 人是商务成功的关键因素。

(2) 最为重要的是，在任何情况下都要对周围的人礼貌且周到。

(3) 请牢记参会时间并确保准备妥当。

(4) 要紧握双方的手。

(5) 第一印象的影响相当长久。

Culture salon

The importance of business etiquette

In business, the relationships you build are critical. Establishing good rapport is significant if you want to progress your professional future, take on new clients, impress your boss or close that final sale. The way to build positive relationships in the business world is by exercising good etiquette, specifically by exhibiting top-notch communication skills. If others are speaking, give them your full attention and make eye contact to let them know you are engaged in the conversation. This is known as active listening. When it is your turn to speak, be clear and concise, and avoid jargon that your audience would not understand. Add a smile and a handshake so others find you pleasant to work with.

Chapter 5-
Section 1-
Text 1-看译文

Text 2 Customer service etiquette

Lead-in questions:

(1) What should the customer service staff remember when dealing with customers?

(2) What is Candy's job?

(3) What is Dick's problem?

(4) Why the items Dick ordered weren't shipped out?

(5) How did Jim solve the problem and make Dick satisfied?

When dealing with customers, in many cases we can forget that we must leave ourselves out of our interactions and remember that we represent a team/company. Always remember to care about the customer's complete satisfaction. Convey the feeling of caring in all of your interactions with customers. Make sure that the customer realizes that you truly appreciate them. They have to understand that you are just as concerned as they are about their problem or question as they are and you want to help them eliminate it.

Chapter 5-Section 1-Text 2-听课文

Following is an example for practice.

(Candy is a customer service officer in GE Healthcare, and Dick is a customer who has ordered the software packs last week.)

Candy: Good afternoon and thank you for calling GE's customer hotline. My name is Candy. How can I be of assistance?

Dick: Hi Candy. My name is Dick Johnson and I'm calling from New York. My company ordered eight packs of Pro 9 design software from your firm, but it has now been a week past the date the items were expected to arrive!

Candy: Okay sir... do you have the tracking number we sent you by e-mail when you first made the order?

Dick: Yes I do, and I also have the order number. The tracking number is T-159 and the original order number was 96153G.

Candy: Okay, please wait one second while I try to find your order. (1 minute later) Alright, I found your order number. Let me confirm the items with you here again: your order was for eight packs of Pro 9 design software, correct?

Dick: Yes, that's correct.

Candy: Well, it says here that the order was canceled. The items were scheduled to be shipped, but for some reason, a last-minute notation indicates the customer asked to cancel.

Dick: Well, I'm the customer and I certainly didn't ask to cancel. These software packets are very important to our company. We were scheduled to begin operations of a new design department, but because the items haven't arrived, I've got 10 designers sitting around with nothing to do!

Candy: I apologize for the mix-up, sir. Would you like me to reschedule a shipping date?

Dick: Yes, but I think there really should be some form of compensation for this error. I was reading the fine print in your customer contract and it says that if items don't arrive within the scheduled time, compensation will be paid.

Candy: Yes, normally that would be the case, but because the items were canceled, we can't be held liable.

Dick: I don't mean to sound rude, Candy, but the order was not canceled. Can you produce any evidence to indicate that we sent a message asking for a cancellation?

(Dick is a little angry)

Look... if you review our file, you'll find that we are long-term loyal customers who have bought a significant amount of software from your company. Are you really telling me that you're just going to ignore a major mistake? I don't think your attitude is very professional.

Candy: I'm sorry, sir... but I am not authorized to offer any compensation at this time.

Dick: Then who would be authorized to do so?

Candy: You would have to speak with my supervisor.

Dick: Okay then, is your supervisor currently available?

Candy: I believe he is... please hold on...

(1 minute later)

Jim: Hello, this is supervisor Jim Bradley. Candy told me your items did not arrive on time. We're very sorry for the inconvenience.

Dick: I appreciate your apology, but because we did not cancel the order, I believe I'm within my rights to ask for some sort of compensation.

Jim: Let me look over your file one more time... do you mind holding again?

Dick: No, that's fine.

Jim: Dick, I believe I know what the problem is. It seems one of our staff confused your order with another client's canceled order. We deeply regret this error and I will have your items shipped out to you today by express mail at no added charge to you. You should receive your software within 24 hours.

Dick: Well, thank you. I appreciate your prompt service.

Jim: Also in accordance with our compensation policy, we will only charge you for half of the original order. I hope these measures are satisfactory.

Dick: Yes, absolutely. Thank you very much Mr. Bradley. You've been very helpful.

Jim: You're very welcome, and I hope you'll continue working with Global Com.

Dick: We definitely will. Thank you again.

Jim: Not at all. Goodbye, Dick.

Words and phrases

etiquette	[ˈetɪket]	n.	礼仪,礼节
interaction	[ˌɪntərˈækʃn]	n.	互动
represent	[ˌreprɪˈzent]	vt.	代表
satisfaction	[ˌsætɪsˈfækʃn]	n.	满意,满足
convey	[kənˈveɪ]	vt.	传达,表达
appreciate	[əˈpriːʃieɪt]	vt.	感激,欣赏
concerned	[kənˈsɜːnd]	adj.	关心的,担心的
eliminate	[ɪˈlɪmɪneɪt]	vt.	消除,排除
hotline	[ˈhɒtlaɪn]	n.	咨询电话,热线电话
original	[əˈrɪdʒənl]	adj.	最初的
confirm	[kənˈfɜːm]	vt.	确认
cancel	[ˈkænsl]	vt.	取消
notation	[nəʊˈteɪʃn]	n.	符号,记号
indicate	[ˈɪndɪkeɪt]	vt.	表明,指出
reschedule	[ˌriːˈʃedjuːl]	vt.	重新排定
form	[fɔːm]	n.	形式
compensation	[ˌkɒmpenˈseɪʃn]	n.	补偿,赔偿
contract	[ˈkɒntrækt]	n.	合同
liable	[ˈlaɪəbl]	adj.	有责任的
rude	[ruːd]	adj.	粗鲁的,无礼的
evidence	[ˈevɪdəns]	n.	证据
cancellation	[ˌkænsəˈleɪʃn]	n.	取消
loyal	[ˈlɔɪəl]	adj.	忠实的,忠诚的
significant	[sɪɡˈnɪfɪkənt]	adj.	相当的
ignore	[ɪɡˈnɔː(r)]	vt.	忽视
professional	[prəˈfeʃnl]	adj.	专业的,职业的
authorize	[ˈɔːθəraɪz]	vt.	授权,批准
supervisor	[ˈsuːpəvaɪzə(r)]	n.	主管
available	[əˈveɪləbl]	adj.	有空的,可得的
inconvenience	[ˌɪnkənˈviːniəns]	n.	不便
right	[raɪt]	n.	权利
confuse	[kənˈfjuːz]	vt.	混淆
regret	[rɪˈɡret]	vt.	抱歉
added	[ˈædɪd]	adj.	额外的

prompt	[prɒmpt]	adj.	立刻的
measures	[ˈmeʒəz]	n.	措施
satisfactory	[ˌsætɪsˈfæktəri]	adj.	令人满意的
definitely	[ˈdefɪnətli]	adv.	肯定地，明确地
last-minute			最后的，紧急关头的
tracking number			跟踪号，查询号
leave out of			撇开，脱离
for some reason			由于某种原因
sit around			无所事事，闲坐着
mix-up			混乱，杂乱
fine print			细则
long-term			长期的
look over			仔细检查，翻阅
express mail			特快专递
in accordance with			依照，根据

Key sentences

(1) When dealing with customers, in many cases we can forget that we must leave ourselves out of our interactions and remember that we represent a team/company.

与客户打交道时，很多情况下我们应摆正自己的位置，并谨记我们代表着团队或公司。

(2) Convey the feeling of caring in all of your interactions with customers.

与客户交流时，向对方传递你的关心。

(3) My company ordered eight packs of Pro 9 design software from your firm, but it has now been a week past the date the items were expected to arrive!

我公司从你公司订购了8套Pro 9设计软件，但目前已超过预计到货时间一星期了。

(4) The items were scheduled to be shipped, but for some reason, a last-minute notation indicates the customer asked to cancel.

这些物品原是计划发货的，但不知由于什么原因，最后记录显示客户要求取消订单。

(5) I was reading the fine print in your customer contract and it says that if items don't arrive within the scheduled time, compensation will be paid.

我刚刚在阅读贵公司的客户合同细则，上面写到如果货品不能在预定时间内到货，贵公司将支付一定的赔偿。

Learning test

Choose the following words below to fill in the blanks, change the form if necessary.

eliminate	cancel	significant	appreciate	confirm
concerned	liable	etiquette	authorize	Confuse

(1) I here contact you to _____ that you have received our gift for you.

(2) I _____ your suggestion about the affair.

(3) A _____ number of drivers fail to keep to speed limits.

(4) He called the police because he was _____ for Jenny's safety.

(5) He _____ me to act for him while he is abroad.

(6) Feeling down is a part of life and you cannot _____ it totally.

(7) Before you have a dinner date with a beautiful girl, you should learn some eating _____ .

(8) He is _____ for the fault of his son.

(9) He kept asking unnecessary questions which only _____ the issue.

(10) I'm afraid I'll have to _____ our meeting tomorrow.

Translation

1. Translate the following words into Chinese.

(1) cancellation   (2) measures

(3) contract   (4) confirm

(5) satisfaction   (6) available

(7) hotline   (8) interaction

(9) supervisor   (10) professional

2. Translate the following phrases into Chinese.

(1) sit around   (2) tracking number

(3) fine print   (4) express mail

(5) look over   (6) mix-up

Practice

1. Translate the following sentences into Chinese.

(1) Would you like me to reschedule a shipping date?

(2) The tracking number is T-159 and the original order number was 96153G.

(3) I don't think your attitude is very professional.

(4) We deeply regret this error and I will have your items shipped out to you today by express mail at no added charge to you.

(5) Also in accordance with our compensation policy, we will only charge you for half of the original order.

2. Translate the following sentences into English.

(1) 这批订单的货件查询号码是2E960。

(2) 他们通过提高运营效率来提高客户满意度。

(3) 如果证明货品有缺陷,顾客有权索赔。

(4) 这些服务都是可以免费享用的。

(5)我们将竭力为你提供优质的产品和满意的服务!

> **Culture salon**
>
> **Customer satisfaction**
>
> Customer satisfaction is a term frequently used in marketing. It is a measure of how products and services supplied by a company meet or surpass customer expectation. Customer satisfaction is defined as the number of customers whose reported experience with a firm, its products, or its services exceeds specified satisfaction goals. It is seen as a key performance indicator within business and is often part of a Balanced Scorecard. In a competitive marketplace where businesses compete for customers, customer satisfaction is seen as a key differentiator and increasingly has become a key element of business strategy. Firms need reliable and representative measures of satisfaction to effectively manage customer satisfaction.

Chapter 5-Section 1-Text 2-看译文

Chapter 5-Section 1-拓展阅读

Chapter 5-Section 1-习题

Section 2　Marketing

Chapter 5-Section 2-PPT

Text 1　Medical device sales jobs

Lead-in questions:

(1) What is the difference between general medical sales jobs and medical device sales jobs?

(2) What do most medical device positions involve?

(3) What do medical device positions require?

(4) Why selling skills are important for medical devices sales jobs?

(5) What is the job outlook for medical device sales?

1. Overview of medical device sales jobs

The difference in general medical sales jobs and medical device sales jobs is the level of complexity of the product. Most medical device positions involve detailing the product often in an operating room or hospital environment. Medical device sales jobs often involve calling on surgeons, specialists or executive level clients. These positions require a high level of expertise in a given area and a proven sales track record. In contrast, a general sales position in medical sales selling medical products would most likely call on hospitals, alternate care facilities or general practitioners. The products sold in a medical sales position are often a little less complicated. Starting your medical sales career selling medical products may feel less glamorous, however it will help you establish your sales track record and your ability to sell to physicians and healthcare providers. Medical device companies seek to hire successful sales representatives with proven sales numbers and strong connections to the healthcare industry.

Chapter 5-
Section 2-
Text 1-听课文

2. Training and education for a medical device sales job

To get into a medical device sales job, it is important to have already gained a refined level of selling skills. Companies offering medical device sales jobs do not have time to train employees on basic sales skills. Training is focused on product. Consider starting your career in a more general medical sales role by selling disposable medical products. Medical sales selling products might help you gain the experience you need to later move into selling more complicated medical devices.

As a medical device sales representative you may be highly marketable to other device companies. Medical device sales jobs are on the rise in today's employment market. These jobs are not for the squeamish individual and require long hours in the operating room. Most medical device sales representatives cover several states and travel 50% of the time.

3. Opportunities in medical device sales

For years medical device companies have enjoyed strong gross margins and top-line growth. The list of medical device companies includes companies such as Medtronic, Stryker, Bard, Karl Storz, St Jude Medical, and many others.

4. Salary in medical device sales

Medical device sales jobs offer a competitive base salary and the ability to earn exceptional commissions. Successful medical device sales representatives are high-energy people and are not afraid of challenges.

5. Job outlook for medical device sales

The future of medical device sales is promising as technology and medicine continue to advance. Regardless of the economy, jobs in medical devices will continue to remain strong due to the constant need for health care.

Let's take the following recruitment as an example.

Medical device marketing manager

Job responsibilities:

(1) Defining marketing strategies based on geographical market analysis, coordinating marketing activities, determining medical imaging product definition and pricing, and defining Go-To-Market strategy for delivering competitive product offering with customer satisfaction and business results.

(2) Performing market analysis and gaining thorough understanding on regional marketing, sales and

hospital user demand, competitor analysis, industry insight, macroeconomic indicators, government policy, and service needs.

(3) Identifying and analyzing trends, creating long-term plans and alternative strategic scenarios to identify growth opportunities for company's business.

(4) Conducting market segmentation and product positioning.

(5) Implementing marketing intelligence processes to have up-to-date information on competitive marketing, communicating customer needs and competitiveness with company's management team, and translating them into product features.

(6) Coordinating with lead design engineers in defining product features and functions.

(7) Identifying expert physicians, and establishing collaboration with universities and reference clinical sites.

(8) Leading market promotion activities such as exhibitions, seminars, user group meetings and sales training.

(9) Leading company profile, web construction, cultural publicity.

Job qualifications:

(1) Bachelor degree or higher majoring in technology, engineering, and medicine, MBA degree is preferred.

(2) 5+ years' working experience in medical device industry as business, technical or clinical role.

(3) 2+ years' experiences in X-ray related products such as DR and CT.

(4) Excellent technical knowledge of medical equipment.

(5) Strong business development skills to enter new product and market segment.

(6) Excellent project management skills with strong results-orientated working style.

(7) Excellent communication skills and ability to collaborate in a team environment as well as working independently.

Words and phrases

marketing	[ˈmɑːkɪtɪŋ]	n.	市场营销
sales	[seɪlz]	n.	销售
complexity	[kəmˈpleksəti]	n.	复杂性
position	[pəˈzɪʃn]	n.	职位
involve	[ɪnˈvɒlv]	vt.	包含,涉及
detail	[ˈdiːteɪl]	vt.	提供细节,详述
surgeon	[ˈsɜːdʒən]	n.	外科医生
specialist	[ˈspeʃəlɪst]	n.	专家,专科医生
executive	[ɪgˈzekjətɪv]	n.	经理,主管
expertise	[ˌekspɜːˈtiːz]	n.	专长,专业技术

Chapter 5-Section 2-Text 1-听词汇

facility	[fəˈsɪləti]	n.	机构
complicated	[ˈkɒmplɪkeɪtɪd]	adj.	复杂的
glamorous	[ˈglæmərəs]	adj.	精彩,富有魅力的
representative	[ˌreprɪˈzentətɪv]	n.	代表
training	[ˈtreɪnɪŋ]	n.	培训
refined	[rɪˈfaɪnd]	adj.	熟练的,精练的
employee	[ɪmˈplɔɪi:]	n.	雇员,员工
disposable	[dɪˈspəʊzəbl]	adj.	一次性的
marketable	[ˈmɑ:kɪtəbl]	adj.	有市场的,有价的
squeamish	[ˈskwi:mɪʃ]	adj.	娇气的
individual	[ˌɪndɪˈvɪdʒuəl]	n.	个人
salary	[ˈsæləri]	n.	薪水,薪金
competitive	[kəmˈpetətɪv]	adj.	有竞争力的
exceptional	[ɪkˈsepʃənl]	adj.	额外的,优越的
commission	[kəˈmɪʃn]	n.	佣金
challenge	[ˈtʃæləndʒ]	n.	挑战
promising	[ˈprɒmɪsɪŋ]	adj.	有前途的
recruitment	[rɪˈkru:tmənt]	n.	招聘
responsibility	[rɪˌspɒnsəˈbɪləti]	n.	责任,职责
geographical	[ˌdʒi:əˈgræfɪkl]	adj.	区域的
coordinate	[kəʊˈɔ:dɪneɪt]	vt.	协调
satisfaction	[ˌsætɪsˈfækʃn]	n.	满意(度)
macroeconomic	[ˌmækrəʊˌi:kəˈnɒmɪk]	adj.	宏观经济的
indicator	[ˈɪndɪkeɪtə(r)]	n.	指标
scenario	[səˈnɑ:riəʊ]	n.	方案
conduct	[kənˈdʌkt]	vt.	管理,实施,执行
segmentation	[ˌsegmenˈteɪʃn]	n.	分割
implement	[ˈɪmplɪment]	vt.	贯彻,执行
intelligence	[ɪnˈtelɪdʒəns]	n.	情报,信息
feature	[ˈfi:tʃə(r)]	n.	特色,特征
profile	[ˈprəʊfaɪl]	n.	概况,简介,形象
qualification	[ˌkwɒlɪfɪˈkeɪʃn]	n.	资格
major	[ˈmeɪdʒə(r)]	vi.	主修,专攻
track record			追踪记录
call on			拜访,访问

on the rise	在增加,在增长
gross margin	毛利
top-line growth	顶线增长,盈利增长
base salary	基本工资
regardless of	不管
up-to-date	最新的
high-energy	精力充沛的
go-to-market	进入市场
bachelor degree	学士学位
company profile	公司简介,公司概况
results-orientated	以结果为导向

Key sentences

(1) Most medical device positions involve detailing the product often in an operating room or hospital environment.

大部分医疗器械的职位都需要在手术室或医院环境中详细说明产品。

(2) These positions require a high level of expertise in a given area and a proven sales track record.

这些职位需要在某一领域拥有高水平的专业知识并能提供令人信服的业绩记录。

(3) To get into a medical device sales job, it is important to have already gained a refined level of selling skills.

熟练的销售技能是获得医疗器械销售工作的重要因素。

(4) Medical device sales jobs offer a competitive base salary and the ability to earn exceptional commissions.

医疗器械销售工作提供有竞争力的薪酬和赚取额外佣金的能力。

(5) The future of medical device sales is promising as technology and medicine continue to advance.

随着科学技术和医学的不断进步,医疗设备行业发展前景光明。

Learning test

Choose the following words below to fill in the blanks, change the form if necessary.

executive	training	representative	competitive	promising
involve	complicated	disposable	commission	coordinate

(1) You must _____ the movements of your arms and legs when swimming.

(2) Like any successful _____, you should seek out the best help you can get.

(3) Part of your conversation might _____ telling parents how you feel.

(4) Dry hands with an air dryer, _____ paper towel or personal towel.

(5) The disposable medical device market is quite _____ in Guangdong with more than 3000 manufactures.

(6) He quit the job and ruined his _____ career.

(7) The sales _____ is calculated based on the quantity of sales in each month.

(8) The company provides technical and safety _____ once a year for the entire employee.

(9) This problem is very _____ and we don't know how to solve it yet.

(10) A sales _____ of medical devices should be high-energy.

Translation

1. Translate the following words into Chinese.

(1) feature   (2) conduct

(3) indicator   (4) major

(5) challenge   (6) profile

(7) recruitment   (8) facility

(9) qualification (10) specialist

2. Translate the following phrases into Chinese.

(1) base salary   (2) company profile

(3) track record   (4) on the rise

(5) gross margin   (6) up-to-date

Practice

1. Translate the following sentences into Chinese.

(1) Performing market analysis and gaining thorough understanding on regional marketing, sales and hospital user demand.

(2) Conducting market segmentation and product positioning.

(3) Coordinating with lead design engineers in defining product features and functions.

(4) Identifying expert physicians, and establishing collaboration with universities and reference clinical sites.

(5) Leading company profile, web construction, and cultural publicity.

2. Translate the following sentences into English.

(1) 调整营销策略和计划满足不断变化的市场需求。

(2) 根据公司发展状况,拟定产品年度销售计划。

(3) 参加此类产品的各项市场营销活动。

(4) 与国内专家保持良好的关系。

(5) 你至少要有2年的医疗器械产品营销的经历。

Chapter 5　Medical device marketing and management

> **Culture salon**
>
> ### Challenges of medical device sales jobs
>
> Frequent travel can be a challenge in terms of managing your schedule as a medical device sales representative. In addition, meeting sales goals, or "quotas" also can be a stressful part of the job, depending on the circumstances and how realistic or reasonably your goals were set. While medical device sales can often be very lucrative, your income is typically tied directly to your sales volume and percentage of goal achieved. Therefore, if you have a weak month in sales, you will also have a smaller paycheck that month. Some people do not like the variable paychecks that salespeople often receive. However, successful device sales reps find the perks of the job to be worth the long hours, extensive travel, and pressure to meet goals.

Chapter 5-
Section 2-
Text 1-看译文

Text 2　How medical device companies can strategically sell products to customers?

Lead-in questions:

(1) What are the factors influencing the health system?

(2) What is the most important approach for medtech sales teams to combat the economic pressure?

(3) What type of service pattern does the commercial customer beg to enjoy?

(4) What are critical steps for the companies to overcome the hurdles?

(5) How to recognize the key customer?

Economic pressures on the health system have been mounting for years, pushed by demographics, shrinking reimbursements, increasingly expensive therapies, and, ultimately, payer consolidation and reforms. Hospitals are the fulcrum and the pinch point for these pressures. Medtech companies no longer can afford to focus on clinical benefits without seeking other avenues to drive value for their hospital customers. This has led to the key account management imperative: Find a way to create value beyond the product alone for stakeholders beyond traditional clinical buyers.

Chapter 5-
Section 2-
Text 2-听课文

Key account management is critical to hone in on the most important customers—a pivotal approach for medtech sales teams to combat the aforementioned economic pressures and find areas of mutual value. Nobody doubts the need, but key account management and strategic selling are clouded by all sorts of practicalities in execution. Many multidivisional companies that stand to gain the most from a key account management and strategic selling capability face the biggest challenges in implementation. Divisions are structured around products and are autonomous with separate management structures, and sometimes these multidivisional companies haven't even fully integrated themselves from previous acquisitions. The practical questions are as follows:

- If we agree on the need, which businesses should "pay" for a key account management capability?
- Which roles are needed in account management?
- Where should these key account management roles report?
- Who actually "owns" the common customers in the end, as well as the final decisions?
- What should be done to compensate when the strength of one business is needed to subsidize the success of another?

These questions are deeply charged with financial importance and polarized by big company politics. When it comes to reporting, many multidivisional companies do not work across divisions until at the very top of the organization. Creating a mandate is the critical step, but many companies struggle to do so.

And yet the commercial customer is begging to be served in a more consolidated and coordinated fashion. "One face to the customer" is a common refrain, but it is so difficult for sales teams to deliver—especially in a multidivisional company, where the customer likely is served by a myriad of business units and service lines.

To overcome these hurdles, companies need to structure their sales approach around the customer, rather than around the offerings. There are three critical steps:

1. Examine the portfolio

Understand the value of the portfolio as a whole. Is it based on economics and scale or on a more integrated value that comes from using all of the component parts? Sometimes the value is in the combined offer to a department or a stakeholder. Occasionally, a portfolio is built around a site of care or even a specific type of patient. Conduct a critical examination of the value of the portfolio, with an emphasis on "critical." It can't be a theoretical value or an ambition. It needs to be a true value that comes from the portfolio. And it needs to be distinct from the value of your portfolio in comparison to your competition.

2. Examine the customer

Naturally, there isn't just a single customer. Department stakeholders, hospital stakeholders, and health system stakeholders must all be understood. In the end, there will be a customer who cares deeply about the portfolio that you offer, and this customer likely sits at some point below the very top of the organization—within cardiology, radiology, general surgery, biomedical engineering, etc. This step often isn't as clear-cut as it seems because no two customers are organized exactly the same, but there will be a point where your company's offering has a particular heft and resonance, where the common customer has a strong stake in the offerings. Another key ingredient is that this customer has leverage. This customer can influence the usage and brand decision. It is nice to have someone who cares, but this customer also needs to be able to drive action—be its use or standardization. This is your key customer, or your key account on which to focus.

3. Assess how the customer wants to interact

After choosing the key customer leverage point, the final step is to understand how this customer wants to be sold to. Is the customer looking for value-added services, programs, and outcomes or simply deals and contracts? Think about what your company currently offers and also where your portfolio is headed.

These three steps are imperative for understanding where the value is in your portfolio and how to operationalize it for a strategic customer. While it can be speculative, the outcome is a strategic selling capability that is aligned to customer needs.

Words and phrases

mount	[maʊnt]	vi.	增加,上升
demographics	[ˌdeməˈɡræfɪks]	n.	人口特征,人口结构
shrink	[ʃrɪŋkɪŋ]	vi.	收缩
reimbursement	[ˌriːɪmˈbɜːsmənt]	n.	报销,赔偿
therapy	[ˈθerəpi]	n.	治疗
ultimately	[ˈʌltɪmətli]	adv.	最后
consolidation	[kənˌsɒlɪˈdeɪʃn]	n.	合并,巩固
reform	[rɪˈfɔːm]	n.	改革
fulcrum	[ˈfʊlkrəm]	n.	支点
imperative	[ɪmˈperətɪv]	adj.	必要的,势在必行的
stakeholder	[ˈsteɪkhəʊldə(r)]	n.	利益相关者
hone	[həʊn]	vt.	磨炼,训练,提高
pivotal	[ˈpɪvətl]	adj.	关键的
combat	[ˈkɒmbæt]	vt.	与……战斗
aforementioned	[əˌfɔːˈmenʃənd]	adj.	上述的
mutual	[ˈmjuːtʃuəl]	adj.	相互的
cloud	[klaʊd]	vt.	混淆,混乱
practicality	[ˌpræktɪˈkæləti]	n.	实用性,实例
multidivisional	[ˈmʌltɪ dəˈvɪʒənl]	n.	多部门的,多分部的
implementation	[ˌɪmplɪmenˈteɪʃn]	n.	实施,履行
autonomous	[ɔːˈtɒnəməs]	adj.	自治的,有自主权的
acquisition	[ˌækwɪˈzɪʃn]	n.	收购,获得
capability	[ˌkeɪpəˈbɪləti]	n.	才能,能力
compensate	[ˈkɒmpenseɪt]	vt.	补偿,弥补
subsidize	[ˈsʌbsɪdaɪz]	vt.	资助,补贴
polarize	[ˈpəʊləraɪz]	vt.	分化,使两极分化
division	[dɪˈvɪʒn]	n.	部门

mandate	[ˈmændeɪt]	n.	授权,命令,指令
refrain	[rɪˈfreɪn]	n.	说法
myriad	[ˈmɪriəd]	n.	无数,大量
hurdle	[ˈhɜːdl]	n.	障碍
offering	[ˈɒfərɪŋ]	n.	产品
portfolio	[pɔːtˈfəʊliəʊ]	n.	投资组合
scale	[skeɪl]	n.	规模,比例
theoretical	[ˌθɪəˈretɪkl]	adj.	理论上的
comparison	[kəmˈpærɪsn]	n.	比较
heft	[heft]	n.	重要性,分量
resonance	[ˈrezənəns]	n.	共鸣,反响
ingredient	[ɪnˈgriːdiənt]	n.	要素,组成部分
leverage	[ˈliːvərɪdʒ]	n.	手段,影响力
standardization	[ˌstændədaɪˈzeɪʃn]	n.	标准化
operationalize	[ɒpəˈreɪʃənlaɪz]	vt.	实施
speculative	[ˈspekjələtɪv]	adj.	推测的
aligned	[əˈlaɪnd]	adj.	一致的
no longer			不再
seek to			追求,争取
mutual value			共同价值
hone in on			锁定目标
all sorts of			各种各样的
agree on			对……取得一致意见
in the end			终于,最后
as well as			也,和……一样
the top of			在……的顶部
clear-cut			清晰的,明确
value-added			增值的

Key sentences

(1) Medtech companies no longer can afford to focus on clinical benefits without seeking other avenues to drive value for their hospital customers.

医疗技术公司不能只专注于产品的临床效益,也应寻找其他途径为医院客户带来更大的价值。

(2) Key account management is critical to hone in on the most important customers—a pivotal approach formedtech sales teams to combat the aforementioned economic pressures and find areas of mutual value.

关键客户管理是锁定最重要客户的关键，这是医疗技术销售团队克服上述的经济压力，找到双方共同价值领域的最有效的途径。

(3) To overcome these hurdles, companies need to structure their sales approach around the customer, rather than around the offerings.

若要克服这些障碍，公司需要围绕客户构建其销售方式，而不是围绕产品。

(4) Naturally, there isn't just a single customer. Department stakeholders, hospital stakeholders, and health system stakeholders must all be understood.

当然，没有单一的客户。部门利益相关者、医院利益相关者和卫生系统利益相关方都必须了解。

(5) After choosing the key customer leverage point, the final step is to understand how this customer wants to be sold to.

在选择了关键的客户杠杆点之后，最后一步就是了解客户想要如何将产品出售给他。

Learning test

Choose the following words below to fill in the blanks, change the form if necessary.

reimbursement   shrink   avenue   compensate   imperative

capability   division   hurdle   ingredient   mutual

(1) The company has established a marketing _____ of 100 employees.

(2) Friendship is based on good feeling and _____ understanding.

(3) Nothing can _____ for the loss of one's health.

(4) The president wants to explore every _____ towards peace in the region.

(5) Don't forget to keep all your receipts, as you'll need them for _____.

(6) Investment in new product development is an essential _____ of corporate success.

(7) No one ever questioned her working _____.

(8) Washing wool in hot water will make it _____.

(9) Open communication between the jewellery designer and the client is _____.

(10) Next time you face a _____ to success, find a quite setting, relax, and close your eyes.

Translation

1. Translate the following words into Chinese.

(1) critical   (2) multidivisional

(3) myriad   (4) reimbursement

(5) mandate   (6) acquisition

(7) offering   (8) pivotal

(9) standardization   (10) subsidize

2. Translate the following phrases into Chinese.

(1) value-added   (2) agree on

(3) as well as (4) hone in on
(5) clear-cut (6) all sorts of

Practice

1. Translate the following sentences into Chinese.

(1) Understand the value of the portfolio as a whole.

(2) This step often isn't as clear-cut as it seems because no two customers are organized exactly the same.

(3) It is nice to have someone who cares, but this customer also needs to be able to drive action—be it use or standardization.

(4) These three steps are imperative for understanding where the value is in your portfolio and how to operationalize it for a strategic customer.

(5) Is the customer looking for value-added services, programs, and outcomes or simply deals and contracts?

2. Translate the following sentences into English.

(1) 制订销售计划是一个好的开始。

(2) 成功的销售是深思熟虑的行动。

(3) 你需要确定目标市场。

(4) 销售，是团队合作而非单枪匹马。

(5) 做一个值得信赖的产品顾问而非产品促销者。

Culture salon

Sales strategy

A sales strategy consists of a plan that positions a company's brand or product to gain a competitive advantage. Successful strategies help the sales force focus on target market customers and communicate with them in relevant, meaningful ways. Planning and creating an effective sales strategy requires looking at long-term sales goals and analyzing the business sales cycle, as well as meeting with sales people about their personal career goals. After creating the long-term sales strategy based on long-term goals, sales managers should create monthly and weekly sales strategies based on the long-term strategy.

Chapter 5-Section 2-Text 2-看译文

Chapter 5-Section 2-习题

Section 3 Business negotiation

Chapter 5-
Section 3-PPT

Text 1 How to win price negotiations?

Lead-in questions:

(1) What is the economic relationship between seller and buyer?

(2) What is the goal of the win-win sales situation?

(3) What sources of the product could you list for reducing value?

(4) Why do the buyers always expect lowered prices from suppliers?

(5) Why negotiation strategy is important in the price negotiation?

Chapter 5-
Section 3-
Text 1-听课文

Think of the economic relationship between seller and buyer as a balance between benefits and price. On one side are the benefits delivered by your solution. On the other side is the price the customer must pay to realize these benefits.

The goal of any win-win sales situation is to maintain a fair and even balance between price and benefits. If benefits are increased, they will outweigh the price and disrupt the balance. Likewise, an increase in price will outweigh the benefits and disrupt the balance.

When a buyer asks for a lower price, he or she is effectively disrupting the balance in value. A lower price essentially tilts the value in favor of the buyer.

In order to maintain balance in the value equation, when the buyer asks for a lower price, we can respond with a resounding "Yes!" …with a caveat. We must remove something from the benefit side of the equation to maintain the balance in value. Essentially, we are reducing the benefits of the product for the lower price.

Look for product features, services, delivery schedules, warranties, payment terms or lower-priced product alternatives as sources for reducing value. Remove or reduce these value features to decrease the delivered benefits. From a negotiation standpoint, the trade for value does not necessarily need to be completely equitable. You just need to communicate to the buyer that there is a "price" to be paid for a lower price. Lower prices are not free.

Buyers often expect lowered prices from suppliers because the buyer has historically trained the seller to agree to lower prices without concessions. Your job is to train the buyer to expect a trade in the benefits delivered in return for a lower price.

Smart salespeople will be prepared for this type of value-trading discussion. Identify items in your proposal that you can remove if the buyer pushes back on price. Be prepared to tell the buyer what items can be removed from the proposal at the moment the buyer asks for a discounted price.

We find, in many cases, that the buyer is not willing to sacrifice any component of the proposal and agrees to the full price.

Empowering your sales team with this negotiation strategy helps you strengthen the perceived value of your brand as you ensure healthier margins. Once buyer and seller agree to place value over price, you will discover how easy it is to hold the line on price and build more profitable business relationships.

Following part is a dialogue for practice.

(Robert Liu is a sales manager of ABC company in China. Dan Smith is a representative of the company in America. They are negotiating the prices of the products.)

Dan: I'd like to get the ball rolling by talking about prices.

Robert: Shoot. I'd be happy to answer any questions you may have.

Dan: Your products are very good. But I'm a little worried about the prices you're asking.

Robert: You think we are asking for more?

Dan: That's not exactly what I had in mind. I know your research costs are high, but what I'd like is a 25% discount.

Robert: That seems to be a little high, Mr. Smith. I don't know how we can make a profit with those numbers.

Dan: Please call me Dan. Well, if we promise future business, volume sales, that will slash your costs, right?

Robert: Yes, but it's hard to see how you can place such large orders. How could you turn over so many? We need a guarantee of future business, not just a promise.

Dan: We said we wanted 1000 pieces over a six-month period. What if we place orders for twelve months, with a guarantee?

Robert: If you can guarantee that on paper, I think we can discuss this further.

(Three days later, they sit down again talking about the prices)

Robert: Even with volume sales, our costs won't go down much.

Dan: Just what are you proposing?

Robert: We could take a cut on the price. But 25% would slash our profit margin. We suggest a compromise 10%.

Dan: That's a big change from 25%! 10% is beyond my negotiating limit. Any other ideas?

Robert: I don't think I can change it right now.

Dan: I understand. We propose a structured deal. For the first six months, we get a discount of 20%, and the next six months we get 15%.

Robert: How about 15% the first six months, and the second six months at 12%, with a guarantee of 3000 units?

Dan: That's a lot to sell, with very low profit margins.

Robert: It's about the best we can do, Dan. We need to hammer something out today. If I go back empty-handed, I may be coming back to you soon to ask for a job.

Dan: OK, 17% the first six months, 14% for the second?!

Robert: Good, done. Dan, this deal promises big returns for both sides. Let's hope it's the beginning of a long and prosperous relationship.

Words and phrases

negotiation	[nɪˌɡəʊʃiˈeɪʃn]	n.	谈判
fair	[feə(r)]	adj.	公平的
price	[praɪs]	n.	价格
benefit	[ˈbenɪfɪt]	n.	利益,利润
outweigh	[ˌaʊtˈweɪ]	vt.	超过,胜过
disrupt	[dɪsˈrʌpt]	vt.	破坏,扰乱
likewise	[ˈlaɪkwaɪz]	adv.	同样地,也
essentially	[ɪˈsenʃəli]	adv.	本质上
tilt	[tɪlt]	vt.	倾斜,倾向
equation	[ɪˈkweɪʒn]	n.	等式,方程
resounding	[rɪˈzaʊndɪŋ]	adj.	响亮的
caveat	[ˈkæviæt]	n.	警告,告诫
warranty	[ˈwɒrənti]	n.	保修,保修期,保证
alternative	[ɔːlˈtɜːnətɪv]	n.	供替代的选择,替换物,替代品
standpoint	[ˈstændpɔɪnt]	n.	立场,观点
equitable	[ˈekwɪtəbl]	adj.	公平的,公正的
concession	[kənˈseʃn]	n.	让步
salespeople	[ˈseɪlzˌpiːpl]	n.	销售人员
discount	[ˈdɪskaʊnt]	vt. & n.	打折,减价;折扣
sacrifice	[ˈsækrɪfaɪs]	vt.	牺牲
empower	[ɪmˈpaʊə(r)]	vt.	授权,允许
perceive	[pəˈsiːv]	vt.	感知,感觉,察觉
margin	[ˈmɑːdʒɪn]	n.	利润,盈余
profitable	[ˈprɒfɪtəbl]	adj.	赚钱的,合算的
volume	[ˈvɒljuːm]	adj.	大量的,批量的
slash	[slæʃ]	vt.	(大幅)削减
guarantee	[ˌɡærənˈtiː]	vt. & n.	保证
propose	[prəˈpəʊz]	vt.	建议,打算
compromise	[ˈkɒmprəmaɪz]	n.	妥协,折中,让步

structured	[ˈstrʌktʃəd]	adj.	结构性的
deal	[diːl]	n.	交易
prosperous	[ˈprɒspərəs]	adj.	良好的,繁荣的
win-win			双赢的
ask for			请求,要求
in favor of			有利于,支持
payment term			付款期限,付款方式
in return			作为回报
at the moment			此刻,目前
get the ball rolling			开始
volume sale			批量销售,大量销售
turn over			营业额达到
take a cut			削减
hammer out			敲定,做出决定
empty-handed			空手地,两手空空

Key Sentences

(1) Think of the economic relationship between seller and buyer as a balance between benefits and price.

把买家和卖家之间的经济关系看作是利益和价格之间的平衡关系。

(2) The goal of any win-win sales situation is to maintain a fair and even balance between price and benefits.

任何双赢销售模式的目标是在价格和利益之间保持公平,寻求平衡。

(3) When a buyer asks for a lower price, he or she is effectively disrupting the balance in value.

当买家要求一个更低的价格,他或她实际上是破坏了价值等式的平衡。

(4) Buyers often expect lowered prices from suppliers because the buyer has historically trained the seller to agree to lower prices without concessions.

买家通常期望从供应商处获得更低价格,因为历来买方都是训练卖方无条件地同意买方提出的更低价格。

(5) Empowering your sales team with this negotiation strategy helps you strengthen the perceived value of your brand as you ensure healthier margins.

放手让销售团队实施谈判策略,有助于你在确保公司的良性利润的情况下增强公司品牌的感知价值。

Learning test

Choose the following words below to fill in the blanks, change the form if necessary.

prosperous	alternative	slash	outweigh	guarantee
concession	profitable	standpoint	empower	warranty

(1) The company has to _____ the price to increase the sales volume.

(2) I wish you a _____ and happy new year!

(3) He said there is no _____ for him but to maintain order under any circumstances.

(4) It is foolish to buy a car without a _____ .

(5) I _____ that he will come on time.

(6) Our cost of production allows us to be _____ everywhere, not only in Europe but also in China.

(7) It's really impossible for us to make any _____ by allowing you any commission.

(8) The advantages of this deal largely _____ the disadvantages.

(9) From my _____ , this thing is just ridiculous.

(10) I _____ my agent to make the deal for me.

Translation

1. Translate the following words into Chinese.

(1) perceive   (2) caveat

(3) guarantee   (4) profitable

(5) compromise   (6) propose

(7) equitable   (8) discount

(9) deal   (10) sacrifice

2. Translate the following phrases into Chinese.

(1) in favor of   (2) payment term

(3) in return   (4) get the ball rolling

(5) win-win   (6) volume sale

Practice

1. Translate the following sentences into Chinese.

(1) Look for product features, services, delivery schedules, warranties, payment terms or lower-priced product alternatives as sources for reducing value.

(2) From a negotiation standpoint, the trade for value does not necessarily need to be completely equitable.

(3) If we promise future business, volume sales, that will slash your costs, right?

(4) We propose a structured deal. For the first six months, we get a discount of 20%, and the next six months we get 15%.

(5) This deal promises big returns for both sides. Let's hope it's the beginning of a long and prosperous relationship.

2. Translate the following sentences into English.

(1) 即使是大量销售,我们的生产成本仍无法降低太多。

(2) 我们提出一项阶段式的协议。

(3) 去年几个季度里,毛利润保持在20%左右。

(4) 这么大的购买数量,贵方应给我们打折扣。

(5) 我们真诚地希望与客户携手合作,建立双赢的关系。

Culture salon

What is a "profit margin"

Profit margin is part of a category of profitability ratios calculated as net income divided by revenue, or net profits divided by sales. Net income or net profit may be determined by subtracting all of a company's expenses, including operating costs, material costs (including raw materials) and tax costs, from its total revenue. Profit margins are expressed as a percentage and, in effect, measure how much out of every dollar of sales a company actually keeps in earnings. A 20% profit margin, then, means the company has a net income of $0.20 for each dollar of total revenue earned.

Text 2 Meet goals in negotiation

Lead-in questions:

(1) What is the crucial skill when a smaller business tries to further develop?

(2) What are the guidelines for businessmen to help them meet goals in negotiation?

(3) Why does a businessman need to set the personal ethical standard?

(4) What do the tempted negotiations tend to do?

(5) What is the first step to improve your negotiating power?

No matter how much you may hate to negotiate yourself into a deal—or even out of one—negotiating is a very legitimate business skill to acquire.

It's even more crucial if you have a smaller business trying to get off the ground. You will have to make your arguments against much bigger, more powerful entities, so it's essential that you know the science behind negotiation skills and how they affect the other party's psyche.

Following are four guidelines will help you meet these goals in the business negotiations.

1. Set a personal standard

Before entering a negotiation, set a personal ethical standard for your behavior.

How ethical do you want to be?

Determining in advance which behaviors are off-limits should help you recognize ethical dilemmas when they arise and make decisions that meet your standard.

Moreover, make a plan to address specific ethical dilemmas that you may encounter. The well-prepared job candidate would be ready respond to strategically yet ethically to questions about your other offers, perhaps by pointing out the benefits of remaining focused on hammering out an offer that satisfies both sides.

2. Question your perceptions

The more tempted negotiators are to lie, the more likely they are to believe their opponents will lie to them. Recognize that your perceptions of your counterpart's ethics may be inaccurate, driven by your own desire to behave unethically.

3. Enhance your power

If powerlessness motivates deception, it stands to reason that you should work hard to increase your negotiating power.

Exploring your outside alternatives is an obvious first step. Roger Fisher, co-founder of the program on negotiation at Harvard Law School, has identified several other sources of power: your skills, your knowledge, a strong relationship with your counterpart, and even the generation of an elegant solution.

Thinking creatively about sources of power helps you avoid making unethical statements. Even thinking about a negotiation in which you had more power can enhance your sense of power.

4. Personalize your opponent

When negotiating with a group, strive to view each member as an individual. Our research revealed that negotiators were more likely to lie to groups than to individuals, but simply providing the names of group members diminished that tendency. Getting to know opposing group members will help you adhere to your ethical standards.

Following part is a dialogue for practice.

(Mark Davis is the representative of Botany Bay, a medical device company. The Medic-Disk is a new product in Botany Bay widely used in hospitals and nursing homes. Wang Lin, the representative of Pacer, is negotiating with Mark Davis to get the sole agency of the product in China market.)

Mark: Mr. Wang, total sales of the Medic-Disk were U. S. $ 100,000 last year, through our agent in Hong Kong.

Wang: Our research shows that most of your sales are made in the Mainland China.

Mark: True, but we are happy with the sales. It's a new product. How could you do better?

Wang: We have already well-established medical products business. The Medic-Disk would be a good addition to our product range.

Mark: What kind of distribution capabilities do you have?

Wang: We have salespeople in ten major areas around the nation, selling directly to customers.

Mark: What about your sales?

Wang: In terms of unit sales, 55 percent are still from North China and East China. The rest comes from central China, south China and northeast China. That's a great deal of untapped market potential, Mr. Davis.

Mark: Mr. Wang, what kinds of sales do you think you could get?

Wang: Well, to begin with, we'd have to insist on sole agency in North China and East China. We believe we could spike sales by 28% to 38% in the first year. But certain conditions would have to be met.

Mark: What kinds of conditions?

Wang: We'd need your full technical and marketing support.

Mark: Could you explain what you mean by that?

Wang: We'd like you to give training to our technical staff; we'd also like you to pay a fee for after-sales service.

Mark: It's no problem with the training. As for service support, we usually pay a yearly fee, pegged to total sales.

Wang: Sounds OK. As for marketing support, we would like you to assume 50% of all costs.

Mark: We'd prefer 40%. Many customers learn about our products through international magazines, trade shows, and so on. We pick up the tab for that, but you get the sales in North China and East China.

Wang: We'll think about it, and talk more tomorrow.

Mark: Fine. We'd like you to tell us about your marketing plans.

Words and phrases

legitimate	[lɪˈdʒɪtɪmət]	adj.	合法的,正当的
crucial	[ˈkruːʃl]	adj.	重要的
argument	[ˈɑːgjumənt]	n.	论据,论点,争论
entity	[ˈentəti]	n.	实体
science	[ˈsaɪəns]	n.	科学
affect	[əˈfekt]	vt.	影响
party	[ˈpɑːti]	n.	一方,当事人
psyche	[ˈsaɪki]	n.	心灵,心理,心智
ethical	[ˈeθɪkl]	adj.	道德的,伦理的
behavior	[bɪˈheɪvjə]	n.	行为,举止
off-limits	[ɔːfˈlɪmɪts]	adj.	禁止的
recognize	[ˈrekəgnaɪz]	vi.	确认
dilemma	[dɪˈlemə]	n.	困境,窘境,进退两难
moreover	[mɔːrˈəʊvə(r)]	adv.	此外,而且
specific	[spəˈsɪfɪk]	adj	具体的,特别的
encounter	[ɪnˈkaʊntə(r)]	vt.	遭遇,遇到
candidate	[ˈkændɪdət]	n.	候选人
satisfy	[ˈsætɪsfaɪ]	vt.	使满意
perception	[pəˈsepʃn]	n.	看法
tempt	[tempt]	vt.	诱惑,吸引

opponent	[ə'pəʊnənt]	n.	对手
counterpart	['kaʊntəpɑːt]	n.	对手方,对方
behave	[bɪ'heɪv]	n.	表现
powerlessness	['paʊələsnəs]	n.	无力,无能为力
motivate	['məʊtɪveɪt]	vt.	激励,刺激,使有动机
deception	[dɪ'sepʃn]	n.	欺骗,欺诈
explore	[ɪk'splɔː(r)]	vt.	探索
statement	['steɪtmənt]	n.	陈述,声明
enhance	[ɪn'hɑːns]	vt.	提高,加强
personalize	['pɜːsənəlaɪz]	vt.	个人化,个性化
strive	[straɪv]	vi.	努力,力求
reveal	[rɪ'viːl]	vt.	显示,透露
diminish	[dɪ'mɪnɪʃ]	vi.	减少,缩小
tendency	['tendənsi]	n.	倾向,趋势
Mainland	['meɪnlænd]	n.	大陆
get off the ground			起步,取得进展
in advance			预先,提前
point out			指出,指明
focus on			集中于
co-founder			共同创立人
adhere to			坚持
nursing home			疗养院
distribution capability			分销能力
in terms of			在……方面
sole agency			独家代理
after-sales service			售后服务
pegged to			与……挂钩
pick up thetab			承担费用,替人付账
marketing plan			营销计划

Key sentences

(1) No matter how much you may hate to negotiate yourself into a deal—or even out of one—negotiating is a very legitimate business skill to acquire.

不管你多么讨厌交易谈判或者甚至没有谈判经历,谈判都是生意场上必须掌握的正当商业技巧。

(2) Before entering a negotiation, set a personal ethical standard for your behavior.

谈判之前,设定个人行为的道德标准。

(3) The more tempted negotiators are to lie, the more likely they are to believe their opponents will lie to them.

谈判者越想撒谎,他们越有可能相信对方会欺骗他们。

(4) If powerlessness motivates deception, it stands to reason that you should work hard to increase your negotiating power.

如果你的弱势助长了对方行骗,那么,你应该努力提高谈判能力。

(5) Getting to know opposing group members will help you adhere to your ethical standards.

了解对方团队的成员,可以帮助你坚持自己的道德标准。

Learning test

Choose the following words below to fill in the blanks, change the form if necessary.

| legitimate | opponent | satisfy | explore | dilemma |
| psyche | specific | motivate | deception | argument |

(1) We should make a concrete analysis of each _____ question.

(2) There have been a lot of _____ about who was responsible for the accident.

(3) I'm going to _____ the possibility of a part-time job.

(4) We need to _____ client expectations.

(5) How do you _____ your employees to work hard and efficiently?

(6) A characteristic of the feminine _____ is to seek approval from others.

(7) She didn't have the courage to admit to her _____ .

(8) Their business operations are perfectly _____ .

(9) He is admired even by his political _____ .

(10) They try to find a way out of their _____ .

Translation

1. Translate the following words into Chinese.

(1) reveal   (2) encounter

(3) motivate   (4) enhance

(5) tendency   (6) candidate

(7) counterpart   (8) negotiation

(9) entity   (10) argument

2. Translate the following phrases into Chinese.

(1) in advance   (2) distribution capability

(3) in terms of   (4) get off the ground

(5) sole agency   (6) after-sales service

Practice

1. Translate the following sentences into Chinese.

(1) It's even more crucial if you have a smaller business trying to get off the ground.

(2) Determining in advance which behaviors are off-limits should help you recognize ethical dilemmas when they arise and make decisions that meet your standard.

(3) Exploring your outside alternatives is an obvious first step.

(4) We have salespeople in ten major areas around the nation, selling directly to customers.

(5) We'd need your full technical and marketing support.

2. Translate the following sentences into English.

(1) 良好的分销能力对公司的发展很重要。

(2) 我们来访的原因是想和你们签订一项为期两年的独家代理协议。

(3) 我们的产品和售后服务都是一流的。

(4) 该公司还缺乏战略性营销计划。

(5) 他们的销售员对自己的产品充满热情,并决定开发新市场。

Culture salon

Distribution methods

Direct sales let you do your market research and choose your own customers while setting the selling price. Direct selling is a good match for a marketing plan that has identified, researched and segmented the final customers. Wholesale distribution leaves the selling to wholesalers and retailers specialized in retail sales. This choice is especially valid if your potential customers are widely dispersed or located far from your facilities. Mail order is a low-cost distribution channel that is convenient for the customer. You can use mail order by buying mailing lists or placing ads in a suitable publication. Online selling features the removal of intermediaries while still reaches large groups of potential customers. Harnessing social networks, online ad campaigns and message boards to spread the word, you can achieve substantial sales volumes quickly.

Chapter 5-Section 3-Text 2-看译文

Chapter 5-Section 3-习题

Chapter 6

Industrial communication

Learning objective

In this chapter, students should learn some skills of writing and communication in job hunting, such as resume, job hunting letter and review. Read the following passage, you should meet the following requirements:

(1) Grasp basic concepts, key points, related words and phrases.

(2) Acquire methods of writing resumes, job-hunting letters and skills of the interview.

(3) Understand responsibilities of medical devices sales representatives, and ways of contacting customers.

Section 1 Job hunting

Text 1 Resume for employment

Lead-in questions:

(1) What is a resume?

(2) What's the main information included in one's resume?

(3) What are the two aspects regarding the applicant's experience?

(4) What's the requirement of the resume layout?

(5) How to avoid wrong grammars when writing the resume?

A resume for employment usually refers to forms that an individual seeking employment must fill out as part of the process of informing an employer of the applicant's availability and desire to be employed, and persuading the employer to offer the applicant employment.

A resume usually asks the applicant at the minimum for their name, phone number, and address. In addition, it also asks for previous employment information, educational background, emergency contacts, references, as well as any special skills the applicant might have. The following categories are the main aspects appearing in the formal resume:

• Contact information

- Physical characteristics
- Experience
- Socio-environmental qualifications
- Skills and accomplishments
- Photograph

1. Contact information

Make sure you feature your name prominently (usually centered and bolded) on the top of the page. Below your name, list your phone number, mailing address and e-mail address.

2. Physical characteristics

If the company has a bona fide occupational qualification (BFOQ) to ask regarding a physical condition, they may ask questions about it, for example: The job requires a lot of physical labor. Do you have any physical problems that may interfere with this job?

3. Experience

Experience requirements can be separated into two groups on an application, educational background and work experience. Educational background is important to companies because by evaluating applicants' performance in school tells them what their personality is like as well as their intelligence. Work experience is important to companies because it will inform the company if the applicant meets their requirements. Companies are usually interested when the applicants were unemployed and when/why the applicant left their previous job.

4. Socio-environmental qualifications

Companies are interested in the applicant's social environment because it can inform them of their personality, interests, and qualities. If they are extremely active within an organization, that may demonstrate their ability to communicate well with others. Being in management may demonstrate their leadership ability as well as their determination and so on.

5. Skills and accomplishments

Summarize your special skills and significant achievements using the chronological experience format. For example, the ability to work in a team, manage people, customer service skills, or specific IT skills.

6. Photograph

Customs vary internationally when it comes to the inclusion or non-inclusion of a photograph of the applicant. In the English-speaking countries, notably the United States, this is not customary, and books or websites giving recommendations about how to design a resume typically advise against it unless explicitly requested by the employer. In other countries, for instance Germany, the inclusion of a photograph of the applicant is still common, and many employers would consider a resume incomplete without it.

There are many ways to create an exceptional resume, but for a solid foundation, concentrate on four main points as below:

(1) Grammar

There should be no mistakes in your resume. Use a spell checker and enlist a second pair of eyes to check over the text. Try and include as many active words as possible to increase the impact of your resume. Use active verbs to replace passive verbs and nouns wherever possible. For example, you could include targeted words like 'created', 'analyzed'; and 'devised' to present yourself as a person that shows initiative.

(2) Layout

Place your most attractive skills and talents towards the top of your resume to boost your chances of impressing an employer. The same rule applies to listing grades-always place your highest grade first.

(3) Presentation

Keep your resume neat and make sure it is easy on the eye. Bullet points should be used to tidy up any lists. Your choice of font can have more impact than you might think. The University of Kent careers service suggest using 10 pointVerdana or Lucida Sans with a larger typeface for headings and sub-headings. You should always avoid Comic Sans.

(4) Style

There a various types of resumes you can employ. Think carefully about what style will suit your needs. Choose the formal written English to complete your resume.

From the employer's perspective, the resume serves a number of purposes. These vary depending on the nature of the job and the preferences of the person responsible for hiring, as "each organization should have an application form that reflects its own environment". A resume usually requires the applicant to provide information sufficient to demonstrate that he or she is legally permitted to be employed. The typical resume also requires the applicant to provide information regarding relevant skills, education, experience and so on. The resume itself is a minor test of the applicant's literacy, penmanship, and communication skills-a careless job applicant might disqualify themselves with a poorly filled-out application, even the smallest grammatical mistake will foil your hard work, by contrast, an effect one might help you successfully apply for your position.

Chapter 6- Section 1- Text 1-听词汇

When you're constructing your resume, you need to be able to brag about yourself without exaggerating. Another point is making your resume stand out without descending to tricks and designs that merely frustrate the resume reader.

Words and phrases

resume	[rɪˈzjuːm]	n.	简历
individual	[ˌɪndɪˈvɪdʒuəl]	n.	个人
applicant	[ˈæplɪkənt]	n.	申请人
desire	[dɪˈzaɪə(r)]	n.	渴望,愿望
persuade	[pəˈsweɪd]	vt.	说服

previous	[ˈpriːviəs]	adj.	以前的
employment	[ɪmˈplɔɪmənt]	n.	就业,雇用
background	[ˈbækɡraʊnd]	n.	背景
emergency	[iˈmɜːdʒənsi]	n.	突发事件,紧急情况
reference	[ˈrefrəns]	n.	推荐信,介绍信
qualification	[ˌkwɒlɪfɪˈkeɪʃn]	n.	资格
accomplishment	[əˈkʌmplɪʃmənt]	n.	成就
feature	[ˈfiːtʃə(r)]	vt.	突出,特写
evaluate	[ɪˈvæljueɪt]	vt.	评估,评价
personality	[ˌpɜːsəˈnæləti]	n.	个性,人格
intelligence	[ɪnˈtelɪdʒəns]	n.	智力,理解力
inform	[ɪnˈfɔːm]	vt.	告诉,通知
organization	[ˌɔːɡənaɪˈzeɪʃn]	n.	机构,组织
leadership	[ˈliːdəʃɪp]	n.	领导,领导才干
significant	[sɪɡˈnɪfɪkənt]	adj.	重要的,有意义的
achievement	[əˈtʃiːvmənt]	n.	成就,完成
format	[ˈfɔːmæt]	n.	版式,格式,形式
customer	[ˈkʌstəmə(r)]	n.	顾客,主顾
specific	[spəˈsɪfɪk]	adj.	特殊的
inclusion	[ɪnˈkluːʒn]	n.	包含
recommendation	[ˌrekəmenˈdeɪʃn]	n.	推荐,建议
design	[dɪˈzaɪn]	vt.	设计
common	[ˈkɒmən]	adj.	常见的
solid	[ˈsɒlɪd]	adj.	坚实的,坚固的
foundation	[faʊnˈdeɪʃn]	n.	根基,基础
concentrate	[ˈkɒnsntreɪt]	vt.	集中,专注于
impact	[ˈɪmpækt]	n.	影响,冲击力
replace	[rɪˈpleɪs]	vt.	取代,替换
initiative	[ɪˈnɪʃətɪv]	n.	主动性,能动性
layout	[ˈleɪaʊt]	n.	布局,安排,设计
talent	[ˈtælənt]	n.	才能,天赋
impress	[ɪmˈpres]	vt.	使……有印象,影响
presentation	[ˌpreznˈteɪʃn]	n.	外观,样式
neat	[niːt]	adj.	整洁的,简洁的
perspective	[pəˈspektɪv]	n.	观点,看法

responsible	[rɪˈspɒnsəbl]	*adj.*	有责任的,负责的
reflect	[rɪˈflekt]	*vt.*	反映,反射
sufficient	[səˈfɪʃnt]	*adj.*	足够的,充分的
penmanship	[ˈpenmənʃɪp]	*n.*	写作,书写技巧
exaggerate	[ɪɡˈzædʒəreɪt]	*vi.*	夸张,夸大
frustrate	[frʌˈstreɪt]	*vt.*	挫败,击败

refer to	涉及,指的是
educational background	教育背景
emergency contact	紧急联络人
contact information	联系方式,联系信息
physical characteristic	体型特征
socio-environmental qualification	社会环境条件
interfere with	影响,妨碍
communicate with	与……交流
as well as	也,还,和……一样
for instance	例如
concentrate on	集中于,专注于
apply to	适用于,运用于
depend on	依靠,取决于
stand out	脱颖而出

Key sentences

(1) A resume for employment usually refers to forms that an individual seeking employment must fill out as part of the process of informing an employer of the applicant's availability and desire to be employed, and persuading the employer to offer the applicant employment.

求职简历通常是指求职者必须填写的表格,作为向雇主通报申请人的就业意愿的一部分,并说服雇主为申请人提供就业机会。

(2) Experience requirements can be separated into two groups on an application, educational background and work experience. Educational background is important to companies because by evaluating applicants' performance in school tells them what their personality is like as well as their intelligence.

工作经历可分为两部分:教育背景及工作经历。教育背景对公司很重要,因为公司可通过评估申请人在学校的表现,得知他们的性格和智力。

(3) Summarize your special skills and significant achievements using the chronological experience format.

按时间顺序总结你的特殊技能和重要成就。

(4) Place your most attractive skills and talents towards the top of your resume to boost your chances of

impressing an employer. The same rule applies to listing grades-always place your highest grade first.

把你最具吸引力的技能和才华放在简历的顶部，以增加你给雇主留下深刻印象的机会。同样的规则也适用于排列成绩—总是把你的最高成绩放在第一位。

(5) When you're constructing your resume, you need to be able to brag about yourself without exaggerating. Another point is making your resume stand out without descending to tricks and designs that merely frustrate the resume reader.

当你写简历的时候，你需要能够自夸，但不能夸大其辞。另一点是，让你的简历脱颖而出，而不是沦落到依靠小伎俩和设计，那些只会挫败简历读者。

Learning test

Choose the following words below to fill in the blanks, change the form if necessary.

| impact | emergency | previous | individual | significant |
| replace | desire | evaluate | responsible | impress |

(1) Nothing can _____ a mother's love.

(2) We need to discuss the _____ on the matter.

(3) You are supposed to be _____ for breaking the rule.

(4) She has her own _____ style of doing things.

(5) Everyone has the _____ to live in a world free of worries and pains.

(6) Ring the bell in an _____ to contact me.

(7) Please _____ the candidate's resume and make a decision.

(8) His _____ attempts had been unsuccessful.

(9) I'm afraid that I don't have any _____ achievements to put on my resume.

(10) She _____ me as a scholar.

Translation

1. Translate the following words into Chinese.

(1) background   (2) responsible

(3) resume   (4) previous

(5) reflect   (6) emergency

(7) exaggerate   (8) talent

(9) impact   (10) achievement

2. Translate the following phrases into Chinese.

(1) contact information   (2) for instance

(3) physical characteristic   (4) educational background

(5) stand out   (6) refer to

Practice

1. Translate the following sentences into Chinese.

(1) A resume usually asks the applicant at the minimum for their name, phone number, and address.

(2) Work experience is important to companies because it will inform the company if the applicant meets their requirements.

(3) Companies are interested in the applicant's social environment because it can inform them of their personality, interests, and qualities.

(4) Choose the formal written English to complete your resume.

(5) Keep your resume neat and make sure it is easy on the eye. Bullet points should be used to tidy up any lists.

2. Translate the following sentences into English.

（1）我们已经面试了 26 个求职者了。

（2）你带简历了吗？

（3）他以前没有从事这种工作的经验。

（4）面试官通常只花 30 秒看一份简历。

（5）一个好的简历可以帮助你成功申请职位。

Culture salon

Resume career highlights section

A career highlights/qualifications section of a resume, known as a resume summary, is an optional customized section of a resume that lists key achievements, skills, traits, and experience relevant to the position for which you are applying. Typically, the career highlights or resume summary section can be found at the top of a resume. The objective of this profile is to concisely describe your background and the strengths you look to gain with this new opportunity. Even those who have a lot of experience in the field should attach importance to a resume summary. It is a way to highlight your best skills and encourage the hiring manager to read more.

Chapter 6-Section 1-Text 1-看译文

Text 2 Job-hunting letter

Lead-in questions:

(1) What's the use of a job-hunting letter?

(2) How do I write a job-hunting letter?

(3) How should I address my job-hunting letter?

(4) What should be included in the first paragraph of a job-hunting letter?

(5) What should we write in the last paragraph of a job-hunting letter?

Chapter 6-Section 1-Text 2-听课文

A job-hunting letter should always accompany your resume unless you are told otherwise. It allows you to personalize an application and highlight key areas of your resume in more depth. A speculative application can

305

sometimes be an effective method of creating a career opening, especially in highly competitive industries.

Also, carry out some research on the job and the company you are sending your application to. Timing is everything when it comes to the creative career search.

1. How do I write a job-hunting letter?

Keep your job-hunting letter brief, while making sure it emphasizes your suitability for the job. It can be broken down into the following three sections.

(1) First paragraph-the opening statement should set out why you are writing the letter. Begin by stating the position you're applying for, where you saw it advertised and when you are available to start.

(2) Middle paragraphs-you should use the next two or three paragraphs to explain: what attracted you to this vacancy and type of work, why you're interested in working for the company, and what you can offer to the organization. Demonstrate how your skills match the specific requirements of the job description.

(3) Last paragraph-use the closing paragraph to indicate your desire for a personal interview, while mentioning any unavailable dates. Finish by thanking the employer and say how you are looking forward to receiving a response.

2. How should I address my job-hunting letter?

Always try and address your cover letter directly to the person who will be reading it. Bear in mind that you are more likely to receive a reply if you send it to the right person.

Advertised positions will usually include a contact name, but if not, it is worth taking the time to find out who the letter should be addressed to.

3. 11 rules of successful job-hunting letters

With employers often receiving huge volumes of applications for each vacancy, you need to ensure that your cover letter makes a lasting impression.

Here are some rules you'll need to stick to if you want to increase your chances of success:

(1) be concise and to the point-keep it to one side of A4;

(2) use the same quality plain white paper you used to print your job-hunting;

(3) include a named contact whenever possible to show you have sent it to them personally;

(4) your skills to the job advert and make a case for why the employer should want to meet with you;

(5) proofread-always double-check your spelling and grammar without relying on a computer spell check program;

(6) target the company by tailoring your cover letter for each application;

(7) page layout should be easy on the eye, set out with the reader in mind;

(8) check to make sure you've got the company name and other key details right;

(9) read it and cut out any unnecessary words or sentences;

(10) if sending electronically, put the text in the body of the email rather than as an attachment to avoid it being detected by spam filters;

(11) stick to your own words, avoiding jargon and formal cliches.

4. Seven of the worst cover letter mistakes

Your covering letter is an opportunity to show employers how well you express yourself and it should entice them to read your CV. If you want to ensure it is as effective as possible, avoid these common mistakes:

(1) failing to address the letter to a named individual at the company;

(2) repeating what is written in your CV;

(3) forgetting to proofread your letter and sending it full of mistakes;

(4) spilling over onto a second page;

(5) sharing unnecessary personal details and giving rambling explanations;

(6) concentrating too much on your qualifications rather than your skills and experience;

(7) failing to target your letter to the specific job you're applying for.

5. How do I apply for a job abroad?

Many overseas job offers are dependent on the applicant already beinglegally allowed to work and live in the country.

If you're planning on working overseas, you'll need to go through the correct visa procedures and obtain any work permits that are applicable to the country you wish to work in. For more details, seeworking abroad.

Most countries will recognize UK qualifications, but you should check to see if there are any comparisons you might need to refer to in your cover letter.

Do some research when writing your cover letter and CV, to ensure that you include everything required by employers in the region of the world where you'd like to work.

If you want to work abroad, take a look at the cover letter of an international student applying for a job in the UK and apply these principles to the country of your choice.

Words and phrases

accompany	[əˈkʌmpəni]	vt.	伴随,与…共存
application	[ˌæplɪˈkeɪʃn]	n.	申请
highlight	[ˈhaɪlaɪt]	vt.	强调,突出
speculative	[ˈspekjələtɪv]	adj	投机的
competitive	[kəmˈpetətɪv]	adj.	竞争的,有竞争力的
research	[rɪˈsɜːtʃ]	n.	研究,调查
creative	[kriˈeɪtɪv]	adj.	创造的,创造性的
career	[kəˈrɪə(r)]	n.	生涯,职业
brief	[briːf]	adj.	简洁的,简明的
emphasize	[ˈemfəsaɪz]	vt.	强调,着重
vacancy	[ˈveɪkənsi]	n.	(职位)空缺,空位

英文	音标	词性	中文
demonstrate	[ˈdemənstreɪt]	vt.	证明,演示
match	[mætʃ]	v.	相配,匹配
personal	[ˈpɜːsənl]	adj.	私人的,个人的
interview	[ˈɪntəvjuː]	n.	面谈,面试
mention	[ˈmenʃn]	vt.	提到,说起
response	[rɪˈspɒns]	n.	反应,回答,答复
reply	[rɪˈplaɪ]	n.	回复,回答,答复
chance	[tʃɑːns]	n.	可能性,机会
proofread	[ˈpruːfriːd]	vt.	校对
tailor	[ˈteɪlə(r)]	vt.	定制,制作
layout	[ˈleɪaʊt]	n.	布局,安排,设计
detail	[ˈdiːteɪl]	n.	细节,琐事
attachment	[əˈtætʃmənt]	n.	附件
avoid	[əˈvɔɪd]	vt.	避免
opportunity	[ˌɒpəˈtjuːnəti]	n.	机会,时机
jargon	[ˈdʒɑːgən]	n.	行话,行业术语
cliche	[ˈkliːʃeɪ]	n.	陈词滥调
rambling	[ˈræmblɪŋ]	adj.	长篇大论
legally	[ˈliːgəli]	adv.	合法地,法律上地
procedure	[prəˈsiːdʒə(r)]	n.	程序,手续
obtain	[əbˈteɪn]	vt.	获得,得到
permit	[pəˈmɪt]	v.	允许,许可
applicable	[əˈplɪkəbl]	adj.	适用的,合适的
region	[ˈriːdʒən]	n.	地区,范围
job-hunting			求职,找工作
carry out			施行,实现
cover letter			附函,求职信
stick to			遵守,坚持
plain white			纯白色
double-check			复核
make sure			确信,证实
break down			分解,划分
apply for			申请
applicable to			适用于
look forward to			期待

rely on 依赖

Key sentences

(1) A job-hunting letter should always accompany your resume unless you are told otherwise.

除非另有说明,否则求职信应该伴随着简历发出。

(2) Keep your job-hunting letter brief, while making sure it emphasizes your suitability for the job.

求职信要简短,并确保要强调你适合这份工作。

(3) With employers often receiving huge volumes of applications for each vacancy, you need to ensure that your cover letter makes a lasting impression.

由于雇主经常因每个职位空缺收到大量的申请,你需要确保你的求职信给人留下深刻的印象。

(4) Your covering letter is an opportunity to show employers how well you express yourself and it should entice them to read your CV.

你的求职信是向雇主展示你自己表现得如何好的机会,应该要能吸引雇主阅读你的简历。

(5) Many overseas job offers are dependent on the applicant already being legally allowed to work and live in the country.

许多海外工作机会取决于申请人是否已被依法允许在该国工作和生活。

Learning test

Choose the following words below to fill in the blanks, change the form if necessary.

| match | permit | interview | mention | vacancy |
| career | detail | application | obtain | research |

(1) Her nails were painted bright red to _____ her dress.

(2) Please tell us immediately after you have_____the import license.

(3) May I make a few preliminary remarks before we start the _____?

(4) Every _____ of her dress is perfect.

(5) The Museum guards _____ me to bring my camera.

(6) I am very interested in the position and want to apply for the _____.

(7) The manager received twenty _____ for the position.

(8) She didn't _____ her final decision in the meeting.

(9) Her dancing _____ was over because of the accident.

(10) We all support his scientific _____.

Translation

1. Translate the following words into Chinese.

(1) research   (2) opportunity

(3) personal   (4) reply

(5) layout   (6) competitive

(7) brief   (8) application

(9) detail (10) attachment

2. Translate the following phrases into Chinese.

(1) job-hunting (2) carry out

(3) plain white (4) apply for

(5) make sure (6) cover letter

Practice

1. Translate the following sentences into Chinese.

(1) The opening statement should set out why you are writing the letter.

(2) Demonstrate how your skills match the specific requirements of the job description.

(3) Advertised positions will usually include a contact name, but if not, it is worth taking the time to find out who the letter should be addressed to.

(4) If you want to ensure it is as effective as possible, avoid these common mistakes.

(5) If you're planning on working overseas, you'll need to go through the correct visa procedures and obtain any work permits that are applicable to the country you wish to work in.

2. Translate the following sentences into English.

(1) 求职信在找工作中起着重要的作用。

(2) 求职信的书写要规范,内容要真实。

(3) 求职信应包括哪些内容?

(4) 求职信中要有联系人及地址。

(5) 求职信中要简要列出个人的工作经历。

Culture salon

Job research

In today's job market, it's not uncommon to submit applications for many positions. That involves lots of time, and lots to keep track of. You don't want to squander those precious hours by missing important application deadlines, garbling companies and positions, confusing interview times.

Accordingly, properly managing your job search is just as important as identifying job opportunities and submitting your application.

Here are some effective and convenient ways to keep track of your job applications and stay on top of the job search process. By using local newspaper, a website, your job search site and so on, you could reach a lot of job vacancies. You should choose the most suitable job for you.

Chapter 6-
Section 1-
Text 2-看译文

Text 3 Interview skills

Lead-in questions:

(1) What is the definition of the interview?
(2) What are the more accurate predictors of which applicants will make good employees?
(3) How to prepare for an interview?
(4) What should we plan before an interview?
(5) How to perform in an interview?

Chapter 6-Section 1-Text 3-听课文

A job interview is a one-on-one interview consisting of a conversation between a job applicant and a representative of an employer which is conducted to assess whether the applicant should be hired. Interviews are one of the most popularly used methods for employee selection. Interviews vary in the extent to which the questions are structured, from a totally unstructured and free-wheeling conversation, to a structured interview in which an applicant is asked a predetermined list of questions in a specified order, structured interviews are usually more accurate predictors of which applicants will make good employees.

A job interview typically precedes thehiring decision. The interview is usually preceded by the evaluation of submitted resumes from interested candidates, possibly by examining job applications or reading many resumes. Next, after this screening, a small number of candidates for interviews are selected.

Multiple rounds of job interviews and/or other candidate selection methods may be used where there are many candidates or the job is particularly challenging or desirable. Earlier rounds sometimes called "screening interviews" may involve fewer staff from the employers and will typically be much shorter and less in-depth. An increasingly common initial interview approach is the telephone interview. This is especially common when the candidates do not live near the employer and has the advantage of keeping costs low for both sides. Since 2003, interviews have been held through video conferencing software, such as Skype. Once all candidates have been interviewed, the employer typically selects the most desirable candidate(s) and begins the negotiation of a job offer.

No matter how your resume and talents are, if you mess up a job interview, you won't get that position. How well you do in the interview often makes the difference in whether you will get a job offer or not. Interview skills will immeasurably improve your chances of winning those hard-to-get jobs and contribute to a rewarding career.

1. Prepare

Do as much research as possible on the company and the position before a job interview. If the company has a web site, read it thoroughly and you can read their annual report to find out as much as you can about the company. You can also learn about the company's products, services and growth from a variety of publications available in public libraries.

2. Practice

Read through your resume again and be ready to talk about everything that you mentioned in your resume. Refresh your memory on the facts and figures of your last or present employer. Think about the questions the interviewer might ask you and rehearse answers to common questions. You should also think about some questions to ask the interviewer. Be aware of any weaknesses you have and be ready to deal with them in the interview.

3. Plan

Plan what you are going to wear, and make sure it is appropriate for the company you might work for to make the most of the opportunity to create a good impression. Know the exact time and the exact location of the interview and aim to arrive 5~10 minutes early, especially taking traffic conditions at the time the interview is planned into consideration.

4. Perform

It is very important that you give out the right signals during the interview. So in the interview, try to be at ease and friendly. Be confident but not arrogant when answering questions and look the interviewer in the eye. Be positive in your comments, outlook and attitude. Be enthusiastic about the position but don't talk too much. Relax, and smile.

Here are some "Dos and Don'ts" that will dramatically increase your chances: from wearing the right expression, to knowing what not to say, to never ever breaking a sweat.

(1) Do plan to arrive on time or a few minutes early. Late arrival for a job interview is never excusable.

(2) Do look at a prospective employer in the eyes while speaking.

(3) Do show enthusiasm. If you are interested in the opportunity, enthusiastic feedback can enhance your chances of being further considered.

(4) Don't be petty. Asking the location of the lunchroom or meeting room will clue the interviewer into your lack of preparation and initiative. Don't inquire about salary, vacations, bonuses, retirement, etc. in the initial interview unless you are sure the employer is interested in hiring you.

Chapter 6-
Section 1-
Text 3-听词汇

(5) Don't answer with a simple "yes" or "no". Explain whenever possible. Describe the things about yourself which relate to the situation.

(6) Don't be a liar. Lying won't get you a job. In a job interview even a slight exaggeration is lying. Don't lie. Never stretch your resume or embellish accomplishments. One lie can ruin your entire interview, and the skilled interviewer will spot the lie and show you the door.

Potential job interview opportunities also include networking events and career fairs. The job interview is considered one of the most useful tools for evaluating potential employees. It also demands significant resources from the employer, yet has been demonstrated to be notoriously unreliable in identifying the optimal person for the job. An interview also allows the candidate to assess the corporate culture and demands of the job.

Words and phrases

interview	[ˈɪntəvjuː]	n. & v.	面试
conversation	[ˌkɒnvəˈseɪʃn]	n.	谈话, 会话
representative	[ˌreprɪˈzentətɪv]	n.	代表
conduct	[kənˈdʌkt]	v.	实施, 执行
assess	[əˈses]	v.	评估, 评定
hire	[ˈhaɪə(r)]	v.	聘用, 雇用
extent	[ɪkˈstent]	n.	程度, 广度, 宽度
predetermine	[ˌpriːdɪˈtɜːmɪn]	v.	预先决定, 预定
accurate	[ˈækjərət]	adj.	准确的, 精确的
predictor	[prɪˈdɪktə(r)]	n.	预测
typically	[ˈtɪpɪkli]	adv.	典型地, 代表性地
decision	[dɪˈsɪʒn]	n.	决定, 决心
precede	[prɪˈsiːd]	vt.	先于, 在……之前
evaluation	[ɪˌvæljuˈeɪʃn]	n.	评估, 评价
submit	[səbˈmɪt]	vt.	提交
candidate	[ˈkændɪdət]	n.	候选人
multiple	[ˈmʌltɪpl]	adj.	多的, 多(个、重、种)的
challenging	[ˈtʃælɪndʒɪŋ]	adj.	挑战性的
involve	[ɪnˈvɒlv]	vt.	包含, 涉及
staff	[stɑːf]	n.	工作人员, 员工
approach	[əˈprəʊtʃ]	n.	方法
advantage	[ədˈvɑːntɪdʒ]	n.	优点
conference	[ˈkɒnfərəns]	vi.	会议, (举行或参加)会议
negotiation	[nɪˌɡəʊʃiˈeɪʃn]	n.	谈判, 协商
annual	[ˈænjuəl]	adj.	每年的, 年度的
product	[ˈprɒdʌkt]	n.	产品
available	[əˈveɪləbl]	adj.	可得到的, 能找到的
rehearse	[rɪˈhɜːs]	vt.	排练, 预演
weakness	[ˈwiːknəs]	n.	弱点, 缺点
appropriate	[əˈprəʊpriət]	adj.	合适的, 适当的
exact	[ɪɡˈzækt]	adj.	准确的, 确切的

impression	[ɪmˈpreʃn]	n.	印象
consideration	[kənˌsɪdəˈreɪʃn]	n.	考虑
signal	[ˈsɪgnəl]	n.	信号
confident	[ˈkɒnfɪdənt]	adj.	自信的,有信心的
arrogant	[ˈærəgənt]	adj.	傲慢的,自大的
positive	[ˈpɒzətɪv]	adj.	正面的,积极的
comment	[ˈkɒment]	n.	评论,意见
enthusiastic	[ɪnˌθjuːziˈæstɪk]	adj.	热情的,热心的
excusable	[ɪkˈskjuːzəbl]	adj.	可原谅的,可容许的
feedback	[ˈfiːdbæk]	n.	反馈,反馈意见
inquire	[ɪnˈkwaɪə(r)]	vi.	询问
clue	[kluː]	vt.	暗示,提示
bonus	[ˈbəʊnəs]	n.	奖金,额外津贴
ruin	[ˈruːɪn]	vt.	毁灭,毁坏
potential	[pəˈtenʃl]	adj.	潜在的,可能的
one-on-one			一对一的
consist of			由……组成
free-wheeling			随心所欲的
mess up			搞砸,弄糟
job offer			工作机会
contribute to			为……作贡献,有助于
deal with			处理,对付
dos and don'ts			注意事项,行为准则
aim to			目的在于,旨在
at ease			放松,自然
relate to			有关,涉及

Key sentences

(1) A job interview is a one-on-one interview consisting of a conversation between a job applicant and a representative of an employer which is conducted to assess whether the applicant should be hired.

工作面试是应聘者和雇主代表之间的一对一的面谈,进行评估应聘者是否应聘用。

(2) Multiple rounds of job interviews and/or other candidate selection methods may be used where there are many candidates or the job is particularly challenging or desirable.

在候选人比较多或者工作特别具有挑战性或吸引力的情况下,可以使用多轮面试和/或其他选择方法。

(3) Do as much research as possible on the company and the position before a job interview.

在面试之前尽可能对公司及招聘职位多做一些调查。

(4) It is very important that you give out the right signals during the interview. So in the interview, try to be at ease and friendly.

在面试过程中发出正确的信号是非常重要的。所以在面试中,尽量放松和友好。

(5) Potential job interview opportunities also include networking events and career fairs. The job interview is considered one of the most useful tools for evaluating potential employees.

潜在的面试机会还包括网络活动和招聘会。求职面试被认为是评估潜在员工最有用的工具之一。

Learning test

Choose the following words below to fill in the blanks, change the form if necessary.

| hire | representative | confident | advantage | impression |
| potential | feedback | candidate | positive | involve |

(1) Your encouragement made me more _____ of my next job.

(2) She had a stressful job as a sales _____ .

(3) These _____ effects must be considered carefully.

(4) We should take _____ of all educational opportunities.

(5) We don't want to _____ an inexperienced and lazy person.

(6) The students are expecting comments and _____ from their supervisor.

(7) His speech left a strong _____ on the audience.

(8) He has a _____ attitude towards life.

(9) Don't _____ other people in your trouble.

(10) We all agree that John is the best _____ for the position.

Translation

1. Translate the following words into Chinese.

(1) conversation   (2) decision

(3) advantage   (4) staff

(5) hire   (6) inquire

(7) comment   (8) positive

(9) candidate   (10) confident

2. Translate the following phrases into Chinese.

(1) deal with   (2) consist of

(3) job offer   (4) free-wheeling

(5) contribute to   (6) mess up

Practice

1. Translate the following sentences into Chinese.

(1) Interviews are one of the most popularly used methods for employee selection.

(2) Once all candidates have been interviewed, the employer typically selects the most desirable candidate(s) and begins the negotiation of a job offer.

(3) No matter how your resume and talents are, if you mess up a job interview, you won't get that position.

(4) If the company has a web site, read it thoroughly and you can read their annual report to find out as much as you can about the company.

(5) Plan what you are going to wear, and make sure it is appropriate for the company you might work for to make the most of the opportunity to create a good impression.

2. Translate the following sentences into English.

(1) 面试时一定要表现得自信。

(2) 面试前先思考可能会问哪些问题。

(3) 面试时不适合穿太休闲的衣服。

(4) 我在这次面试中表现很好。

(5) 我非常荣幸参加这次面试。

Culture salon

How to dress for an interview

The first impression you make on a potential employer is incredibly meaningful and important. When you first meet, employers will form an assessment based on what you're wearing and how you carry yourself. Regardless of the work environment, it's important to dress professionally for a job interview. This will help guarantee that you make a great first impression. In general, the candidate dressed in a suit and tie is going to make a much better impression than the candidate dressed in scruffy jeans and a T-shirt. That said, different industries do have varying expectations for what candidates and employees should wear. The appropriate dress code can vary greatly depending on company and industry.

Chapter 6-Section 1-Text 3-看译文

Chapter 6-Section 1-拓展阅读

Chapter 6-Section 1-习题

Section 2 Technical communication

Chapter 6-
Section 2-
PPT

Text 1 How to become a successful medical device sales representative?

Lead-in questions:

(1) Who do medical device salespeople work for?

(2) What are the requirements for a sales representative?

(3) What are the responsibilities for a sales representative?

(4) What is vital to landing a sales position in the medical device industry?

(5) Is it important for medical device sales representatives to possess the product knowledge?

Imagine competing against 375 candidates with the same qualifications as you for a single job opening. For those interested in medical device equipment sales, that is the reality of trying to land a position in this highly competitive field. The competition doesn't stop there--once you have a job, you have to perform well, or else there are hundreds of candidates who would be happy to take your place. Although being successful in the field can depend on a variety of factors, developing skills in certain areas can pave the way to a successful career in medical device equipment sales.

Chapter 6-
Section 2-
Text 1-听课文

1. Career overview

Medical device salespeople work for manufacturers or wholesalers and sell goods to hospitals, doctor offices, clinics, government facilities and nursing homes. They are responsible for contacting potential customers, explaining the features and benefits of their products, answering questions and negotiating the final sales price. The route to a medical device sales representative position involves education, training, experience and networking.

2. Get sales experience

While you're in school, get as much hands-on sales training as possible because medical device companies are not interested in providing basic sales training techniques to recruits, according to Global Edge Recruiting. A new employer may provide you with in-depth product training to ensure you're familiar with the extremely technical aspects of the medical devices and how they're used, but they won't be teaching you how to overcome objections or how to close a sale. Sales experience is vital to landing a sales position in the medical device industry.

3. Research potential employers

To successfully land a sales position with a medical device company, you need to know what kind of equipment a company sells, who are its primary customers and where you'll find prospects. Tailor your job search to those companies that sell products you are most familiar with. For example, if you are trying to get into a company that sells surgical supplies, you can rely on your stint as a receptionist in a surgeon's office or the fact that you did marketing for a plastic surgery practice. Look for companies that sell products that relate to your previous experience. Medical device companies rely on representatives to sell in large geographic territories, too, so look for companies that operate in areas where you have built a reputation with local health care providers.

4. Medical and product knowledge

Because medical device equipment sales representatives are usually selling directly to surgeons and other specialists, it's critical that you cannot only "talk the talk" but also have advanced knowledge of medical procedures, anatomy and other medical knowledge. Once you have a device to sell, you must know everything about that product. The pros and cons, how it works, competitors' products—you need to know it all. It is common for medical device sales representatives to be present when surgeons or other physicians are actually using the product on patients, so be prepared to give good instructions and answer difficult medical-related questions.

5. Build industry relationships

Even while you're still in school, you can start building industry relationships through your professors and professional organizations that cater to your industry. Join a professional association such as the National Association of Medical Sales Representatives or the Independent Medical Distributors Association to take advantage of the networking opportunities at conventions and seminars. Sit on a committee and seek career services they might offer. You could find a lead for a good job while keeping up with industry changes and garnering support from your peers and industry leaders.

6. Communication skills

Good interpersonal communication skills can be the difference between an average and a great medical device equipment salesperson. Listening to the needs of the client is the key to truly establishing a relationship and gaining their trust. Establishing positive relationships with clients can also lead to referrals, which can also help you to be successful in this field.

7. Sales skills

Most companies prefer to hire experienced medical device sales representatives, so the training can focus on the complex products the sales representative will be selling. Different types of sales may require you to have different skill sets. For example, if you are selling high-priced devices, you should have more aggressive sales techniques. On the other hand, if you are selling to the same groups regularly, you may need stronger relationship-building skills. Knowledge of a variety of approaches to sales is best, as you may have to use a variety of techniques throughout your career.

The requirements for the job vary depending on the company, but sales representatives for technical products typically must possess certain knowledge, such engineering or biology, which can be useful to sell the highly technical equipment and understand its medical applications. In addition, you could enrich your relative experience of medical equipment sales that will increase your knowledge of the field, highlight your commitment to your career.

Chapter 6-Section 2-Text 1-听词汇

Words and phrases

imagine	[ɪˈmædʒɪn]	vt.	想象,设想
equipment	[ɪˈkwɪpmənt]	n.	设备,装备
sales	[seɪls]	n.	销售
reality	[riˈæləti]	n.	现实,真实
factor	[ˈfæktə(r)]	n.	因素
pave	[peɪv]	vt.	铺路,铺设
manufacturer	[ˌmænjuˈfæktʃərə(r)]	n.	制造商,厂商
wholesaler	[ˈhəʊlseɪlə(r)]	n.	批发商
facility	[fəˈsɪləti]	n.	机构
feature	[ˈfiːtʃə(r)]	n.	特征,特色
benefit	[ˈbenɪfɪt]	n.	优点,益处
negotiate	[nɪˈgəʊʃieɪt]	vt.	谈判,协商
recruit	[rɪˈkruːt]	n.	新员工
overcome	[ˌəʊvəˈkʌm]	vt.	战胜,克服
objection	[əbˈdʒekʃn]	n.	异议
prospect	[ˈprɒspekt]	n.	展望,前景
tailor	[ˈteɪlə(r)]	vt.	定制
surgical	[ˈsɜːdʒɪkl]	adj.	外科的,外科医生的
supply	[səˈplaɪ]	n.	供应品,供给物
reputation	[ˌrepjuˈteɪʃn]	n.	名声,声誉
advanced	[ədˈvɑːnst]	adj.	高等的,先进的
anatomy	[əˈnætəmi]	n.	解剖(学)
instruction	[ɪnˈstrʌkʃn]	n.	(用法)说明
professional	[prəˈfeʃənl]	adj.	专业的,职业的
cater	[ˈkeɪtə(r)]	vt.	面向,适合,迎合
association	[əˌsəʊʃiˈeɪʃn]	n.	(行业)协会
convention	[kənˈvenʃn]	n.	会议,大会

seminar	[ˈsemɪnɑː(r)]	n.	研讨会
committee	[kəˈmɪti]	n.	委员会
garner	[ˈgɑːnə(r)]	vt.	获得,积累
referral	[rɪˈfɜːrəl]	n.	介绍
establish	[ɪˈstæblɪʃ]	vt.	建立,确立
gain	[geɪn]	vt.	获得
aggressive	[əˈgresɪv]	adj.	激进的,进攻性的
possess	[pəˈzes]	vt.	拥有,持有
highlight	[ˈhaɪlaɪt]	vt.	突出,强调
according to			按照,根据
job opening			职务空缺
hands-on			实践的,动手的,实操的
close a sale			达成买卖
pros and cons			优缺点
professional association			专业协会,行业协会
take advantage of			利用
cater to			面向
lead to			导致,引起
in addition			另外,此外

Key sentences

(1) Although being successful in the field can depend on a variety of factors, developing skills in certain areas can pave the way to a successful career in medical device equipment sales.

尽管在这一领域取得成功取决于很多因素,但在某些领域提高一定技能可以为医疗器械设备销售的成功铺平道路。

(2) Medical device salespeople work for manufacturers or wholesalers and sell goods to hospitals, doctor offices, clinics, government facilities and nursing homes.

医疗器械销售人员为制造商或批发商工作,并向医院、医生办公室、诊所、政府机构和养老机构出售医疗器械商品。

(3) Sales experience is vital to landing a sales position in the medical device industry.

销售经验对于在医疗器械行业取得销售职位至关重要。

(4) Once you have a device to sell, you must know everything about that product. The pros and cons, how it works, competitors' products--you need to know it all.

一旦你有了一个销售设备,你就必须了解产品的一切信息。优缺点、工作原理、竞争对手的产品——你需要知道这一切。

(5) The requirements for the job vary depending on the company, but sales representatives for technical

products typically must possess certain knowledge, such engineering or biology, which can be useful to sell the highly technical equipment and understand its medical applications.

对工作的要求因公司而异,但技术产品的销售代表通常必须具备一定的知识,如工程或生物学,可用于销售高技术设备并了解其医疗应用。

Learning test

Choose the following words below to fill in the blanks, change the form if necessary.

| wholesaler | recruit | aggressive | prospect | factor |
| supply | imagine | facility | feature | possess |

(1) The support from the government is an important _____ in the success of the project.

(2) A large number of applicants can't explain the _____ of the products.

(3) Many manufactures produce low-priced disposable _____ in China.

(4) He is respected as a very _____ leader.

(5) _____ you are an interviewer, what do you want to know about the candidates?

(6) Different workers _____ different skills.

(7) Retailers buy products directly from the manufacturer or from a _____ .

(8) Common medical _____ include hospitals, clinics and nursing homes.

(9) I chose to work abroad to improve my career _____ .

(10) The _____ will have a 20-day training course before they start to work.

Translation

1. Translate the following words into Chinese.

(1) equipment   (2) overcome

(3) wholesaler   (4) factor

(5) recruit   (6) benefit

(7) facility   (8) highlight

(9) establish   (10) prospect

2. Translate the following phrases into Chinese.

(1) professional association   (2) in addition

(3) job opening   (4) cater to

(5) close a sale   (6) pros and cons

Practice

1. Translate the following sentences into Chinese.

(1) The competition doesn't stop there—once you have a job, you have to perform well, or else there are hundreds of candidates who would be happy to take your place.

(2) They are responsible for contacting potential customers, explaining the features and benefits of their products, answering questions and negotiating the final sales price.

(3) Even while you're still in school, you can start building industry relationships through your professors and professional organizations that cater to your industry.

(4) Listening to the needs of the client is the key to truly establishing a relationship and gaining their trust.

(5) Most companies prefer to hire experienced medical device sales representatives, so the training can focus on the complex products the sales representative will be selling.

2. Translate the following sentences into English.

（1）医疗器械是一类特殊的高科技产品。

（2）医疗器械销售需要具备一定的医疗知识。

（3）销售经验和良好的专业知识是一名成功医疗器械销售必须具备的。

（4）医疗器械销售人员要对所销售的医疗设备工作原理有所了解。

（5）销售经理想开辟一个新市场。

Culture salon

A day as a medical device sales representative

As in any sales career, there are not many "typical" days. Every day is new when you're dealing with new clients and existing clients. Some companies may divide the responsibilities, but many medical device representatives are responsible for both new and existing clients.

A typical day would primarily include time on the phone with clients and prospects, as well as meetings and appointments with clients and prospects. Remaining time would entail administrative duties such as replying to emails, setting appointments, sending out proposals or quotes, and other deskwork.

Often, medical device sales jobs are "outside" sales jobs, meaning as a sales representative or territory manager, you will be traveling out "in the field" to meet with clients in person.

Chapter 6-Section 2-Text 1-看译文

Text 2 The Internet-where the customers are?

Lead-in questions:

（1）Why is e-mail so popular?

（2）Why is e-mail cheap?

（3）When can I send a message to everyone on my list?

（4）What is SPAM?

（5）How often can the salespeople send a message to everyone about their new products?

Chapter 6-Section 2-Text 2-听课文

E-mail can be a very effective tool for marketing your products to buyers around the world if you know what you are doing and follow certain rules. Some of the companies on the database list of USA Importers and Wholesalers have included either their e-mail address or their web site address.

Why is e-mail so popular? First of all it's easy to get. When you register your URL web address with the internet service provider(ISP) as mentioned above, you automatically get several e-mail addresses. It is important that you have at least three basic e-mail addresses. Buyers are looking for them and if you don't have them, it raises suspicion about your company.

The second reason that e-mail is so popular is because it is cheap. It costs nothing more than your time. No paper expense, no postage, no envelope expense.

The third reason it's popular is that sending e-mail messages is relatively easy. You must keep a database and make sure it is up-to-date. You must compose the message. But anyone who can use a word processor can also write an e-mail message.

There is a right way to market your products by e-mail. Let's discuss the right way.

You should always have the person's permission before you send them a sales e-mail message. The common question is "How can I ask them if I can contact them until I contact them?" Go through all the e-mails of prospective buyers that have contacted you in the past. Send each of them an e-mail asking if they would like to be on your "mailing list". Many of them will say yes.

Secondly, there are e-mail address lists that you can buy. These are e-mail addresses of companies that might be interested in your product. But you should contact them first and ask for permission to send them a sales message.

The third way to get permission to send prospective buyers an e-mail is have them "opt-in" for your list. On your web site you will announce that you have periodic announcements of your products and if the viewer wants to get on(or opt-in) the mailing list that they should contact you.

If you have a new product that you think the buyers might be interested in, you can send a message to everyone on your list. However, do not send too many messages. One message every two or three months is enough. Less than this, they will forget you. More than that, they may think you are sending too many.

Make the messages you send as interesting as possible. In addition to introducing your new product you should also include some interesting information. Buyers will read it if they think they will learn something. They may or may not read it if it is only a sales message.

It's this simple. Sending e-mail messages to buyers who have asked for it can be very effective. Sending SPAM e-mail message to people who have not asked for is ineffective and destroys your company's reputation.

One thing that you do not want to do, never ever, is send a sales e-mail message to someone who has not asked for it. It is one of the worst things you can do.

It's very tempting to do. You find the e-mail address of a company that might be interested in your product and you say to yourself, "I'll send them an e-mail. What do I have to lose?" The answer is that you have a lot to lose.

Sending a sales e-mail message to someone who did not ask for it is called "SPAM". It is considered a very unprofessional and rude thing to do. SPAM messages are hated by everyone. It is the most offensive and at the same time least effective way to market on the internet.

In short, sending SPAM e-mail messages is not an effective way to sell your products. It is a waste of your time and that of the person who gets your unwanted message.

Following are Dos and Don'ts of an e-mail message:

(1) Do make sure you have permission to send that person the message.

(2) Do make the message interesting.

(3) Do have a message at the end of the message telling the buyer what to do if he/she does not want to get your messages anymore.

(4) Do keep your e-mail message short. No one wants to read a long message. Give the highlights and direct the buyer to your web site for more details.

(5) Do answer all e-mail messages in one business day or less.

(6) Don't ever send SPAM.

(7) Don't use all capital letters. It looks like you are shouting or yelling and is rude.

(8) Don't use abbreviations. It's considered lazy. Buyers do not like to do business with lazy people.

(9) Don't use different colors and different font sizes on your message. It's considered childish and unprofessional. Use one size and all black.

(10) Don't use a colored background. Again, it's considered unprofessional. In addition, they can be hard to read.

Words and phrases

effective	[ɪˈfektɪv]	adj.	有效的,生效的
market	[ˈmɑːkɪt]	vt.	营销,推销
database	[ˈdeɪtəbeɪs]	n.	数据库
importer	[ɪmˈpɔːtə(r)]	n.	进口商
wholesaler	[ˈhəʊlseɪlə(r)]	n.	批发商
include	[ɪnˈkluːd]	vt.	包括
register	[ˈredʒɪstə(r)]	vt.	登记,注册
internet	[ˈɪntənet]	n.	互联网
service	[ˈsɜːvɪs]	n.	服务
mention	[ˈmenʃn]	vt.	提到
automatically	[ˌɔːtəˈmætɪkli]	adv.	自动地
suspicion	[səˈspɪʃn]	n.	猜疑,怀疑

Chapter 6-
Section 2-
Text 2-听词汇

expense	[ɪkˈspens]	n.	费用，花费
postage	[ˈpəʊstɪdʒ]	n.	邮费
envelope	[ˈenvələʊp]	n.	信封
relatively	[ˈrelətɪvli]	adv.	相对地，比较地
permission	[pəˈmɪʃn]	n.	许可，允许
compose	[kəmˈpəʊz]	vt.	编写，写作
prospective	[prəˈspektɪv]	adj.	可能的，有希望的
announce	[əˈnaʊns]	vt.	宣布，宣告
periodic	[ˌpɪəriˈɒdɪk]	adj.	周期的，定期的
destroy	[dɪˈstrɔɪ]	vt.	破坏
reputation	[ˌrepjuˈteɪʃn]	n.	名声，声誉
tempting	[ˈtemptɪŋ]	adj.	吸引人的，诱惑人的
consider	[kənˈsɪdə(r)]	v.	认为
unprofessional	[ˌʌnprəˈfeʃənl]	adj.	不专业的，外行的
offensive	[əˈfensɪv]	adj.	冒犯的，无礼的
waste	[weɪst]	n.	浪费
rude	[ruːd]	adj.	粗鲁的，无礼的
yell	[jel]	vi.	叫喊
abbreviation	[əˌbriːviˈeɪʃn]	n.	缩写
size	[ˈsaɪz]	n.	大小，尺寸
childish	[ˈtʃaɪldɪʃ]	adj.	幼稚的，孩子气的
at least			至少
web address			网址
internet service			互联网服务
word processor			文字处理器
up-to-date			最新的
go through			查看，浏览
in addition to			除……之外
make sure			确保

Key sentences

(1) E-mail can be a very effective tool for marketing your products to buyers around the world if you know what you are doing and follow certain rules.

如果你知道你在做什么，并遵循一定的规则，那么电子邮件就可以成为一个非常有效的工具将你的产品推销给全世界的买家。

(2) It is important that you have at least three basic e-mail addresses. Buyers are looking for them and

if you don't have them, it raises suspicion about your company.

重要的是你至少要有三个基本的电子邮件地址。买家会寻找你邮件地址,如果没有,就会对你的公司产生怀疑。

(3) You should always have the person's permission before you send them a sales e-mail message.

在给客户发销售电子邮件之前,应先得到客户的允许。

(4) If you have a new product that you think the buyers might be interested in you can send a message to everyone on your list.

如果你有一个新产品,认为买家可能感兴趣,你可以给你名单上的每个人发个信息。

(5) In short, sending SPAM e-mail messages is not an effective way to sell your products. It is a waste of your time and that of the person who gets your unwanted message.

简而言之,发送垃圾邮件不是销售产品的有效方法。这是浪费你的时间和那个不想收到这条消息的人的时间。

Learning test

Choose the following words below to fill in the blanks, change the form if necessary.

| childish | consider | rude | service | mention |
| relatively | suspicion | reputation | expense | include |

(1) Our aim is to provide the best _____ at the lower price.

(2) It is _____ to run into danger for nothing.

(3) He has a very good _____ in the matter of honesty.

(4) You should _____ other people's feeling before you act.

(5) _____ and jealousy can ruin a marriage.

(6) Cut the unnecessary _____ to reduce your living cost.

(7) Don't be so _____ to your parents!

(8) The exam was _____ easy.

(9) Does the price _____ the postage?

(10) Speaking of Macao, it is necessary to _____ lotus flowers.

Translation

1. Translate the following words into Chinese.

(1) suspicion (2) importer

(3) consider (4) compose

(5) permission (6) database

(7) rude (8) waste

(9) reputation (10) offensive

2. Translate the following phrases into Chinese.

(1) at least (2) up-to-date

(3) web address (4) go through

(5) make sure (6) in addition to

Practice

1. Translate the following sentences into Chinese.

(1) The second reason that e-mail is so popular is because it is cheap.

(2) You must keep a database and make sure it is up-to-date.

(3) But you should contact them first and ask for permission to send them a sales message.

(4) It is considered a very unprofessional and rude thing to do.

(5) In addition to introducing your new product you should also include some interesting information.

2. Translate the following sentences into English.

(1) 现在,利用网络做生意很普遍。

(2) 网络可以给商家和客户提供更大的交流平台。

(3) 电子邮件是商业交流的一种有效手段。

(4) 用电子邮件发送信息大大提高了工作效率。

(5) 根据客户的需求发送产品信息。

Culture salon

How to contact your customers?

A business cannot survive without conducting ongoing efforts to better understand customer needs. In order to discover whether your product or service is having a positive effect and whether customers are satisfied with your products, you should take time to ascertain your customer's emotional and material needs, and then offer valuable incentives for remaining loyal to your company. Acquiring customer feedback is difficult; from creating simple email surveys, you can obtain the information and can learn a lot about what customers want just by asking and listening. Also to learn more, you could even ask them for a little bit of time just to fill out a questionnaire.

Chapter 6-Section 2-Text 2-看译文

Chapter 6-Section 2-习题

参考文献

[1] Seeram Ramakrishna, Wee-Eong Teo, Lingling Tian, et al. Medical Devices: Regulations, Standards and Practices. Cambridge: Woodhead Publishing Ltd, 2015.

[2] Ferdiansyah Mahyudin, Hendra Hermawan. Biomaterials and medical devices. Switzerland: Springer International Publishing, 2016.

[3] 刘金龙,谷青松.科技英语阅读与翻译.北京:国防工业出版社,2013.

[4] 汪丽.新编科技英语阅读.广州:华南理工大学出版社,2015.

[5] 刘芳.医学检验专业英语.北京:人民卫生出版社,2003.

[6] 师丽华.医疗器械专业英语.北京:人民卫生出版社,2011.

[7] 周剑涛.检验技术专业英语.北京:人民卫生出版社,2015.

目标检测参考答案（Learning Test 部分）

Chapter 1　Fundamentals of medical devices

Section 1　Basic introduction to medical devices

Text 1　Introduction to medical devices

(1) revenue

(2) reached

(3) instrument

(4) disposable

(5) reagent

(6) diagnosis

(7) prevention

(8) injury

(9) innovative

(10) vary

Text 2　FDA: definition and classification of medical devices

(1) treatment

(2) electronic

(3) complex

(4) assure

(5) include

(6) affects

(7) assist

(8) consumer

(9) effective

(10) regulated

Text 3　Global medical device market

(1) highlight

(2) protein

(3) ambition

(4) expenditure

(5) market

(6) healthcare

(7) imaging

(8) invasive

(9) dynamic

(10) demand

Section 2 Fundamentals of medical devices

Text 1 A brief introduction to electronic basics

(1) contain

(2) value

(3) electronic

(4) current

(5) transform

(6) passive

(7) technology

(8) digital

(9) fundamentals

(10) resistor

Text 2 A brief introduction to biological basics

(1) propel

(2) genes

(3) field

(4) phenomenon

(5) microscopic

(6) tissue

(7) respond

(8) blood

(9) constituent

(10) inorganic

Text 3 A brief introduction to microorganisms

(1) invisible

(2) tiny

(3) observe

(4) virus

(5) cold

(6) discover

(7) nutrient

(8) micrometers

(9) vomited

(10) lethal

Chapter 2　Introduction to typical products

Section 1　Consumer products

Text 1　Contact lenses

(1) germs

(2) Saliva

(3) irritation

(4) deposits

(5) routine

(6) primarily

(7) oxygen

(8) involve

(9) disorder

(10) Remove

Text 2　Blood pressure meters

(1) heartbeat

(2) damaged

(3) periodically

(4) meter

(5) appropriate

(6) automatic

(7) accurate

(8) valve

(9) measure

(10) diastolic

Text 3　Glucose meters

(1) check

(2) diabetes

(3) complications

(4) diet

(5) urgent

(6) solution

(7) recommend

(8) manual

(9) glucose

(10) lancet

Text 4　Hearing aids

(1) situation

(2) magnify

(3) throat

(4) tumor

(5) participate

(6) plastic

(7) substances

(8) benefit

(9) amplifier

(10) profound

Section 2　Electrocardiograph (ECG) monitors

Text 1　Introduction to electrocardiograph (ECG) monitors

(1) instantaneously

(2) dizziness

(3) congenital

(4) inflammation

(5) breath

(6) cardiac

(7) elicit

(8) visualize

(9) Allergy

(10) consumption

Text 2　How an ECG monitor works

(1) electrodes

(2) gel

(3) Silver

(4) precedes

(5) positive

(6) electrolyte

(7) horizontal

(8) arise

(9) circular

(10) fibers

Section 3　Ultrasound imaging

Text 1　Introduction to ultrasound imaging

(1) dynamics

(2) audible

(3) visualize

(4) capture

(5) attenuate

(6) biological

(7) consequences

(8) prudently

(9) exposure

(10) untrained

Text 2　Fundamentals of ultrasound imaging

(1) diagnostics

(2) portable

(3) vibrated

(4) transform

(5) transmit

(6) penetrate

(7) pitch

(8) spread

(9) intensity

(10) approaching

Section 4　X-ray imaging

Text 1　Introduction to medical X-ray imaging

(1) intervention

(2) sensitive

(3) wavelength

(4) principle

(5) slices

(6) appropriate

(7) undergo

(8) injected

(9) outweigh

(10) radiation

Text 2　Radiography (Plain X-rays)

(1) disorder

(2) scattered

(3) sensitive

(4) restrict

(5) density

(6) availability

(7) structure

(8) preview

(9) evaluation

(10) Shield

Text 3　Computed tomography (CT)

(1) prominent

(2) rotation

(3) reconstruct

(4) motorized

(5) assembly

(6) register

(7) fraction

(8) preparation

(9) therapy

(10) manifested

Section 5 Magnetic resonance imaging (MRI)

Text 1 Introduction to magnetic resonance imaging (MRI)

(1) intravenous

(2) diagnostic

(3) identify

(4) tissues

(5) magnetic

(6) spinal

(7) scanner

(8) rivals

(9) utilize

(10) abnormality

Text 2 Physics of magnetic resonance imaging (MRI)

(1) undergo

(2) physical

(3) pregnant

(4) converted

(5) absorb

(6) apply

(7) phenomenon

(8) Symptoms

(9) radiation

(10) complementary

Section 6 Clinical laboratory equipment

Text 1 Hematology analyzer

(1) toxicity

(2) quantify

(3) classified

(4) analyze

(5) anemia

(6) indicates

(7) variation

(8) Hematology

(9) hormone

(10) bacterial

Text 2　Biochemistry analyzer

(1) guarantee

(2) permanent

(3) discard

(4) qualitatively

(5) hardware

(6) biochemistry

(7) pale

(8) expendable

(9) absorbance

(10) calibration

Text 3　Urine analyzer

(1) urinalysis

(2) clerical

(3) emit

(4) culture

(5) bacteria

(6) optimum

(7) concentration

(8) eliminate

(9) reflect

(10) proportionally

Section 7　Biomaterials

Text 1　Medical dressing materials

(1) crucial

(2) infection

(3) plasma

(4) ease

(5) expedite

(6) moist

(7) Desiccation

(8) wound

(9) chronic

(10) venous

Text 2　Dental materials

(1) dental

(2) fabricate

(3) solid

(4) contamination

(5) alloy

(6) interim

(7) oral

(8) loosen

(9) Silicone

(10) durable

Text 3　Metallic biomaterials in orthopaedic surgery

(1) reconstruction

(2) resect

(3) reliability

(4) defective

(5) hardness

(6) resist

(7) element

(8) hazard

(9) Microstructure

(10) Electrochemical

Chapter 3　Medical devices regulations

Section 1　Global medical device regulatory harmonization

Text 1　Global regulations of medical devices

(1) Electronic

(2) function

(3) regulation

(4) requirement

(5) risk

(6) performance

(7) objective

(8) submission

(9) certificate

(10) category

Text 2　Global harmonization of medical devices

(1) organization

(2) industry

(3) trade

(4) representative

(5) cost

(6) registration

(7) purpose

(8) increase

(9) disease

(10) standard

Section 2　US FDA regulations

Text 1　How to market your medical devices in U.S.?

(1) specialty

(2) report

(3) staff

(4) market

(5) receive

(6) information

(7) authorize

(8) document

(9) clinical

(10) hardcopy

Text 2　FDA: Premarket notification(PMN) 510(k)

(1) factor

(2) drug

(3) affect

(4) draft

(5) design

(6) source

(7) chemical

(8) diagnose

(9) modify

(10) inspect

Section 3　NMPA regulations

Text 1　Regulations on supervisory management of medical devices

(1) Credit

(2) manner

(3) local

(4) scope

(5) operation

(6) Production

(7) deleted

(8) promote

(9) association

(10) catalogue

Text 2　Provisions for medical device registration

(1) coordinate

(2) department

(3) applicant

(4) procedure

(5) principles

(6) entrust

(7) issued

(8) progress

(9) permission

(10) designated

Section 4　Quality Management System (QMS) for medical device

Text 1　Quality Management System (QMS) for medical devices

(1) purchased

(2) vary

(3) determine

(4) permits

(5) meets

(6) assure

(7) training

(8) follow

(9) framework

(10) adopt

Text 2 ISO 13485: 2003 Medical devices-quality management systems-Requirements for regulatory purposes

(1) guidance

(2) maintain

(3) service

(4) relationship

(5) resource

(6) influence

(7) size

(8) monitor

(9) criteria

(10) convenience

Chapter 4 Medical device standards and practices

Section 1 Medical device standards

Text 1 ISO 10993-5: Test for in vitro cytotoxicity

(1) qualitative

(2) test

(3) proliferation

(4) cytotoxicity

(5) means

(6) Numerous

(7) simulate

(8) documenting

(9) principle

(10) dilution

Text 2　Heavy metals impurities test methods in United States pharmacopeia

(1) prepare

(2) content

(3) specified

(4) dissolve

(5) Dilute

(6) downward

(7) comparison

(8) pipet

(9) exceed

(10) percentage

Text 3　IEC 60601-1-2-2014: Medical electrical equipment-general requirements for basic safety and essential performance

(1) electromagnetic

(2) scope

(3) phenomenon

(4) wireless

(5) addressed

(6) foreseeable

(7) interference

(8) guidance

(9) meet

(10) prohibited

Section 2　Application cases

Text 1　Classes and types of medical electrical equipment

(1) protection

(2) insulation

(3) fault

(4) symbol

(5) deem

(6) disconnect

(7) operative

(8) short

目标检测参考答案（Learning Test 部分）

(9) conductive

(10) mandatory

Text 2　The AK 96 dialysis machine

(1) reverse

(2) parameters

(3) check

(4) precaution

(5) sensor

(6) osmosis

(7) input

(8) coagulation

(9) deviation

(10) venous

Text 3　Maintenance manual of the AK 96 dialysis machine

(1) technician

(2) validate

(3) inactive

(4) manual

(5) schedule

(6) rinse

(7) drainage

(8) inlet

(9) procedure

(10) hygiene

Chapter 5　Medical device marketing and management

Section 1　Business etiquette

Text 1　Business etiquette tips everyone should follow

(1) etiquette

(2) distracted

(3) region

(4) productive

(5) gratitude

(6) focus

(7) concise

(8) available

(9) guideline

(10) complex

Text 2　Customer service etiquette

(1) confirm

(2) appreciate

(3) significant

(4) concerned

(5) authorized

(6) eliminate

(7) etiquette

(8) liable

(9) confused

(10) cancel

Section 2　Marketing

Text 1　Medical device sales jobs

(1) coordinate

(2) executive

(3) involve

(4) disposable

(5) competitive

(6) promising

(7) commission

(8) training

(9) complicated

(10) representative

Text 2　How medical device companies can strategically sell products to customers?

(1) division

(2) mutual

(3) compensate

(4) avenue

(5) reimbursement

(6) ingredient

(7) capability

(8) shrink

(9) imperative

(10) hurdle

Section 3　Business negotiation

Text 1　How to win price negotiations?

(1) slash

(2) prosperous

(3) alternative

(4) warranty

(5) guarantee

(6) profitable

(7) concession

(8) outweigh

(9) standpoint

(10) empower

Text 2　Meet goals in negotiation

(1) specific

(2) arguments

(3) explore

(4) satisfy

(5) motivate

(6) psyche

(7) deception

(8) legitimate

(9) opponents

(10) dilemma

Chapter 6　Industrial communication

Section 1　Job hunting

Text 1　Resume for employment

(1) replace

(2) impact

(3) responsible

(4) individual

(5) desires

(6) emergency

(7) evaluate

(8) previous

(9) significant

(10) impressed

Text 2　Job-hunting letter

(1) match

(2) obtained

(3) interview

(4) detail

(5) permitted

(6) vacancy

(7) applications

(8) mention

(9) career

(10) research

Text 3　Interview skills

(1) confident

(2) representative

(3) potential

(4) advantage

(5) hire

(6) feedback

(7) impression

(8) positive

(9) involve

(10) candidate

Section 2　Technical communication

Text 1　How to become a successful medical device sales representative?

(1) factor

(2) feature

(3) supply

(4) aggressive

(5) Imagine

(6) possess

(7) wholesaler

(8) facilities

(9) prospects

(10) recruits

Text 2 The Internet-where the customers are?

(1) service

(2) childish

(3) reputation

(4) consider

(5) Suspicion

(6) expense

(7) rude

(8) relatively

(9) include

(10) mention

医疗器械专业英语课程标准

（供医疗器械类专业用）